The Technology and Business of Mobile Communications

The Technology and Business of Mobile Communications

An Introduction

Mythri Hunukumbure
Samsung R&D Institute, UK

Justin P. Coon
Oxford University and Oriel College, UK

Ben Allen
University of Oxford, UK

Tony Vernon
Vector Sum Consulting Ltd, UK

This edition first published 2022
© 2022 John Wiley & Sons, Ltd.

The right of Mythri Hunukumbure, Justin P. Coon, Ben Allen, and Tony Vernon to be identified as the authors of this work has been asserted in accordance with law.

Registered Offices
John Wiley & Sons, Inc., 111 River Street, Hoboken, NJ 07030, USA
John Wiley & Sons Ltd, The Atrium, Southern Gate, Chichester, West Sussex, PO19 8SQ, UK

Editorial Office
The Atrium, Southern Gate, Chichester, West Sussex, PO19 8SQ, UK

For details of our global editorial offices, customer services, and more information about Wiley products visit us at www.wiley.com.

Wiley also publishes its books in a variety of electronic formats and by print-on-demand. Some content that appears in standard print versions of this book may not be available in other formats.

Library of Congress Cataloging-in-Publication Data
Names: Hunukumbure, Mythri, 1972- author.
Title: The technology and business of mobile telecommunications : an introduction / Mythri
    Hunukumbure, Justin P. Coon, Ben Allen, Tony Vernon.
Description: Hoboken, NJ : John Wiley & Sons, 2022. | Includes bibliographical references and index.
Identifiers: LCCN 2021020796 (print) | LCCN 2021020797 (ebook) | ISBN 9781119130291 (hardback) |
    ISBN 9781119130307 (pdf) | ISBN 9781119130314 (epub) | ISBN 9781119130345 (ebook)
Subjects: LCSH: Mobile communication systems. | Cell phone services industry.
Classification: LCC TK5103.2 .H85 2022 (print) | LCC TK5103.2 (ebook) | DDC 621.384--dc23
LC record available at https://lccn.loc.gov/2021020796
LC ebook record available at https://lccn.loc.gov/2021020797

Cover image: © Tony Vernon
Cover design by Wiley

Set in 9.5/12.5pt STIXTwoText by Integra Software Services Pvt. Ltd, Pondicherry, India
Printed and bound by CPI Group (UK) Ltd, Croydon, CR0 4YY
C9781119130291_151121

To Dilshani, Kaveesha and Nethaya and to my parents Leela and Jayatissa

–Mythri Hunukumbure

To my mother, Jennifer Hill, and my father, Bob Coon

–Justin P. Coon

To Louisa, Nicholas and Bethany and my whole family, friends and colleagues past and present

–Ben Allen

For Christopher, Marcus and Caitlin

–Tony Vernon

Contents

Foreword *xv*
Preface *xvii*
About the Authors *xix*
Acknowledgements *xxi*
List of Abbreviations *xxiii*

1 **A Technology that Changed the World** *1*
1.1 Social and Economic Impact of Mobile Communications *2*
1.1.1 Social Impact *3*
1.1.2 Economic Impact *5*
1.2 A Brief History of Mobile (Cellular) Communications *8*
1.3 The Journey of Mobile Communications as Seen from User and Operator Perspectives *18*
References *20*

2 **The Mobile Telecoms Ecosystem** *23*
2.1 Introduction *23*
2.2 Telecommunications Ecosystem *24*
2.3 Regulation and Spectrum *26*
2.3 Standardisation *27*
2.4 Research *28*
2.5 End Users *30*
2.6 The Role of Operators (Carriers) *30*
2.7 The Role of Vendors/Manufacturers *31*
2.8 The Role of Standard Bodies and Regulators *31*
2.9 Telecoms Ecosystem Dynamics and Behaviour *32*
2.10 5G Ecosystem *35*
2.10.1 Datacentres *36*
2.10.2 RF Chip and Component Manufacturers *36*
2.10.3 Telecom Operators (Carriers) *36*
2.10.4 Infrastructure Service Providers *36*
2.10.5 Gaming *37*
2.10.6 Over The Top (OTT) *37*

2.10.7   Low-Cost Processing Unit Manufacturer   *37*
2.10.8   Investors   *38*
2.10.9   Potential Disruptions in the 5G EcoSystem   *38*
2.11   Summary   *41*
           References   *41*

3   The Business of a Mobile Operator   43
3.1   Business Challenges Faced by Operators   *43*
3.1.1   Third-Party Costs   *43*
3.1.2   Radio Access Network Costs   *45*
3.1.3   Transmission Costs   *49*
3.1.4   Physical Locations   *53*
3.1.5   Power Costs for Multiple Technologies   *54*
3.2   MVNOs – Mobile Virtual Network Operators   *56*
3.2.1   Economics of an MVNO   *57*
3.2.2   Modelling MVNOs and SPs   *59*
3.3   Operator Business around International Roaming   *63*
3.3.1   The EU Roaming Regulation 'Roam like at Home'   *64*
3.3.2   Covid-19 Impact on Roaming Revenues   *66*
3.4   The Likely Operator Business Models in 5G   *66*
3.5   Conclusion   *69*
           References   *69*

4   Why Standards Matter   73
4.1   The Creation of a New 'G'   *74*
4.1.1   Research   *74*
4.1.2   Standardisation   *75*
4.1.3   Commercialisation   *77*
4.1.4   Continued Innovation   *79*
4.1.5   Intellectual Property as a Metric and Political Currency   *81*
4.2   Shifting Political Power and the Making of an Ecosystem   *81*
4.2.1   2G GSM – Europe Leads   *82*
4.2.2   3G UMTS – Universal (Except Not Quite)   *85*
4.2.3   4G EPS – Avoiding Old Mistakes (and Making New Ones?)   *89*
4.2.4   5G NR – New World Order?   *94*
4.3   Future Standards   *97*
           References   *99*

5   The Mobile Network   101
5.1   Mobile Network Architecture   *101*
5.2   The Radio Access Network (RAN)   *103*
5.2.1   Synchronisation   *104*
5.2.2   Broadcast Messages   *104*
5.2.3   Paging   *104*

5.2.4   Random Access   *105*

5.2.5   Scheduling   *105*

5.2.6   Power Control   *106*

5.2.7   Handover   *106*

5.2.8   Link Adaptation   *108*

5.2.9   HARQ, Error Correction   *108*

5.2.10   MIMO Techniques   *109*

5.2.11   The Control/data Channels and Reference Signals   *109*

5.3   The Core Network (CN)   *110*

5.3.1   Circuit Switching and Packet Switching Networks   *110*

5.3.2   Tunnelling and Encapsulation   *111*

5.4   The Protocol Stack   *112*

5.4.1   The OSI Model of 7 Layer Protocol Stack   *113*

5.4.2   Protocol Stacks for Mobile Communications   *115*

5.5   The 2G Network   *118*

5.5.1   The Network Architecture of 2G   *118*

5.5.2   The GSM Frame Structure   *120*

5.5.3   GSM (And GPRS) RAN Features   *122*

5.5.4   2G Evolutions   *124*

5.6   The 3G Network   *124*

5.6.1   The UMTS Terrestrial Radio Access Network (UTRAN)   *125*

5.6.2   UTRAN Features   *129*

5.6.3   The IP Multimedia Subsystem (IMS)   *130*

5.6.4   Issues with the UMTS Air Interface   *131*

5.6.5   3G Evolution to HSPA   *132*

5.7   The 4G Network   *133*

5.7.1   LTE System Architecture   *134*

5.7.2   LTE Protocol Layers   *136*

5.7.3   LTE Multiple Access Schemes   *139*

5.7.4   LTE Frame Structures   *142*

5.7.5   LTE Reference Signals   *144*

5.7.6   LTE main RAN procedures   *144*

5.7.7   Main Features of Subsequent LTE Releases   *148*

5.8   The 5G Network   *150*

5.8.1   5G-NR Deployment Options   *152*

5.8.2   5G-NR System Architecture   *153*

5.8.3   Spectrum Options for 5G-NR   *154*

5.8.4   5G-NR Protocol Layers   *155*

5.8.5   The 5G-NR Air Interface   *158*

5.8.6   5G-NR RAN procedures   *160*

5.8.7   5G-NR Reference Signals   *161*

5.8.8   5G Core – Concepts and Functionalities   *162*

5.9   The Centralisation and Virtualisation of the Mobile Network   *163*

5.9.1   The Centralised RAN (C-RAN)   *164*

5.9.2   NFV (Virtualised Network Functions) and SDN (Software Defined Networking) Concepts   *166*
5.10   Conclusions   *169*
         References   *170*

6   Basics of Network Dimensioning and Planning   *173*
6.1   Properties of Signal Strength, Noise and Interference   *174*
6.2   The Link Budget and Coverage Dimensioning   *178*
6.2.1   The Transmit Power   *178*
6.2.2   The Antenna Gains   *178*
6.2.3   Transmit and Receive Diversity Gains   *179*
6.2.4   The EIRP   *179*
6.2.5   Modelling the Path Loss   *180*
6.2.6   Modelling the Log Normal Fade Margin   *183*
6.2.7   The FFM   *184*
6.2.8   Building Penetration Loss   *185*
6.2.9   Building the Link Budget   *185*
6.3   Capacity Dimensioning   *187*
6.3.1   The Capacity Demand Estimation Process   *188*
6.3.2   Capacity Demand Estimation – Worked Example   *189*
6.3.3   Resource Provision – Worked Example   *194*
6.4   The Dimensioning of Backhaul Links   *199*
6.4.1   LTE Backhaul Provision – General Aspects   *200*
6.4.2   LTE Backhaul Provision – Capacity Aspects   *201*
6.4.3   New Developments in Backhaul/fronthaul Provision   *207*
6.5   The Network Planning Process   *208*
6.5.1   The Network Area Maps   *208*
6.5.2   Site Placement and Antenna Radiation Patterns   *209*
6.5.3   Traffic Modelling and Capacity Provision Information   *210*
6.5.4   Fine Tuning and Optimisation   *212*
6.6   A Look at 5G Networks   *213*
         References   *216*

7   Spectrum – The Life Blood of Radio Communications   *219*
7.1   Introduction   *219*
7.2   Spectrum Management and Its Objectives   *219*
7.2.1   The Role of the ITU   *220*
7.2.2   Regional Bodies   *221*
7.2.3   National Regulators and Their Roles   *222*
7.2.4   The Spectrum Management Process   *223*
7.3   Spectrum Allocations   *225*
7.4   Spectrum Assignment   *225*
7.4.1   Administrative Assignments   *226*
7.4.2   Market Based Mechanisms   *226*
7.4.3   Beauty Contests   *227*

7.5 Spectrum Licensing *228*
7.5.1 Spectrum for Mobile Services *228*
7.5.2 Dimensions of Spectrum Sharing *233*
7.6 Spectrum Bands Considered for 5G *235*
7.6.1 Example Illustration of Spectrum Deployment Strategy for MNOs *236*
7.6.2 Local Access Spectrum *237*
References *238*

8 Fundamentals of Digital Communication *241*
8.1 Basic Digital Communication System Overview *241*
8.2 Encoding Information *243*
8.2.1 Sampling *243*
8.2.2 Source Coding *245*
8.2.3 Channel Coding *246*
8.3 Signal Representation and Modulation *251*
8.3.1 Mapping Bits to Signals *253*
8.3.2 Signal Spectrum *256*
8.4 Signal Demodulation and Detection *257*
8.4.1 System Model and Sources of Noise *257*
8.4.2 Demodulation *258*
8.4.3 Detection *260*
8.5 Performance Analysis *260*
8.5.1 Capacity *260*
8.5.2 Bit-error Rate and Symbol-error Rate *262*
8.6 Communication Through Dispersive Channels *264*
8.6.1 Time-domain Equalization and Detection *264*
8.6.2 Frequency-domain Equalisation *267*
8.7 Multiple Access: A Second Look *272*
8.7.1 CDMA and 3G *272*
8.7.2 OFDMA/SC-FDMA and 4G *275*
8.7.3 NOMA and 5G *277*
8.8 System Impairments *278*
8.8.1 Carrier Phase Estimation *279*
8.8.2 Timing Recovery *280*
8.8.3 Channel Estimation *280*
8.9 Further Reading *282*
Notes *282*
References *283*

9 Early Technical Challenges and Innovative Solutions *285*
9.1 Wireless Channels: The Challenge *285*
9.1.1 Propagation *285*
9.1.2 Fading and Multipath *287*
9.1.3 Signal-to-Noise Ratio in Fading Channels *293*
9.2 Multicarrier Modulation: A Second Look *295*

9.2.1   Coded OFDM *295*

9.2.2   Capacity and Adaptive Modulation *295*

9.3   Diversity *297*

9.3.1   Macro Diversity *297*

9.3.2   Time Diversity *298*

9.3.3   Frequency Diversity *300*

9.3.4   Spatial Diversity *300*

9.4   Multiple Input Multiple Output (MIMO) *307*

9.4.1   Capacity *308*

9.4.2   MIMO Transmission Techniques *309*

9.4.3   MIMO Reception Techniques *311*

9.4.4   MIMO vs Multicarrier *312*

9.4.5   Multi-User and Massive MIMO *313*

    References *315*

10   **Small Cells – an Evolution or a Revolution?** *317*

10.1   Introduction *317*

10.2   Small Cells Concept Formation *319*

10.3   Multi-tier Cellular Networks/HetNets Architecture *320*

10.3.1   Interference Management *320*

10.3.2   Mobility Management *321*

10.3.3   Backhaul *322*

10.4   Interference Management and Modelling in Small cell/HetNets *322*

10.4.1   Interference Management *322*

10.4.2   Interference Modelling *325*

10.5   Mobility Management *329*

10.6   Backhaul *332*

10.7   Small-Cell Deployment *335*

10.8   Future Evolution of Small Cells *339*

10.9   Conclusion *342*

    References *342*

11   **Today's and Tomorrow's Challenges** *345*

11.1   The Capacity Crunch *345*

11.1.1   A Historical Perspective *345*

11.1.2   Methods for Capacity Enhancement *346*

11.1.3   Impact on Transport and Core Networks *349*

11.1.4   Complementary Technologies *352*

11.2   Increasing Network Complexity *354*

11.2.1   The Self-Organising Networks *355*

11.2.2   Network Automation in 5G *359*

11.2.3   The Business Rationale for Network Automation *361*

11.3   The Need for Greener and Lower EMF Networks *362*

11.3.1   Greener Mobile Networks *362*

11.3.3   Green Manufacturing and Recycling *364*

11.3.4   Applications of Mobile Networks for Energy Reduction   *364*

11.3.5   Electromagnetic Field Exposure and Mobile Networks   *365*

11.4   Covering the Unserved and Under-served Regions   *368*

11.4.1   New Access Technologies   *368*

11.4.2   Initiatives Driven by Government Funding and Policy   *371*

Reference   *373*

12   The Changing Face of Mobile Communications   *377*

12.1   Changes with Centralisation and Virtualisation of the Mobile Network   *377*

12.2   Supporting Multiple Vertical Industries through 5G   *380*

12.2.1   Automotive Sector   *380*

12.2.2   Smart City   *383*

12.2.3   Industry 4.0   *386*

12.2.4   Critical Communications Sector   *388*

12.2.5   Other Vertical Areas under Development   *391*

12.3   The Continuous Evolution of the Mobile Device   *393*

12.4   What Will 6G Look Like?   *395*

12.4.1   Machine Learning and Artificial Intelligence   *395*

12.4.2   Blockchain and the Internet of Things   *396*

12.4.3   Evolutions in Cloud and Edge Computing   *397*

12.4.4   Advanced Hybrid Beamforming   *398*

12.4.5   New Modulation Schemes   *399*

12.4.6   Tera-Hertz (Thz) Communications   *399*

12.4.7   Orbital Angular Momentum   *401*

12.4.8   Unmanned Aerial Vehicles   *401*

12.4.9   Quantum Technology   *401*

References   *402*

Index   *407*

Foreword

I am thrilled to have been invited to write the foreword to this text as the authors are former doctoral students of mine. This book expertly draws together the technology and business of mobile communications. Whilst both the technical and business elements are thought by many to be highly complex in their own right, and may only be grasped by 'experts', this text renders both as readily accessible to those with an understanding of the basic concepts and an interest in the subject matter. This text is further commendable since it is rare that a single volume seamlessly tackles both technical and business elements in such a way that the reader with a background in either is readily drawn to appreciate the other.

Cellular or mobile communication technologies and their capabilities have evolved rapidly over the last 40 years, with 5 generations (the'G') of research, standardisation, product development, and roll-out now serving more than 14 billion devices. With each transition between the generations, users have benefited from an evolution in connectivity performance and new service offerings. This rapid growth is not solely driven by innovation, but, as a sound economic case is a key requirement given the level of investment required.

The authors of this book, with their doctoral research contributing to wireless communications and enriched by their industrial and academic contributions, have brought together this textbook as a one-stop reference of the journey through the 5 generations of cellular, as well as what is to follow. The breadth of knowledge and experience from the team provides a rich narrative on the mobile telecommunications ecosystem, including spectrum, standards, costing and disruptors, alongside network architecture design. An accessible explanation of the Radio Access Network (RAN) and the mechanisms of call setup, control and mobility complements the former. This is followed by highlights of numerous technology evolutions, including antenna arrays, interference management, network planning, electromagnetic field safety and development of greener technology. The final chapter takes a look forwards in terms of applications in the vertical industrials as well as the application of machine learning (ML) and artificial intelligence (AI), and higher frequency bands (Tera-Hertz) as we enter the era of 6G R&D.

At the time of writing, 5G mobile technologies are being rolled out and becoming embedded in mobile communications networks in many countries. This text provides a gateway into 5G technologies and networks such as the 'core' and RAN and also details forthcoming network technologies such a Cloud-RAN, Network Function Virtualisation (NFV) and Software Defined Networks (SDN). An in-depth look into techno-economic aspects of the emerging 5G verticals is also covered, where the related-use cases underpin the economic outlook for 5G networks and associated businesses.

It is also an exciting and pivotal time for mobile communications research. As 5G networks are being established, research into 6G is underway with increasing intensity. This includes identifying key challenges and technologies for addressing them, hence providing momentum for further research and development. This book also provides further details of some of the key potential technological enablers for 6G.

For those entering the field of mobile communications, this book provides a valuable resource to capture four decades of cellular development. For those already working in this domain, a handy reference.

I commend this book and sincerely hope you enjoy the read, as I have done.

Mark
Professor Mark Beach
Professor of Radio Systems Engineering
Department of Electrical and Electronic Engineering
University of Bristol

Preface

Mobile communications is one of the most rapidly evolving technical fields, with a new generation of mobile communications appearing every 10 years or so. For a new entrant to this field, be it as a post-graduate student or an industry employee, keeping pace with the current developments in their specialisation becomes the priority. While this is critically important, it can obscure or delay development of an overall understanding of the different components of this wide ecosystem and their interactions. Having this overall view of how the different pieces of this complex 'jigsaw puzzle' fit together can be really useful when they want to navigate beyond their specialised area.

The intention of this book is to provide such an overall view of the Mobile ecosystem and show the interactions of individual components, their evolutions and future directions in an introductory manner. The authors and the specialised chapter contributors use their knowledge and wide experience in industry, academia and regulatory domains to provide a broad encapsulation of all the critical components in this field. Special care has been taken to acknowledge that all readers of this book will not have a deep technical background or the requirement to master all the technical details of this field. A separate chapter sequence for following this book while omitting the deep technical details and focussing more on the business aspects has been suggested in the introduction (Chapter 1).

In a book covering a broad array of topics and a target audience with a wide span of interests in mobile communications, it is not possible to delve deeper into a single topic. Ample references, which provide in-depth coverage of particular topics, have been provided particularly in deeply technical chapters. All efforts have been taken to capture the latest developments in each of the topics at the time of writing. However, in such a dynamic field as mobile communications, the lead time in bringing written content to print can mean that the 'latest' has moved a few notches forward. We have also captured many web resources as citations, especially in chapters looking at the latest and future technology trends, where printed and published citation options are somewhat in short supply. We have reviewed the chapter contents several times to remove any errors or inconsistencies but if a reader finds any such inconsistencies we would be very happy to be notified of such. This will help us to improve any subsequent editions or prints of this book.

We hope that this book will become a useful resource to aspiring practitioners in this highly dynamic and exciting field of mobile communications.

Mythri, Justin, Ben and Tony.

About the Authors

Mythri Hunukumbure

Dr. Mythri Hunukumbure obtained his BSc (Eng) degree with first-class honours from the University of Moratuwa, Sri Lanka in 1998 and MSc and PhD degrees in Telecommunication Engineering from the University of Bristol in 2000 and 2004 respectively. He is currently a Principal Research Engineer and a Project Lead at the Samsung Electronics R&D Institute, UK. In an industry career spanning over 15 years, he has contributed to, and later led, mobile communication research, standardisation and product development activities. Whilst at Samsung, he has participated in flagship EU projects mmMAGIC, ONE5G and 5G LOCUS as a work package leader. He is also actively contributing to 3GPP RAN1 and SA2 standardisation topics, securing vital IPR. He has filed around 50 patents to date and has also published extensively in leading conferences and journals, receiving the best paper award at the World Telecommunications Congress (WTC) in 2012.

Justin P. Coon

Professor Justin P. Coon received a BSc degree (with distinction) in electrical engineering from the Calhoun Honours College, Clemson University, USA and a PhD in communications from the University of Bristol, UK in 2000 and 2005, respectively. From 2004 until 2013, he held various technical and management positions at Toshiba Research Europe Ltd. (TREL). Prof. Coon also held a Reader position in the Department of Electrical and Electronic Engineering at the University of Bristol from 2012 until 2013. In 2013, he took a faculty position at Oxford University with a Tutorial Fellowship at Oriel College. Prof. Coon is a Fellow of the Institute of Mathematics and Its Applications (FIMA) and a Senior Member of the Institute of Electrical and Electronics Engineers (IEEE). He is also a regular consultant to industry.

Ben Allen

Dr. Ben Allen completed his MSc and PhD degrees at the University of Bristol in 1997 and 2001 respectively. His career has spanned academia and industry, most recently as a Royal Society Industry Fellow with the University of Oxford and Network Rail. He has been the lead for several R&D activities involving telecoms for railways, several of which exhibited state-of-the-art advances. He has published numerous papers and several books on radio and telecommunications research developments. Dr. Allen is a Chartered Engineer, Fellow of the Institution of Engineering & Technology, Institute of Telecommunications Professionals and the Higher Education Academy.

Tony Vernon

Dr. Vernon graduated from the University of Glasgow in 1987 with a Joint Honours in Electronic Engineering with Physics. After a few years in the cellular industry, he obtained chartered status and, in 2002, received a PhD in Mobile Telecoms from the University of Bristol. His main interests and career contributions lie in the planning and optimisation of digital mobile networks ranging from the dawn of 2G in 1991 to 5G in 2021. With 6G on the horizon, Dr. Vernon's focus has moved to the vehicular channel (V2X) and the future use of soon-to-be-ubiquitous mobile broadband networks for national and public-access broadcasting. He is based on the Scottish Outer Hebridean island of South Uist and is thus passionate about expanding 4G and 5G mobile broadband connectivity to rural and remote areas.

Acknowledgements

This book has truly been a collective effort, with the authors and contributors providing expertise and resources from a wide array of specialisations within mobile communications. The authors would like to jointly thank Wiley publishers, especially Sandra Grayson, Juliet Booker and Britta Ramaraj, for their continued support throughout this project. The authors also acknowledge the three specialised chapters provided by the contributors, which really widen the scope of this book:

Daniel Warren (Samsung Research UK): Chapter 4: Why Standards Matter?

Abhaya Sumanasena (Real Wireless): Chapter 7: Spectrum – the life blood of radio communications

Jie Zhang (University of Sheffield and Ranplan) and **Haonan Hu** (Chongqing University of Posts and Telecommunications): Chapter 10: Small Cells – an evolution or a revolution?

The authors fondly reminisce of the opportunity they had to pursue further studies at the Electrical and Electronics Department, the University of Bristol, all these years ago. They gratefully recall the guidance they received from their common PhD supervisor Dr. Mark Beach and also thank him for providing an insightful foreword to this book.

Mythri Hunukumbure wishes to acknowledge Samsung Electronics R&D Institute UK and thanks his colleagues for their fellowship and support. He also acknowledges the European Commission funded 5G research projects mmMAGIC, ONE5G and 5G LOCUS where the cutting-edge research conducted has helped him to grow his expertise and knowledge in this field. He wishes to thank Fujitsu Labs of Europe and the University of Bristol where he previously worked. Last, but not least, he is so grateful for the support he has received from his family members and friends here in the UK and in Sri Lanka without which this book would not have been possible.

J. P. Coon would like to acknowledge Oxford University and, in particular, the Department of Engineering Science, for providing an atmosphere of academic freedom that has allowed him to follow his curiosities over the past decade. He also wishes to thank Oriel College, in its 695th year of existence, for continuing to be such a welcoming and collegiate seat of learning and fellowship. Additionally, he gratefully acknowledges the support of EPSRC, the US ARO, the John Fell Fund, Toshiba Europe Ltd, and Moogsoft, which has enabled him to explore many exciting areas of research during the writing of this book.

Ben Allen particularly wishes to thank Dr. Mythri Hunukumbure for organising this book, as well as Drs. Justin Coon and Tony Vernon for their substantial contributions and fellowship over the years. He also gratefully acknowledges those who have contributed both directly and indirectly to this text, without which this book would not have come to fruition. Finally, he wishes to thank his family for their enduring love and support. Thank you!

Tony Vernon would like to thank, first and foremost, Dr. Mythri Hunukumbure for his unstinting patience, understanding and encouragement over the five-year gestation of this book, and Drs. Justin Coon and Ben Allen for their expert contributions and fellowship during some personally difficult years spent together at Bristol University. Grateful thanks are also due to Professors Mark Beach and Joe McGeehan, who took a gamble on a chap who had failed first-time round, but came good on his second chance. Finally, he pays tribute to all of the great engineering managers he has had over the course of a three-decade career; most notable in this number is Phil Cornforth, currently VP Emerging Technologies with Deutsche Telecom, whose 'servant-not-master' example he has always tried to follow.

List of Abbreviations

6G	6th Generation
5G	5th Generation
4G	4th Generation
3G	3rd Generation
2G	2nd Generation
1G	1st Generation
3GPP	3rd Generation Partnership Project
8-PSK	8-Phase Shift Keying
AAA	Authentication, Authorisation and Accounting
ABS	Almost Blank Sub-frames
AI	Artificial Intelligence
AMF	Access and Mobility Function
AMPS	Advanced Mobile Phone System
ANDSF	Access Network Discovery and Selection Function
ANR	Automatic Neighbour Relation
API	Application Programming Interfaces
AR	Augmented Reality
ARPU	Average Revenue Per User
AUSF	Authentication Server Function
AWGN	Additive White Gaussian Noise
B2B	Business to business
BBU	Baseband Processing Unit
BER	Bit Error Rate
BiTA	Blockchain in Transport Alliance
BJCR	Bahl-Cocke-Jelinek-Raviv
bps	bits per second
BSC	Base Station Controller
BSS	Base Station Subsystem
BTS	Base Transceiver Station
BWP	Bandwidth Parts
CA	Carrier Aggregation
CAGR	Compound Annual Growth Rate
CAPEX	Capital Expenditure

CBRS	Citizen's Broadband Radio Service
CDF	Cumulative Distribution Function
CDMA	Code Division Multiple Access
C-ITS	Co-operative Intelligent Transport Systems
CLSP	Close-loop Spatial Multiplexing
CN	Core Network
CO	Central Office
CoMP	Coordinated Multipoint
COTS	commercial off-the-shelf
CP	Control Plane
CPE	Common Phase Error
CPRI	Common Public Radio Interface
CQI	Channel Quality Indicator
C-RAN	Cloud Radio Access Network
CRC	Cyclic Redundancy Check
CRE	Cell Range Expansion
CRS	Common Reference Signal
CU	Centralised Unit
D2D	Device to Device
DAB	Digital Audio Broadcasting
DAS	Distributed Antenna Systems
DC	Dual Connectivity
DEM	Digital Evaluation Model
DFD	Division Free Duplex
DFE	Decision Feedback Equalizer
DFP	Dynamic Frequency Planning
DRX	Discontinuous Reception Cycle
DSCs	Drone Small Cells
DSP	Digital Signal Processing
DSRC	Dedicated Short Range Communications
DSSS	Direct-sequence spread spectrum
DTM	Digital Terrain Model
DU	Dense Urban
DVB	Digital Video Broadcasting
DWDM	Dense Wavelength Division Multiplexing
EB	Exabyte
ECC	Electronic Communications Committee
EDGE	Enhanced Data rates for GSM Evolution
EEA	European Economic Area
Eicic	Enhanced Inter-cell Interference Coordination
EMF	Electromagnetic Field
EPC	Evolved Packet Core
ESN	Emergency Services Network
ETSI	European Telecommunications Standards Institute
E-UTRAN	Evolved UTRAN

EV-DO	Evolution Data Only
EV-DO	Evolution Data Optimised
FAT	Frequency Allocation Table
FCC	Federal Communications Commission
FD	Full Duplex
FDD	Frequency Division Duplexing
FDMA	Frequency Division Multiple Access
FFM	Fast Fade Margin
FFR	Fractional Frequency Reuse
FOMA	Freedom of Mobile Multimedia Access
FPS	Full Power Sub-frames
FRAND	Fair, Reasonable and Non-Discriminatory
FSPL	Free-space Path Loss
FTP	File Transfer Protocol
FTTH	fibre-to-th-home
FWA	Fixed Wireless Access
GAN	Generic Access Network
Gbps	gigabits per second
GBT	Guaranteed Bit Rate
GCF	Global Certification Forum
GDP	Gross Domestic Product
GGSN	Gateway GPRS Support Node
GMSK	Gaussian Minimum Shift Keying
GPRS	General Packet Radio System
GSM	Groupe Spécial Mobile
HARQ	Hybrid Automatic Repeat Request
HEO	High Elliptical Satellite
GSMA	Global System for Mobile Communications
GSS	GPRS sub-system
HAP	High Altitude Platforms
H-CRAN	Heterogeneous Cloud Radio Access Network
HDD	Hard Disk Drive
HetNet	Heterogeneous Network
HH	Horizontal to Horizontal
HLR	Home Location Register
HeNB	Home eNodeB
HM	Hysteresis Margin
HNB	Home NodeB
HOFs	Hand Over Failures
HSS	Home Subscription Server
HSDPA	High Speed Downlink Packet Access
HSPA	High Speed Packet Access
HSUPA	High Speed Uplink Packet Access
IAB	Integrated Access and Backhaul
ICIC	Inter-cell Interference Coordination

ICNIRP	International Commission on Non-Ionising Radiation Protection
ICO	Intelligent Cell Optimisation
IETF	Internet Engineering Task Force
IFFT	Inverse Fast Fourier Transform
IMS	IP Multimedia Sub-system
IP	Internet Protocol
IPR	Intellectual Property Rights
IoT	Internet of Things
ISD	Inter-site Distance
ISG	Industry Specification Group
ISI	Inter-Symbol Interference
ITU	International Telecommunication Union
kbps	kilobits per second
KPI	Key Performance Indicator
LA	Location Area
LAA	Licensed Assisted Access
LAMM	Linear Angular Momentum Multiplexing
LAPD-m	Link Adaptation Channel D – mobile
LDC	Local Data Centre
LDPC	Low-density parity check
LEO	Low Earth Orbits
LLC	Logical Link Layer
LMR	Land Mobile Radio
LNA	Low-noise amplifier
LNFM	log normal fade margin
LOS	Line of Sight
LSAS	Large-scale Antenna Systems
LTE	Long Term Evolution
MAC	Medium Access Control
MANO	Management and Network Orchestration
MC-PTT	Mission Critical Push To Talk
MCS	Modulation and Coding Scheme
MEC	Multi-access Edge Computing
MIB	Master Information Block
MIMO	Multiple Input Multiple Output
ML	Machine Learning
MLB	Mobility Load Balancing
MM	Mobility Management
MMDS	Multi-point Microwave Distribution Systems
MME	Mobility Management Entities
M2M	Machine to Machine
MMSE	Minimum mean-square error
mMTC	massive Machine Type Communication
MORAN	Multi Operator Radio Access Networks
MRC	Maximum Ratio Combining

MRO	Mobility Robustness Optimisation
MRP	Market Representation Partner
MRRC	Maximum Ratio Receiver Combining
MSC	Mobile Switching Centre
MUD	Multi User Detection
MVNO	Mobile virtual network operator
N3IWF	Non-3GPP Interworking Function
NaaS	Network as a Service
NAS	Non-Access Stratum
NCS	Network Slice Controller
NB-IoT	Narrow Band Internet of Things
NEF	Network Exposure Function
NFV	Network Function Virtualisation
NFVI	Network Function Virtualisation Infra-structure
NLOS	Non-line of Sight
NMT	Nordic Mobile Telephones
NOMA	Non-orthogonal multiple access
NRZ	Non-return-to-zero
NSA	Non-StandAlone
NSS	Network and Switching Sub-System
NTN	Non-Terrestrial Networks
OAI	Open Air Interface
OAM	Orbital Angular Momentum
OBSAI	Open Base Station Architecture Initiative
OFDMA	Orthogonal Frequency Division Multiple Access
O&M	Operations and Maintenance
ONF	Open Networking Foundation
OOK	on off keying
OPEX	Operational Expenditure
OSI	Open Systems Interconnection
OSS	Operations Support Systems
OTT	over-the-top
OVSF	Orthogonal Variable Spreading Factor
OWB	Optical Wireless Broadband
P2P	Point-to-point
P2MP	Point-to-multipoint
PaaS	Platform as a Servide
PAPR	Peak-to-average Power Ratio
PBCH	Physical Broadcast Channel
PCEF	Policy and Charging Enforcement Function
PCF	Policy Control Function
PCI	Physical Cell Identity
PCRF	Policy Control and Charging Rules Function
PDF	Probability Density Function
PDSCH	Physical Downlink Shared Channel

PDU	Protocol Data Unit
PF	Proportional Fair
PL	Path Loss
PLE	Path Loss Exponent
PLL	Phase Locked Loop
PLMN	Public Land Mobile Networks
PN	Pseudo Noise
PNDS	Packet Data Networks
PoP	Point of Presence
PPC	Policy and Charging Control
PRB	Physical Resource Block
PRS	Positioning Reference Signal
P-SCH	Primary Synchronisation Channel
PSK	Phase Shift Keying
PSS	Primary Synchronisation Signal
PSTN	Public Switched Telephone Networks
PUSH	Physical Uplink Shared Channel
QAM	Quadrature Amplitude Modulation
QFI	Quality of Service Flow Identifier
QPSK	Quadrature Phase-shift Keying
QoE	Quality of Experience
QoS	Quality of Service
RA	Routing Area
RACH	Random Access Channel
RAN	radio access network
RDC	Regional Data Centre
RE	resource element
RED	Radio Equipment Directive
RGC	Rural Gigabit Connectivity
RIP	Received Interference Power
RLAH	Roam Like At Home
RLF	Radio Link Failure
RNC	Radio Network Controller
ROI	Return on Investment
RPE	Radiation Pattern Envelope
RPS	Reduced Power Sub-frames
RRH	Remote Radio Head
RRM	Radio Resource Management
RRPS	Ranplan Radio Propagation Simulator
RSPG	Radio Spectrum Policy Group
RSQ	Received Signal Quality
RSS	Received Signal Strengths
RU	Rural
RWP	Random Waypoint
SA	StandAlone

SAE	System Architecture Evolution
SAW	Stop and Wait
SBA	Service-Based Architecture
SC	Small Cell
SC-FDE	Single Carrier with Frequency Domain Equalisation
SC-FDMA	Single Carrier Frequency Division Multiple Access
SCH	Synchronisation Channel
SDN	Software Defined Network
SER	Symbol Error Rate
SGW	Serving Gateways
SGSN	Serving and Gateway Support Nodes
SIB	System Information Block
SIC	Successive Interference Cancellation
SIM	Subscriber Identity Module
SINR	Signal to Noise Ratio
SMF	Session Management Function
SON	Self Organised Networks
SSB	Synchronisation Signal Blocks
SSD	Solid-State Drive
S-SCH	Secondary Synchronisation Channel
SSS	Secondary Synchronisation Signal
SU	Suburban
SVD	Singular Value Decomposition
TACS	Total Access Communication System
TCH	Traffic Channels
TCO	Total Cost of Ownership
TCP	Transmission Control Protocol
TDD	Time Division Duplex
TDE	Time-Domain Equalisation
TDMA	time Division Multiple Access
TD-SCDMA	Time Division – Synchronous CDMA
TEID	Tunnel End-point ID
TETRA	Terrestrial Trunked Radio
TMA	Tower Mount Amplifier
TNGF	Trusted Non-3GPP Gateway Function
TPS	Transmit Power Control
TRAU	Transcoder/Rate Adaptor Unit
TSG	Technical Specification Groups
TTT	Time to Trigger
U	Urban
UAVs	Unmanned Ariel Vehicles
UDP	User Datagram Protocol
UE	User Equipment
UMB	Ultra-Mobile Broadband
UMTS	Universal Mobile Telecommunication System

UP	User Plane
UPF	User Plane Function
UP-SOC	User Plane Service Oriented Core
URLLC	Ultra Reliable and Low Latency Communications
UTRAN	UMTS Terrestrial Radio Access Network
UWB	Ultra-wideband
V2P	Vehicle to Pedestrian
V2V	Vehicle to Vehicle
V2X	vehicle-to-anything
VCO	Voltage Controlled Oscillator
VLC	Visible Light Communication
VLR	Visitor Location Register
VNF	Virtualised Network Functions
VoIP	Voice over IP
VR	Virtual Reality
VU	Victim User
VV	Vertical to Vertical
WAN	Wide Area Network
WCDMA	Wideband Code Division Multiple Access
WiBro	Wireless Broadband
WiMAX	Worldwide Inter-operability for Microwave Access
ZB	Zetabyte
ZC	Zadoff-Chu

1

A Technology that Changed the World

Mobile communications and the internet are widely regarded as the most influential technologies that reshaped human behaviour and interactions over the last 30 years. Today, there are more mobile connections worldwide than the global human population. Mobile networks are functional from the teeming mega cities to the sparsely populated arid lands. From humble beginnings in the early 1980s, the industry has experienced phenomenal growth fuelled by technological advancement. Today, the challenge of providing 'anywhere, anytime' communications and connectivity has mostly been achieved. Yet, the industry is facing a new set of challenges, created largely by its own success.

In this book, we will focus not just on the past successes of the mobile industry, but also on the current challenges it is facing and the likely shape it will take in future. Technologically, the mobile industry, from its inception, has endeavoured to provide faster, more reliable and cheaper connectivity to its customers. We will cover the basics of cellular network deployment and progress to the most advanced technologies used today, with special emphasis on the technical challenges the industry faced and the ingenious solutions it has developed. We will discuss the technological journey, from the first widespread mobile generation (2G) to the current 5th Generation (5G) and look at what likely shape the next wave of deployments, the 6G will take. From a business perspective, the mobile industry has evolved into a complex ecosystem generating huge revenues and touching the lives of billions of people. We will analyse this ecosystem, paying special attention to the business aspects of the mobile operator. Basically, this book is an attempt to introduce not just what happened and what is happening in the industry but to answer why, looking closely at both the technical and business rationale.

We will begin with a brief chapter guide and recommend streams for technical and non-technical reading. We will endeavour to keep the maths to a bare minimum and use it only in the technical chapters, where its usage would help clarify complex concepts. Where mathematical concepts are introduced, we will provide thorough references, so the interested readers can delve deeper into these concepts. As this is meant to be an introductory book, the intention is not to fill readers with in-depth knowledge on a certain topic, but to equip them with an overall understanding of the mobile technology and industry, which hopefully will make them more eager to explore further.

The Technology and Business of Mobile Communications: An Introduction, First Edition.
Mythri Hunukumbure, Justin P. Coon, Ben Allen, and Tony Vernon.
© 2022 John Wiley & Sons Ltd. Published 2022 by John Wiley & Sons Ltd.

The next part of Chapter 1 is dedicated to illustrating the social and economic impact the mobile communications industry is exerting to improve lives globally. Then, we will skim through the history of the industry, spanning from 1G to 5G, also briefly looking at the likely shape it will take in thefuture. In Chapter 2, we will look at the main players in the mobile industry ecosystem, with special emphasis on the interactions amongst them. Chapter 3 is focussing on the business of a mobile operator with the major costs and revenues explained and also looking at the interactions with the other ecosystem players. Being the most prominent part of the industry, we will examine the current challenges faced by mobile operators and the business innovations (backed by technology) prompted to counter these. As a highly dynamic industry, its reliance on research, innovation and standardisation cannot be overemphasised. We dedicate Chapter 4 to discussing the matters related to research and standardisation, especially a historical examination into why certain standards became so successful and others did not. From Chapter 5 until Chapter 10, we delve a little deeper into technical matters. Chapter 5 is a journey through the mobile generations from 2G to 5G, examining the evolution of basic technical features across the generations and network features governing each mobile generation. We will also discuss the successes and limitations in technologies adapted for each generation, giving clues as to why it was necessary to move into newer generations. Chapter 6 provides an introduction to the fundamentals of mobile network planning, with the intention of familiarising the reader with the meticulous planning process carried out before network deployment. Chapter 7 will look at the management of radio spectrum, which is the single most essential component for all radio communications. The basic signal processing techniques (mainly at the receiver), required to extract the wanted signal from a mixture with noise and interference, are presented in Chapter 8. Chapter 9 is an appreciation of the early technical challenges faced by the industry and the innovative solutions developed to tackle them. In Chapter 10, we look at how the cells have evolved to become smaller, how some significant revolutionary aspects were introduced with the small cells in the recent past and the likely contribution they will make in the future. From Chapter 11, we revert back to a non-technical style of writing. Chapter 11 is a look at the main challenges facing the mobile industry today and the nature of the solutions envisaged to tackle these issues. The concluding Chapter 12 is a holistic look at the changing face of mobile telecoms and what could be expected in the future.

For readers who may want to explore both the deeper technical and business aspects of the industry, we suggest they follow the orderly chapter sequence, from Chapters 1 to 12. For readers who may want to omit the deep technical details, we suggest a reading sequence from Chapters 1 to 4, Chapter 7 and Chapters 11 and 12.

1.1 Social and Economic Impact of Mobile Communications

We will first look at the social and economic impact the mobile communications industry is having today, to truly appreciate this potent technology. Mobile communications have made virtually everyone in the world reachable at the touch of a button, whenever or wherever they are. When the power of mobile internet is added to this, truly remarkable capabilities in the social and economic domains have been achieved. We will briefly look through some of the shining examples in these domains. Many of the socio-economic development enablers in developing countries are cited from the supportive work the GSMA (GSM Association – the trade body of the mobile operators) is conducting through their 'Mobile for Development' initiative [1].

1.1.1 Social Impact

Before the mobile communications era, you would always call a 'number' at a certain 'place', with no guarantee of being able to reach the person you intended to connect to. Mobile communications made the 'number' uniquely associated with a 'person' wherever he is, creating a new 'personal communications' phenomenon. The advent of mobile internet and the usage of social media have again revolutionised this personal communications space. Now, you can connect with not just one single person at a time, but simultaneously broadcast yourself to a preferred social group. The so-called 'social media' such as Twitter and Facebook enables you to be in touch with a social group, be it your family, friends or work colleagues at the touch of a button, wherever you are. In this era of 'selfies' and Instagram, sharing personal experiences with your social groups has never been easier. This 'social networking" also carries a significant political and business value. Most of the prominent politicians today operate their own twitter accounts, with millions of followers. They use these to subtly promote their policies and political views, ramping up their communications near election times. Many of the business leaders also have massive twitter followings and these are used to create a 'social aura' about themselves and their businesses. These social networking tools are not just a tool for the rich and powerful but also for the dissenting voices in our society. Many of the demonstrations, rallies and uprisings of today are organised and publicised though these tools, by-passing mainstream media.

The mobile internet era has given rise to hundreds and thousands of 'Mobile Apps', which can make your life easier whilst on the move. These Apps utilise the 'always on' feature of the mobile as the mobile is the first device the user will revert to, whenever or wherever he is. These Apps can help you find a restaurant or a parking space in an unfamiliar city (e.g.: Four square location App), as well as plan your travel route considering live traffic information etc. Many of the Apps (like Waze for travel planning) utilise 'crowd sourcing', to gather, corroborate and analyse data and present it in optimal ways. Data analytics on the data gathered from social Apps (whilst still an evolving science) can achieve far-reaching social, political and commercial objectives. Mobile data connectivity (commonly known as Mobile broadband) has also blurred the boundaries of work and personal spaces and has enabled people to work from anywhere, be it at home, whilst travelling or even from a coffee shop. Whilst some people argue that this has altered the work/life balance negatively, mobile broadband has opened up so many possibilities, giving flexibility and choice of work to employees.

In terms of entertainment, mobile gaming and mobile TV are becoming ever more popular. These applications are again capitalising on the fact that the user can connect to them wherever or whenever he wants to, through the mobile. Many of the video or movie streaming services have the option to include your mobile device to the subscription. With the advancements of the cellular connectivity, many of the games can now be played on your mobile interactively against other users. The lower latencies and higher data rates supported by the newer cellular generations and the video quality of today's Smartphones make this a seamless experience. Streaming music to your mobile has had a phenomenal growth in the last decade, with Apps like Spotify, Apple Music and Deezer becoming the main tools through which the young enjoy their music.

The social impacts of mobile technology extend into the fields of health and education. With ageing populations, developed countries are facing increasing costs in providing elderly care. Countries like Japan are looking for innovative ways of utilising mobile technology so the elderly can be provided with 'assisted living'. The elderly, even with serious conditions, can continue to live in their own homes and their conditions can be monitored with the use of wearable communication devices. In developing countries, providing healthcare to rural communities has always been a challenge with the available limited resources. Mobile technology has stepped in as a means of disseminating basic medical and sanitary information in these countries. A successful scheme in Bangladesh armed 100 000 health volunteers with mobile phones. When they meet a pregnant woman, they enlist her details and her (or a close relative's) mobile number and start to text regular standard advice on prenatal and later, postnatal care [2]. In Pakistan, a mobile and web-based telemedicine scheme 'Sehat Kahani' was launched in 2017, which connects female doctors with patients in underserved areas and communities remotely. There were special walk-in-clinics created with qualified nurses and health workers acting as intermediaries in these remote consultations. Whilst enabling affordable, good quality healthcare to these underserved communities (mainly women) it has also opened the door for female doctors in Pakistan to continue with their profession, where the majority of female medical students do not move into their profession after graduation [3]. This telemedicine solution is now also linked to a Mobile money application, so the patients can book and pay for these services through their mobiles.

The lower rates of birth registration in some developing countries has been a significant barrier for socio-economic development of affected children, with impacts lasting throughout their lives. There have been successful efforts by mobile operators in some of these countries (like Tanzania, Ghana and Bolivia) to partner with government agencies and UNICEF to equip the registry points with mobile devices and software to record births digitally and upload them to a central database [4]. These schemes have increased the availability of registration points by many fold, making long journeys to a central point unnecessary.

There are numerous examples of the use of mobile technology in disaster and emergency situations, to provide warnings efficiently and to coordinate relief efforts. There is a big push in the US and UK to replace traditional emergency communication networks (like TETRA) with cellular LTE-based mobile networks. We will look at this shift in detail in Chapter 12. On a personal level, the mobile is the one communication device that is likely to be with you always, even of you get lost on a mountain or in a jungle. There are numerous location-based apps emerging now (for example, what3words), which can quickly trace a precise location and assist the emergency services. Even after a major disaster, mobile communications tend to be the first line of connectivity that could be restored, providing vital information and advice to affected people. With adverse extreme weather conditions becoming more frequent as a result of climate change, the role mobile communications can play in these support operations will continue to grow.

In education too, the prevalence of mobile coverage is helping to bridge the 'social divide'. The 'English in Action' scheme in Bangladesh, for example, is utilising mobile phones (amongst other methods) as a simple, low-cost learning device and has helped millions of people to take the first steps in improving their English language proficiency. A digital education platform launched by Eneza Education in 2016 in Kenya is now enabling millions of children to access educational resources through SMS, mobile web or android apps [5]. This platform features 'an ask a Teacher' service where students ask real, live questions to teachers and get responses within 15–30 minutes, send Wikipedia content through SMS (Short Message Service) texts, reader boards and other non-curriculum content such as health education. With mobile being the only communication device available for many families in rural and under-privileged communities, these kinds of services go at least some way to bridge the enormous divide in the quality of education available to their children.

1.1.2 Economic Impact

In the last three decades, mobile communications have grown to become major economic pillars in most countries. The industry employs hundreds of thousands of people and contributes significant tax returns to national revenues. In many countries, the mobile spectrum auctions, where frequency bands are licensed for mobile operations, continue to raise billions of pounds for government coffers. The industry is sustaining sizeable research and development activities spread out across its own subsidiaries, universities and other institutes, contributing to develop and sustain knowledge-based economies. All these direct contributions to the macroeconomic growth have been widely acknowledged [6]. For example, the GSMA reported that the mobile industry ecosystem directly and indirectly contributed $3.9 Trillion (or 4.6% of the global GDP) to the global economy in 2018 [7]. Figure 1.1 is an illustration of the various components of this economic contribution.

Mobile communications also contribute in numerous indirect ways to the economic growth of communities. This indirect economic impact is perhaps best felt in developing countries. Traditionally, vast areas of many developing countries, except the main cities

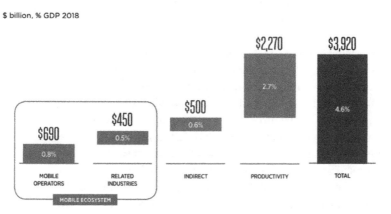

Figure 1.1 The global economic contribution of Mobile Telecoms – 2018. (*Source* [7]).

and towns, are not well connected in either telecommunications or transport infrastructure. The introduction of mobile communications has meant that wireless connectivity for the 'last mile' of the telecommunications link was facilitated. This 'last mile' in rural communities literally extended to hundreds of miles, connecting towns and villages with their main cities and commercial hubs. This 'connectivity' opens up numerous economic opportunities for these rural communities. If this 'revolution' was not to have happened and the normal course of evolutionary infrastructure development was to solve this problem, it would have taken decades to achieve the same level of telecoms service penetration that we see in these countries today.

The advent of mobile communications in these communities was not just a social benefactor, but a potent economic enabler. People soon found numerous economic usages with their mobile phones that would otherwise not have been possible, even with the basic voice and text connectivity. Today, rural farmers in India check the prices for their crops in nearby towns before harvesting and transporting them to the most profitable market. Fishermen in Tanzania and Kerala (in India) check the prices for their catch whilst at sea, to agree on the best deal from any number of local harbour markets within their reach. Small local industries and craftsmen are advertising themselves through text messaging, without solely relying on word of mouth to expand their sales. Small hotel and lodge owners in remote tourist locations can now advertise their businesses to the whole world through mobile Apps, increasing their revenues many fold. The consumers of these Apps and services now have a major say through the feedback they provide and these transparent business models have also improved overall service quality.

Mobile communications in developing countries have even stepped into remedy the deficiencies in the financial and banking infrastructure. Many of these countries, especially in Africa, lack an adequate network of bank branches, ATM machines and financial transaction mechanisms that people in western countries take for granted. Many of their citizens lack sound financial knowledge and do not possess bank accounts, credit cards or other financial tools. Mobile telecommunications have ingeniously been adapted to fill this void and enable a smooth flow of financial activity. Perhaps the best example of this is the use of mobile money, known as M-Pesa in Africa. It originated in Kenya in 2007, enabling users to register accounts through their mobiles, top-up accounts by paying money at numerous agents and conduct financial transactions on these account credits through SMS messaging. This system has become so popular that by 2012, there were over 17 million M-Pesa accounts registered in Kenya. Mobile money systems have now spread to other countries in Africa and Asia and by the end of 2018, there were 866 million registered accounts with transactions over $1.3 billion per day [8]. Mobile money, now as an established payment platform, has enabled account holders to access other financial services and services in healthcare, education, social protection etc. These services have benefitted the most disadvantaged groups in society, including displaced communities due to war and natural disasters. Mobile money has also helped to teach the skills of saving and financial planning to absorb future shocks, by bringing financial literacy to their lives [8].

Another good example of mobile money is the award winning eZ Cash initiative in Sri Lanka [9]. Although Sri Lanka is fast moving towards a middle income country with 70% of adults having access to banking services, there is a significant proportion of lower income earners (mainly in rural areas) who are outside the formal banking scope. Also, the electronic banking tools are not well established even amongst the urban banking population. With more mobile subscriptions than country population and networks covering over 96% of the country, there is significant potential for mobile money solutions. The Central bank of Sri Lanka having realised this potential, opened up the mobile payment market to non-bank institutions in 2011. In June 2012, the largest mobile operator Dialog initiated the eZ Cash service, where any of its mobile subscribers could register and open an account by simply dialling a number. The registered users can send and receive cash directly through their mobiles and carry out payments on utilities and to certain institutions. Within the first month of launching, the service had attracted 300 000 subscribers. By August 2015, with the inclusion of two other operators to this service, the customer base has grown to more than 2 million with over 20 000 agents and merchant points supporting the system.

Some economic applications and disruptions with mobile communications are not limited to developing countries. Mobile banking is fast spreading in developed countries as an effective disruptor to established banking giants. With App-based mobile accounts, Monzo, Revolut and Starling are taking on the banking giants in the UK, for example [10]. Your debit/credit cards are linked to the App with instant updates of the transactions. The user identity verifications and credit checks to open accounts and obtain credit are all done through the Apps, with results obtained in minutes. The minimal operating costs in this model have helped reduce fees to a bare minimum, attracting mainly younger Millennials as tech savvy customers.

The App 'Uber' developed by a start-up in San Francisco, has revolutionised the ride sharing/call taxi-market sector. With the advanced localisation features enabled by Smart phones, the App could immediately identify user location and calculate the fare to the requested destination. The fare is calculated taking into account the demand for taxi services at that time. The transaction goes through the App, without the passenger paying directly to the driver. The reduced cost and wait times have made this App so popular, it is now functional in many cities of developed and developing countries. As a derivative of this concept, similar digital courier services for parcel and restaurant food delivery have sprung up in many large cities. From an employee perspective, these Apps enable you to select jobs convenient to you and plan your work times around other priorities. This is a fundamental shift in the traditional view of a worker and his rights [11], which have come to be questioned and legally challenged many times over recent months.

A similar disruption in the rental property market is spearheaded by the online and mobile App Airbnb. It allows millions of homeowners to list their spare homes or rooms on-line and match these with potential holiday makers. The growth of Airbnb really took off with the launch of their mobile App in 2012. It made listing your spare room or property with a collection of photos taken from your mobile very easy. Also, for consumers looking for the perfect property to rent, the Airbnb app uses deep linking technology [12] to provide the right content and provide personalisation so that users keep coming back to the App for future bookings. Again, the traditional model of listing properties permanently and paying a levy to city councils for this privilege has been disrupted by the Airbnb model, where home owners can now rent out whenever it suits their circumstances.

The social and economic restrictions put in place to help control the Covid-19 pandemic has made the mobile and landline broadband connectivity a lifeline for many people and businesses. Many small businesses have taken to introducing on-line ordering and delivery solutions to keep their businesses afloat. Working from home has become the norm for many industry and business sectors and broadband connectivity has become crucial in this regard. Mobile video-calling Apps have also become an essential tool for many of us, to keep in touch with our families and friends during these difficult and challenging times.

Here, we have only listed a handful of examples to showcase the potent social and economic contributions mobile communications continue to make all across the globe. With the advent of 5G and new disruptive technologies like artificial intelligence (AI), blockchain and quantum computing, the capabilities of today's smart phone will increase many fold in the next few years. This will, undoubtedly, open up new innovative use cases for the Mobile industry as it will continue to re-shape our socio-economic interactions in a highly connected world.

1.2 A Brief History of Mobile (Cellular) Communications

The basis of Mobile communications lies with the wireless connectivity enabled by radio-wave propagation. The exploration of using radiowaves for communication was pioneered by Marconi, during the turn of the twentieth century. Most notable of his experiments was the Trans-Atlantic wireless telegraphy experiments of 1901 [13], where he showed that information embedded onto radio signals could traverse thousands of miles. These discoveries paved the way for many modern day wireless communication modes, including radio, television and Mobile telephony. Various forms of Mobile Telephony systems, employing a single transmission tower and multiple, mobile receiver units were operational from the early 1950s. These towers would provide radio coverage for around a 50-mile radius, enough to cover a large metropolis. It would typically be the government or emergency services such as the police or fire crews who would use these services. Heavy and bulky receiver units would be fitted into the vehicles of these services. Only one-way communication would be possible at a time (we call this half-duplex), with many systems featuring push-to-talk buttons. The coverage would often be patchy and there was no concept of fair radio resource sharing amongst users. The first users to connect would grab the available radio channels of the single transmitter and others would have to wait till radio channels became free. In some cases, a human operator would need to connect these radio channels to the required destination line. Due to these limitations, there was little scope for these systems to evolve as broad-based commercial operations. Some technologists term these pre-commercial mobile systems as 0th Generation or 0G [14]. As we move on to a generic overview of the actual cellular generations, a more technically inclined overview of these early mobile generations (1G to 3G) can be found in [15].

The advent of 1G heralded the first commercial deployments of Mobile communication services, in the early 1980s. Technological evolutions on several fronts made this possible. The advances in semi-conductor and power electronics fields made receiver units portable, although they were still bulky by today's standards. The cellular concept (although first presented in 1947 [14]) was adapted, marking a distinct change from previous Mobile networks. Rather than pumping very high power from a single tower to cover a vast area, the power and height of transmit towers was limited to restrict the coverage to a 'cell' and multiple cells, adjacent to each other, were enacted. This enabled the radio resources to be re-used (we will discuss the radio resources in detail in Chapter 5) and increase the number of simultaneous active users in the network. To support user mobility, basic handover protocols were developed. So, when an active user crossed from one 'cell' to another, the call would be picked up and supported by this new cell. The advances of switching technology meant that automated connectivity to other Mobile or landline fixed networks could also be established.

One of the core features that started with 1G was that governments offered (in many cases auctioned) dedicated spectrum bands for Mobile telephony use. Different regions developed specific standards for 1G operations. Amongst these, the more successful were TACS (Total Access Communication System) and NMT (Nordic Mobile Telephones) in Europe, AMPS (Advanced Mobile Phone System) in the US and TZ-801 in Japan. The first reported 1G network was initiated in Japan, in 1979. Two years later, the Nordic countries started their own 1G networks with the NMT standards. In 1983, the first 1G network in the US was launched in Chicago. The common feature in all these standards and networks was that Analogue techniques were used to modulate (or in simple terms embed) the voice signal onto the radio waves. We will introduce these technical details in Chapter 5. The sole emphasis was on voice transmissions; there was no capability for data transmissions as we see today. Each region operated to their specific standards, without worrying about interoperability or the provision of roaming for international travellers. However, the Nordic countries who deployed NMT standards had the provision for roaming within their region. At the peak of the 1G systems in and around 1990, they were globally supporting subscriber numbers reaching 11 million [16].

The key difference in moving from 1G to 2G Mobile was the introduction of the Digital Signal Processing (DSP) techniques to transform the analogue voice signal to a digital signal and embed into the radio carriers. For the first time, this enabled direct (although basic) data services such as SMS to be included. The digital revolution brought many benefits to mobile communications. The miniaturisation trends in DSP chips and circuitry meant that the Base station equipment and the mobile handsets could be made smaller, cheaper and more power efficient. Many signal processing techniques such as error correction coding and interleaving could be introduced in the digital signal domain, making the mobile signal reception more reliable. We will introduce some of these techniques in Chapter 8. Digital speech compression techniques enabled three times more voice calls to be carried on the radio spectrum as compared to 1G analogue transmissions. The lower costs of the network deployment and operation could be passed onto the customers, making subscriptions cheaper, which inevitably increased customer numbers.

The real success of 2G systems came with standardisation and one standard in particular, GSM (Global System for Mobile Communications or originally Groupe Spécial Mobile), enjoyed phenomenal growth. Started as a European standardisation effort in 1982, it quickly gained the support of many European governments. The EU heads of state endorsed the GSM standardisation project in 1986, making its success a firm priority for the European Commission [17]. The first commercial GSM network was deployed by Radiolinja in Finland in 1991. GSM networks mushroomed in many European countries soon after, enabling a true pan-European roaming capability. The success of these European deployments meant that GSM was set for global expansion. By 1998, GSM networks were operating in more than 100 countries, serving 100 million subscribers. This phenomenal growth continued into the twenty-first century. The global GSM subscriber numbers reached 1 billion in 2004 and in two-year steps it doubled to 2 billion by 2006 and 3 billion by 2008 [18]. Even today, many operators with advanced 3G and 4G networks have retained their GSM networks, mainly to provide fall-back, wide-area coverage. This is a testament to the durability and reliability of GSM technology. The global expansion of GSM networks is illustrated in Figure 1.2, captured as in January 2005 [19].

The huge proliferation of the GSM systems brought about the need for mobile technology to meet the rapidly-growing customer demands. By the turn of this century, the success of the wired internet meant that there was a growing demand to access the internet on the move, i.e. through mobile devices. Technically, this required much faster data rates to be supported by the mobile technology than the nominal GSM data rates of 9.6 kbps (or kilobits per second). GSM however, was designed with a voice communication-centric switching architecture, known as circuit switching. For data communications, a radical shift towards packet-switching systems was needed. (We will cover these aspects in detail in Chapter 5). GPRS or General Packet Radio System was the first successful standardisation move to make this shift. With support for packet switching, it could achieve data rates of up to a theoretical maximum of 172 kbps. In practice, many networks averaged around 100 kbps, which was a significant improvement over GSM. These data rates were capable of supporting basic internet browsing and web e-mail access systems. GPRS standards were still backward compatible with GSM, i.e. GSM voice calls could be still supported by GPRS systems. By the year 2000, the International Telecommunication Union (ITU) was setting the IMT-2000 specifications, an official specification list that true 3G standards should meet [20]. The GPRS data rates were far below these requirements. Thus, GPRS became known as a 2.5G system. The first commercial GPRS system was launched in the year 2000 and some systems are still operational today.

The next significant step in the GSM progression came with the EDGE (Enhanced Data rates for GSM Evolution) technology. This is actually an evolution from GPRS technology, to improve the data rates of packet switched networks. The novelty was a significant increase in spectral efficiency or, in simpler terms, the amount of data that could be carried

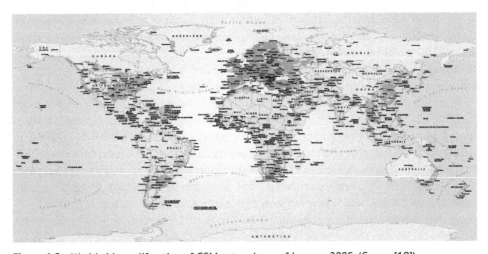

Figure 1.2 Worldwide proliferation of GSM networks as of January 2005. (*Source* [19]).

by a unit of radio spectrum. The process of embedding a data stream onto a radiowave (which is an analogue signal) is called modulation. GSM and GPRS used a simple modulation scheme called GMSK (Gaussian Minimum Shift Keying), where only one data bit per symbol (or radio resource component) could be embedded. Technological improvements in EDGE enabled the more advanced 8-PSK (Phase Shift Keying) modulation scheme to be incorporated. This allowed three data bits per symbol to be carried, which essentially can triple spectrum efficiency. We will cover these technical details in depth in Chapter 5. EDGE systems, in theory, could carry 384 kbps data rates but in practice it was, in most cases, around 200 kbps [21].

Due to the limited bandwidth (200 kHz) of legacy GSM systems used by EDGE, it could never reach the peak data rates stipulated by the IMT-2000 specifications. Thus, it was considered as a 2.5G standard, although some technologists called it 2.9G to differentiate from GPRS. One of the other advantages of EDGE was that the network modifications needed to implement EDGE were also the stepping stones for the all IP/packet-switching based 3G systems. Thus, investments on EDGE infrastructure by the operators were to be future proof for subsequent 3G systems. This kind of evolution in the technology, bringing in massive cost savings, was a key expectation from operators which could make or break a standard. You can read about these intricacies in cellular standards in Chapter 4. The first networks supporting EDGE technology became operational in 2003 and some operators still continue to use this technology.

The other 2G standard which gained traction in the industry was the IS-95 (or Interim Standard 95). It used quite a revolutionary technology (for the mobile industry at that time), known as CDMA (Code Division Multiple Access), which allocates users different codes and allows them to operate in the same frequency and the same time domain. CDMA was by no means a new discovery. It had been in use for decades, especially for military communication systems, where its properties of resilience against jamming and unauthorised detection were very useful. However, the use of it in cellular systems was a novel move, which subsequently paved the way for the 3rd cellular generation (3G) to adapt this. We will again discuss the technical details of this technology in Chapters 5 and 8.

The IS-95 CDMA technology became known as CDMA One. It operated in 1.25 MHz bandwidth and IS-95B standards could achieve 115 kbps data rates. The IS-95 systems were mostly deployed in North America and in some East Asian markets. By 2001, CDMA One and subsequent technologies accounted for roughly 13% of new mobile subscription numbers globally, concentrated mostly in the US [22]. CDMA quickly proved to be a robust technology and led to it being investigated as the prime candidate for 3G systems, which we cover in the next few paragraphs.

The need and demand for mobile data applications such as web browsing, e-mail and file transfer on mobile devices became clear even before the heyday of the 2G systems. As noted before, 2G/GSM was designed as a circuit switching, narrowband system to support voice communications. Although data rates were improved with GPRS and EDGE evolutions, the need for a true high-speed, packet-switching mobile data network was never really satisfied. Backed by extensive research, the cellular industry came to the realisation that much wider bandwidth systems (compared to 2G) were needed to support the high data-rate applications. The CDMA technology offered the possibility to significantly increase user data rates by using wider bands. Technically, 3G standards were required to meet the IMT-2000 set of requirements set by the ITU. The basic data rate requirements set by IMT 2000 are 144 kbps for high mobility, 384 kbps for low mobility and 2 Mbps for stationary environments [19].

Different flavours of CDMA technology were adapted by different 3G standards. By far the most popular 3G standard, known as UMTS (Universal Mobile Telecommunication System) employed the W-CDMA (or Wideband CDMA) technology with a 5 MHz system bandwidth. UMTS was the first mobile standard to be developed with truly global collaboration. A new body called the 3GPP (3rd Generation Partnership Project) was initiated in December 1998 by the European Telecommunications Standards Institute (ETSI), with participation of other national and regional standards organisations of Asia and North America [23]. The 3GPP oversaw development of UMTS as a natural successor to the GSM standard, with the view of providing a truly global standard.

UMTS was intended to be a technology operable in a globally harmonised spectrum. Specific bands in the 1800–2200 MHz range were originally year-marked as a global band for deploying UMTS networks. Having a globally harmonised spectrum was intended to bring down network equipment and handset costs, helped by economies of scale in mass scale manufacturing for the global market. However, the 3G-spectrum costs in many developed countries spiralled out of control, due to anticipated demand for 3G data services. In the 3G-spectrum auction in UK, for example, bidding by incumbent and prospective operators pushed up the total auction price to almost £22.5 billion, which was about four times the expected value the auction would raise. In Germany, the operators dished out £28.5 billion for 12 frequency blocks. This excessive pricing of 3G spectrum had a hugely negative impact on the European cellular industry, with many analysts suggesting that European 3G deployments were set back by a number of years as a result [24].

The first commercial 3G network was launched by NTT DoCoMo in Japan, in October 2001, known as FOMA (Freedom of Mobile Multimedia Access). In Europe, Telenor of Norway launched their 3G network in December 2001. The UMTS-based 3G networks followed an evolution path of HSDPA (High Speed Downlink Packet Access), HSPA (High Speed Packet Access) and HSPA+. The HSPA+ standards, defined as 3GPP Release 7 in the year 2007, could achieve theoretical data speeds of 168 Mbps in the downlink (from Base station to Mobile) and 22 Mbps (from Mobile to Base station) in the uplink. The 3G services were not taken up by consumers in the expected market penetration rates, so this journey through the standard releases happened at a far slower rate in the commercial networks.

A competing 3G standard, known as CDMA 2000, was developed by the 3GPP2 partnership [24], as a continuation of the IS-95 standards stream. The basic standard, known as CDMA 2000 1x, utilised CDMA technology with a bandwidth of 1.25 MHz, being an evolution of the CDMA One technology used for IS-95. This technology achieved 153 kbps data rates in both downlink and uplink, and was recognised as an IMT 2000 technology. This strand of 3G technologies followed a path of CDMA 2000 1x EV-DO, EV-DO Rev B and EV-DO Rev. C. The EV-DO originally stood for Evolution Data Only, where voice calls could be routed through baseline CDMA 2000 or even IS-95 technology. Later EV-DO became known as Evolution Data Optimised. The EV-DO Rev B technology supported Multi-carrier CDMA, meaning more spectrum in the guise of 1.25 MHz chunks could be aggregated to support higher data rates. With a 20 MHz aggregated bandwidth, and with favourable radio channel and mobility conditions, EV-DO Rev B could achieve up to 73.5 Mbps data rates.

As with the 3GPP set of standards, all CDMA-2000 based standards were backward compatible, enabling co-existence. The EV-DO Rev B standard could be enforced with a software upgrade to the existing CDMA 2000 hardware. The EV-DO Rev C development later morphed into UMB (Ultra Mobile Broadband) an early competitor for the 4th Generation (4G) standards, which we will detail later. The first CDMA 2000-based 3G network was deployed in South Korea by SK Telecom in October 2000.

A third 3G standard, known as TD-SCDMA (Time Division – Synchronous CDMA) was developed mainly in China [25]. Whilst using CDMA technology, it also employed different time slots for uplink and downlink transmissions, in the same carrier frequency. The duration of these time slots can be varied, making TDD more suited to support asymmetric traffic like the mobile internet. The term 'synchronous' refers to having all active mobiles in a cell time aligned for transmissions. This makes it much easier to localise the mobiles to enable techniques like beamforming and also simplifies the use of multi-user detection (MUD) schemes. We will touch upon these schemes when we introduce advanced technologies in Chapter 8. The chip rate (or code rate) used in CDMA spreading codes is 1.28 Mcps, which is far lower than for W-CDMA.

The TD-SCDMA standard was developed in the late 1990s and was approved by the ITU-R group in May 2000. In March 2001, China joined the 3GPP group and TD-SCDMA was approved as a 3G standard, becoming part of the UTRA – TDD specifications. The Chinese government announced in 2006 that TD-SCDMA would be China's official standard for 3G communications. The largest Chinese operator, China Mobile, started building a trial network of TD-SCDMA in 2006, but it was not until late 2008 that these trials were completed. The launch of other 3G standards in China was delayed until TD-SCDMA was ready for commercial launch. In 2009, the Chinese government took the unusual step of assigning TD-SCDMA, W-CDMA and CDMA 2000 1xEV-DO licenses separately to the three largest carriers in China. Following on from their trials, China Mobile deployed the commercial TD-SCDMA network. The standard did not find much success elsewhere. There are only a few examples of TD-SCDMA networks being deployed in developing countries.

At this point, it is perhaps worth taking a brief look at the reasons why the mobile industry generates and supports multiple standards within the same generation. There is clearly a regional emphasis on the generation of multiple standards. Traditionally, the US and Europe have favoured different cellular technologies, continuing to develop different standard threads across mobile generations. The standardisation process should not be viewed as purely a technical activity. There is also a significant commercial and political element to it. The commercial interests of the significant industry players in a specific region (mainly operators and equipment vendors) weigh heavily on the decisions for promoting different standards. It should be noted that the business model of mobile vendors depends on their ability to innovate, to protect their innovations through IPR and then to transfer these into a widely used standard. We will discuss this model in detail in Chapter 4. There are also political initiatives to promote the development of standards to secure industry leadership in certain technologies and to reduce dependency on another region's technology. The TD-SCDMA development in China, for example, is viewed as an attempt to give local industries a head start in competing with the then global technology giants. Technically, having different standards can help to evaluate the performance of different technologies in actual deployments. For example, the adaptation of CDMA-based technologies for the 3rd Generation was massively supported by the knowhow gained through 2G CDMA One deployments.

The advent of the iPhone in June 2007 marked a watershed moment in the mobile tele-communication industry. Not only did it kick-off the Smartphone era, it also revitalised mobile operators' revenue streams. Up until then, mobile data consumption was a rather cumbersome experience for the user and it was nowhere near anticipated levels. The iPhone provided a seamless web interface to access the internet, making it as slick an experience for the user as sending a text. Apple iOS, the iPhone operating system, could navigate seamlessly between phone, text and numerous new data applications through the iPhone touch screen. It also revolutionised the mobile applications domain by introducing the Apple App store, where third-party developers could market their certified Apps to Apple device users. This Apps market soon grew into a substantial mini-economy and by mid-2012 there were more than 650 000 Apps available through Apple, who paid out more than $5 billion to the developers [26].

Soon after the launch of the iPhone, Google launched their own operating system for Smartphones. Named Android, it was commercially released for mobile devices in October 2008. It soon became the default choice for smartphone manufacturers competing with the Apple iPhone. By 2014, the Android-based Smartphones were dominating the market with around 85% of the market share [27]. The Android-based Apps market also expanded exponentially. Some estimates put the number of Android Apps available to be over 1.4 million by mid-2014.

Thus, the advent of Smartphones heralded an era of exponential growth in mobile data consumption. The following graph in Figure 1.3 shows how the mobile data traffic grew 350 fold from 2005 to 2010 [28]. At the beginning, mobile operators were delighted with this growth, as the extra capacity they provisioned with the 3.5G and 3.9G systems could be monetised. Many operators started providing packages with unlimited data bundles to spur demand. But soon the consumers' appetite for data was putting these networks under considerable strain and unlimited data bundles mostly vanished. We will discuss the advent of 4G systems under these circumstances, where mobile data was becoming dominant over mobile voice for the first time.

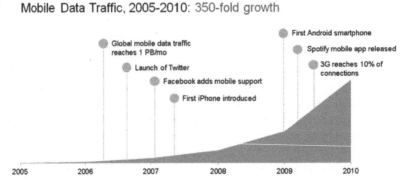

Figure 1.3 Mobile data traffic growth from 2005 to 2010. (*Source* [28]).

The development of 4G standards gathered pace in the mid-2000s, with the aims of providing much higher mobile broadband speeds and all IP (packet) based switching. In the beginning, there were three main strands of potential 4G standards. The 3GPP group were beginning the development of LTE (Long-Term Evolution) standards. One of the aims of LTE standardisation was to provide a smooth transition from 3G UMTS-based networks to LTE, hence it was named as a long-term evolution. The 3GPP2 group initiated work on UMB with a similar aim for transition from EV-DO systems. The IEEE entered the mobile standardisation foray with the development of the 802.16 series, better known as the WiMAX (Worldwide Inter-operability for Microwave Access) standards.

Soon, 3GPP2 decided to discontinue development of UMB, leaving LTE and WiMAX as the competing 4G standards. Both these standards opted for a new radio access technology, known as OFDMA (Orthogonal Frequency Division Multiple Access). Unlike in W-CDMA, the data stream would be split and embedded into a multitude of (mutually orthogonal) frequencies, which together would make up a wideband system. We will explain the technical details of OFDMA in Chapters 5 and 8. Also, both LTE and WiMAX opted for a flatter architecture than 3G, where the base stations (eNodeB as known in LTE) would connect directly to elements of the core network. This would be an all IP network where even the voice would be converted to voice over IP packets (VoIP or VoLTE) and transmitted, moving away from the circuit-switched components of the previous generations.

The WiMAX standardisation had a head-start over LTE, with the 802.16d fixed wireless system being released in 2004. This was intended to provide wireless broadband access to homes and offices, competing with cable and DSL service provisions. The mobile WiMAX standard 802.16e was released one year later in 2005. The Wireless Broadband (WiBro) standard developed by the South Korean Telecom Industry had many similarities to WiMAX and in 2005 WiBro was included in the 802.16e family of standards. The first LTE standard, LTE Release 8, came out in 2008. However, the performance of all these early systems did not reach 4G thresholds set by the ITU-R, through their IMT-Advanced specifications [29]. Under these IMT-Advanced thresholds, the peak downlink data rates for low-mobility users should reach 1 Gbps and, for high-mobility users, should reach 100 Mbps. In this context, the real 4G standards were the WiMAX 2.0 (802.16m) released in 2011 and LTE-Advanced (also known as LTE release 10), released in 2012.

Despite being first to the 4G market, backed by huge investments from industry promoters and being technically sound, WiMAX did not succeed in carving out a market share [30]. The first big mobile WiMAX deployments came out in 2008–09, amidst real optimism for the technology. But by 2011, all of the big WiMAX operators (including Sprint in the US) had announced the transition plans of their networks to LTE. We will analyse the reasons for WiMAX failure in more detail in Chapter 4, but a few salient points can be noted here. First and foremost, WiMAX did not have the support of wider established mobile operators, whilst LTE was their choice. LTE offered a long-term transition path for their 3G UMTS networks, rather than a complete revamp of the networks (with the associated huge cost) needed for WiMAX. WiMAX was mostly supported by new entrants, who saw an opportunity to establish a market presence under the Royalty free IPR policy WiMAX was promoting. Many observers point out that WiMAX did not attempt to build upon the relative success of 802.16d fixed wireless system in the mobile versions [30]. The 802.16e version was not backward compatible with 802.16d, preventing the fixed operators an opportunity to migrate to 802.16e whilst safeguarding their investments. Also, the finalising of the mobile WiMAX standard took far too long, allowing LTE to catch-up and, importantly, fit into the future plans of major operators. From the device perspective, WiMAX was supported by only very few vendors and none of the emerging Smartphones ventured into supporting WiMAX. The mobile WiMAX (being a Time Division – TD version) was perceived to have some advantage in emerging markets, with its lower cost of deployment and ability to utilise narrower spectrum. However, 3GPP also came up with a TD-LTE version, which was preferred to WiMAX in these markets. By 2014, even the WiMAX forum was acknowledging dominance of LTE, announcing plans to make WiMAX Advanced compatible with TD-LTE [31].

So, by 2015, for the first time in the history of mobile communications, the mobile market had only one single standard as the latest flag bearer, LTE. As of July 2015, there were 422 LTE networks deployed in 143 countries, providing services to over 755 million LTE subscribers. Over the next 4 years, LTE continued its impressive growth, becoming the dominant mobile technology accounting for more than half of all global mobile subscriptions by 2019 [32]. The peak data rates that can be achieved on the ground in some of the advanced LTE networks reaches 300 Mbps in the downlink. Evolution of the LTE standards continued at a rapid pace, with LTE Release 10 to Release 14 finalised from 2012 to 2016. The LTE Release 14 was the last of the 3GPP 4G standards and the focus shifted to 5G from Release 15. The later LTE releases have enabled carrier aggregation, with the aim of amalgamating multiple carrier frequencies spread over a wider range, thus providing more frequency resources to carry more data. In some small-cell (SC) deployments (a topic we will cover in Chapter 10) the targeted data rates with carrier aggregation and advanced multi-antenna configurations now extend up to 838Mbps [33]. These data rates were simply unimaginable and may have been deemed as totally excessive in the era before the Smartphones. Although it will have many more years as the dominant cellular technology, 4G LTE will be remembered as the standard that brought the mobile broadband concept to life, after bit of a false start with 3G UMTS.

Around 2012, researchers were already conceptualising how 5G should look, in early research projects like METIS [34]. 5G would encompass a much grander vision than any of the previous generations, which were only focussed on connecting the mobile consumer. A key aim for 5G would be to support different vertical industries, from Smart cities to vehicular communications to Industry 4.0, in addition to the mobile consumer. We will take a detailed look at these vertical areas in Chapter 12 and at 5G in general in Chapter 5. As a brief overview here, it would be suffice to say that the technical capabilities of 5G would span three dimensions. The traditional expansion of data rates would be captured in the eMBB (enhanced Mobile Broadband), where stationary data rates of 20 Gbps and mobile rates exceeding 1 Gbps are targeted. Then, for applications requiring URLLC (Ultra Reliable and Low Latency Communications) reliabilities over 5 '9's (99.999%) and latencies lower than 1 ms are targeted. Then, to facilitate massive deployment of sensor networks, mMTC (massive Machine Type Communication) with up to 1 million sensor devices per square kilometre (1 Mn/km^2) are targeted. The IMT-2020 specifications [35] are evolving to define exactly what requirements should be met by standards to be classified as 5G. An early conceptualisation of these IMT 2020 requirements in the above three and associated dimensions is captured in Figure 1.4, the spider diagram, produced by the ITU [35].

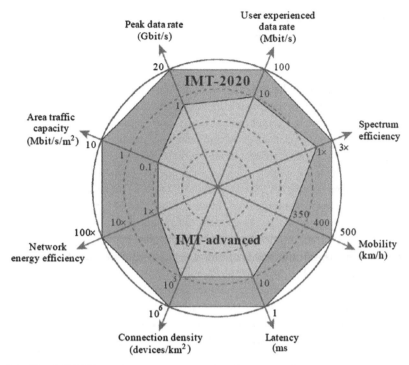

Figure 1.4 The IMT-2020 requirements to be met by 5G systems. (*Source* [35]).

3GPP, now as the sole entity responsible for developing cellular 5G standards, embarked on developing a channel model capturing the frequencies from 6 GHz to 100 GHz (the millimetre wave spectrum [mmWave]) as an enabler for 5G standardisation in 2016. Work on the first 5G version, termed as 5G-NR (5G New Radio) Release 15, started in 2017 and completed in 2018 [36]. This first 5G Release is seen as more of an evolution of LTE, aimed primarily at providing the eMBB capability. The same radio access method as LTE, OFDM, was utilised but now with wider bandwidths (including support for higher mmWave frequencies) and different frequency finger (known as sub-carrier) spacings. At the time of writing this chapter (late 2020), work in Release 16 has come to an end, with the standard frozen in July 2020. This is viewed as phase 2 of 5G-NR standards functionality and both Releases 15 and 16 will be included in the 3GPP submission to ITU-R, for certification as a 5G compliant standard.

The first 5G commercial networks came up in South Korea in April 2019, closely followed by a smaller scale launch in the US [37]. Launched separately by three South Korean mobile operators KT, SK Telecom and LG Uplus, these networks are expected to provide coverage accessible to more than half of the country's population of 52 million at the launch. Soon afterwards, Verizon in the US launched their 5G network in two cities, with a much smaller initial coverage footprint. Within the latter half of 2019, different operators from a number countries (including the UK) joined the 5G network operators' list but most of these have very limited initial coverage, confined to densely populated urban areas. We will see a wider expansion of 5G networks in 2020–21, when more network equipment and 5G enabled user devices will enter the market.

1.3 The Journey of Mobile Communications as Seen from User and Operator Perspectives

Thus, the story of the evolution of mobile communications is a story of continuous technological advancement through the generations 1G to 5G. Technological advancements have enabled a steady stream of new applications to come to the fore in this journey. Looking back from the user-experience perspective, the user devices (mobile phones) tell their own story. As depicted in Figure 1.5, each generation can be characterised by the features of the typical user devices. In 1G, the mobile phones were heavy and bulky with often unreliable connections. The phones were solely used for voice communications. In the 2G digital era, the phones became much smaller, lighter and more battery efficient. In addition to voice, short messaging (or texting) emerged as a killer application. In the 3G era of mobile data provision, there was initially an indifferent take-up of data services. The use of Blackberry type devices with web-mail applications gradually became popular, with their mini keypads to type in e-mails. But the real breakthrough came with the smartphones and their numerous apps. Thus, in the latter part of 3G, and mostly in 4G, the mobile phone transformed into a high-end device, capable of supporting a plethora of third-party applications. Throughout the mobile broadband age of the 4G, the mobile phones continued to grow in capabilities of primarily processing power, memory, screen resolution and camera specification. Displays were becoming sharper and screen sizes getting bigger to support continuing growth in mobile video consumption. As discussed, 5G will spread beyond the provision consumer mobile broadband to different verticals. The plethora of envisaged 5G devices is a testimony to this diversity of applications. These devices include Smartphones, AR/VR devices, IoT sensors, Automated/intelligent vehicles, industrial machines/robots and many more. Thus, 5G consumers will vary from single users to large industry players and service providers, creating a much more complex ecosystem than previous generations. An illustration of the representative devices in each mobile generation is depicted in Figure1.5, from [38].

When looking from the operators' business perspective, 1G was a niche market where the phones and line rentals came at high price points. Technological limitations and costs in the network connection and devices meant that the service was only appealing to a

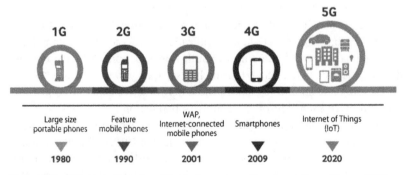

Figure 1.5 The evolution of mobile devices across the generations. (*Source* [38]).

limited customer base, mostly high-end business users. In 2G, the digital technology and ability to mass produce network equipment and devices of a single dominant standard (GSM) pushed prices down. This, in turn, made cellular communications more popular and attracted millions of customers. At the height of 2G success, device prices were pushed down almost to the level of a commodity. Voice call minutes and number of texts were stipulated in the price packages and voice was the main revenue generator for operators. With consumer numbers showing exponential growth in many markets, the operators' revenue growth relied heavily on adding new customers year on year. The market entered 3G with a lot of hype and expectation about mobile data but the lack of nimble mobile data applications meant take up was much slower. Having spent huge amounts on 3G spectrum, and with the subscriber numbers reaching saturation, many operators in developed markets found it difficult to record any business growth. Early 3G phones like the Blackberry did capture traction with business users, but it was the advent of Smartphones that really stimulated growth again. Smartphones gave a unique experience to the user with numerous applications of mobile data and users were willing to pay a premium price for these devices. The sheer magnitude of mobile data demand showed the limitations with 3G technology and paved the way for a quicker transition to 4G. For the mobile operators, who for so long relied on voice as their main source of income and started adapting to the new age of mobile data, new challenges emerged as over-the-top (OTT) Voice over IP (VoIP) packet applications enabled virtually free voice calls and text messages through data networks. OTT applications like Apple Facetime, WhatsApp, Viber, Skype, Wechat in China and Kakao Talk in South Korea made significant dents in mobile operator revenues [38] and, in turn, hastened transition to mobile data oriented business models in 4G.

In many countries now, we are at a stage where 4G networks have really matured and technological advances have driven down the cost per data bit significantly from 3G levels. Mobile data has established itself as the prime income source for operators in many advanced markets. Users now expect GB/month data packages, with the majority of the data consumed for mobile video. However, many advanced markets are seeing consistent drops in ARPU (Average Revenue Per User). An example from European markets is shown below in Figure 1.6 [39]. This is in the context that mobile data is not the first choice of data provision to Smartphones in many indoor environments, where Wi-Fi is predominantly used. Some estimates put that by 2022, only 20% of all IP traffic will be carried by mobile broadband [40].

We are entering the 5G era with many challenges, including falling ARPU facing the mobile operators, as we will detail in Chapter 3. 5G will be an entirely new business proposition for operators, who will have to build collaborative business models with many other vertical industries. Mobile data of many facets and types (eMBB, URLLC and mMTC) will feature in these 5G networks, which will feature partnerships with many stakeholders in both private and public domains. The technical and business aspects of some of these vertical partnerships will be covered in Chapter 12.

The mobile industry has shown a remarkable ability to sustain technological advancement and business vitality throughout its history, from 1G to 4G. It has been highly resilient in challenging times, particularly during the wobbly start to 3G and the shake-up of its voice-oriented business models in the early part of 4G. As 5G dawns, there is a lot of

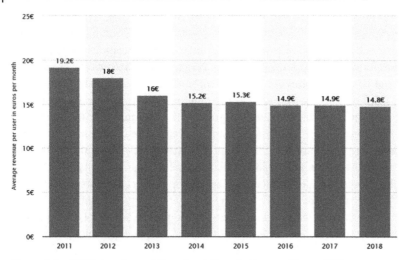

Figure 1.6 ARPU drop for mobile subscriptions in Europe. (*Source* [39]).

excitement in the industry about the wider scope of opportunities in a number of vertical industry areas. There will undoubtedly be a lot of challenges in realising the full potential of 5G, which will require new business partnerships with many of the established vertical industries. However, the mobile industry is well placed, with a single set of 5G-NR standards from 3GPP, progressive spectrum allocations in many countries and active industry bodies in the key verticals working to bridge the gap between ICT and the relevant industry domains. There is a lot to look forward to in this enduring journey of mobile communications.

References

[1] *GSMA web report on '10 years of mobile for development'.* Available at: https://www.gsma.com/mobilefordevelopment/10yearsofm4d

[2] *Article in the economist.* https://www.economist.com/briefing/2012/11/03/the-path-through-the-fields

[3] *GSMA article on Sehat Kahani.* Available at https://www.gsma.com/mobilefordevelopment/wp-content/uploads/2019/05/Sehat-Kahani-Employing-mobile-technology-to-connect-women-doctors-to-underserved-communities-in-Pakistan.pdf

[4] *GSMA article under digital identity.* Available at: https://www.gsma.com/mobilefordevelopment/blog/mapping-access-birth-registration-updates-tanzania

[5] https://www.gsma.com/subsaharanafrica/eneza-education-making-education-reality-working-hand-hand-mobile-operators

[6] Boston consultancy group report. (January 2015). *The Mobile Revolution: How Mobile Technologies Drive a Trillion-Dollar Impact.* Available at: https://www.bcg.com/publications/2015/telecommunications-technology-industries-the-mobile-revolution.aspx

[7] GSMA Intelligence report. *The Mobile Economy 2019*. Available at: https://www.gsmaintelligence.com/research/?file=b9a6e6202ee1d5f787cfebb95d3639c5&download

[8] *GSMA mobile money state of the industry report 2018*. https://www.gsma.com/r/state-of-the-industry-report

[9] GSMA study report. *The Rise of eZ Cash*. Available at: https://www.gsma.com/mobilefordevelopment/wp-content/uploads/2013/07/Enabling-Mobile-Money-Policies-in-Sri-Lanka-GSMA-MMU-Case-Study-July2013.pdf

[10] KPMG insights report. Jan. 2018. *The Rise of the Challenger Banks*. Available at: https://assets.kpmg/content/dam/kpmg/xx/pdf/2018/02/kpmg-rise-of-challenger-banks.pdf

[11] Cramer, J. and Krueger, A.B. (2016). Disruptive change in the taxi business: The case of Uber. *Published in the American Economic Review* 106 (5): May.

[12] *The success of Airbnb app*. Online article, Aug. 2015, available at: https://appsamurai.com/mobile-app-success-story-how-airbnb-did-it

[13] *The original nature journal notes*. On 19.12.1901 available at: http://www.nature.com/physics/looking-back/marconi/marconi.pdf

[14] *Lecture notes on inception of mobile communications*. Sep. 2015. Available at: http://www.cntr.salford.ac.uk/comms/etacs_mobiles.php

[15] Kwok and Lau. (2007). *Wireless Internet and Mobile Computing: Interoperability and Performance*. Chapter 5. Hobeken, NJ, USA: Wiley-IEEE Press.

[16] ITU report. (1999). *World Telecommunication Development Report – Mobile Cellular Executive Summary*. https://www.itu.int/ITU-D/ict/publications/wtdr_99/material/wtdr99s.pdf

[17] Badenoch, A. and Fickers, A. (2010). *Materializing Europe: Transnational Infrastructures and the Project of Europe*. London, UK: Palgrave and Macmillan.

[18] Online article. *Brief History of the GSM and the GSMA*. Available at: https://www.gsma.com/aboutus/history

[19] *Europa Technologies Resource*. https://www.europa.uk.com

[20] ITU online handbook. (June 2003). *Deployment of IMT-2000 Systems*. Available at: https://www.itu.int/dms_pub/itu-r/opb/hdb/R-HDB-60-2003-PDF-E.pdf

[21] Holma, H. and Toskala, A. (2006). *HSDPA/HSUPA for UMTS: High Speed Radio Access for Mobile Communications*. Hoboken, NJ: Wiley.

[22] Melero, J. (2003). 3G Technology strategy and evolution paths. In: *GSM, GPRS and EDGE Performance: Evolution Towards 3G/UMTS*, 2e. (ed. T. Halonen, J. Romero and J. Melero), 543–553. Hoboken, NJ: Wiley.

[23] 3GPP (2021). *The inception of 3GPP*.https://www.3gpp.org/about-3gpp (accessed 14 June 2021).

[24] Camarillo, G. and García-Martín, M.-A. (2011). The history of the IMS standardization. In: *The 3G IP Multimedia Subsystem (IMS): Merging the Internet and the Cellular Worlds* (ed. G. Caramillo and M.-A. García-Martín), 9–23. Hoboken, NJ: Wiley.

[25] Korhonen, J. (2003). History of mobile cellular systems. In: *Introduction to 3G Mobile Communications*, 2e (ed. J. Korhonen), 1–24 . Norwood, MA: Artech House Publishers.

[26] Tech Crunch (2013). Apple's app store hits 50 billion downloads. https://techcrunch.com/2013/06/10/apples-app-store-hits-50-billion-downloads-paid-out-10-billion-to-developers (accessed 14 June 2021).

[27] IP Watchdog (2014). A breif history of Google's android operating system http://www.ipwatchdog.com/2014/11/26/a-brief-history-of-googles-android-operating-system/id=52285 (accessed 14 June 2021).

[28] *Cisco (2016). Major mobile milestones – The last 15 years, and the next five.* https://blogs.cisco.com/sp/mobile-vni-major-mobile-milestones-the-last-15-years-and-the-next-five (accessed 14 June 2021).

[29] ITU (2015). ITU global standard for international mobile telecommunications. https://www.itu.int/en/ITU-R/study-groups/rsg5/rwp5d/imt-adv/Pages/default.aspx (accessed 14 June 2021).

[30] *Web article* https://wirelesstelecom.wordpress.com/2012/05/29/the-rise-and-fall-of-wimax-2/.

[31] WiMAX Forum (2014). WiMAX Forum and Global TD-LTE initiative jointly annouce strategic collaboration. http://wimaxforum.org/Page/News/PR/20140225_WiMAX_Forum_and_Global_TD-LTE_Initiative_Jointly_Announce_Strategic_Collaboration (accessed 14 June 2021).

[32] GSMA (2020). The mobile economy. https://www.gsma.com/r/mobileeconomy (accessed 14 June 2021).

[33] Nakamura, T., Nagata, S., Benjebbour, A. et.al. (2013). Trends in small cell enhancements in LTE advanced. *IEEE Comms Magazine* 51(2): 98–105.

[34] METIS EU project overview. https://www.ericsson.com/en/blog/2015/5/the-metis-5g-system-concept

[35] ITU-R (2015). *IMT Vision – Framework and Overall Objectives of the Future Development of IMT for 2020 and Beyond.* https://www.itu.int/dms_pubrec/itu-r/rec/m/R-REC-M.2083-0-201509-I!!PDF-E.pdf (accessed 14 June 2021).

[36] 3GPP overview of release 15. *Web Page.*

[37] BBC news article. 05/04/2019, *5G:World's First Commercial Services Promise 'Great Leap,* available at:

[38] PHD China (2019). *Voice Led Marketing in China.* https://www.phdmedia.com/china/phd-china-whitepaper-voice-led-marketing-in-china (accessed 14 June 2021).

[39] Statista (2021). Average revenue per user (ARPU) of mobile broadband in Europe from 2011 to 2018 (in euros per month). https://www.statista.com/statistics/691710/mobile-voice-arpu-evolution-in-europe (accessed 14 June 2021).

[40] Cisco (2019). Cisco Annual Internet Report (2018–2023) White Paper. https://www.cisco.com/c/en/us/solutions/collateral/service-provider/visual-networking-index-vni/white-paper-c11-738429.html (accessed 14 June 2021).

2

The Mobile Telecoms Ecosystem

2.1 Introduction

For the first time, 2014 saw more mobile communications devices deployed than the worlds' population which was around 7 billion [1]. This was a significant landmark in that it showed there are numerous devices that are no longer for personal communications use, but for applications reporting sensor data and other command and control applications, often referred to as Machine-to-Machine (M2M) communications as part of the Internet of Things (IoT). M2M and IoT are emerging as a radical extension to the internet. It will enable the internet to encompass connectivity with both people and 'things' [2].

The purpose of this chapter is not, however, to discuss M2M communications, but to mention it because of its growing significance in, and impact on, the mobile communications ecosystem. Substantial growth of both global population and the mobile communications market is predicted for the foreseeable future [3]. Furthermore, modern smart devices have more computing power than the combined computing power of NASAs 1969 space missions, and in a compact pocket-sized device that includes a battery able to provide power for around one day [4]. This technical feat is testament to intensive and detailed engineering, spanning several disciplines that include: electronics, material science, mechanics and chemistry. However, this has only been made possible by the favourable economic backdrop.

Incredible economic and technical growth, the success of smart devices and related infrastructure has accelerated over the last three and a half decades, but it has only been possible because of an ecosystem being formed to sustain the technology, legal, economic, political and social factors of mobile telecommunications. This has, of course, been driven by supply and demand at its core.

Although the past is not always an indication of the future, at the time of writing, the future continues to look bright for mobile telecommunications. Demand is perceived to be higher than ever, with new services emerging almost every day, including the aforementioned opportunities relating to M2M and IoT.

Commentators perceive a change in the dynamics of this ecosystem, stimulating a fundamental change in the way business is conducted through the arrival of new entities such

The Technology and Business of Mobile Communications: An Introduction, First Edition.
Mythri Hunukumbure, Justin P. Coon, Ben Allen, and Tony Vernon.
© 2022 John Wiley & Sons Ltd. Published 2022 by John Wiley & Sons Ltd.

as big data, M2M, and IoT service providers, the development and integration of 'vertical' markets such as health, transportation, energy and entertainment [5].

This chapter explains the major factors that make up the mobile telecommunications ecosystem and how they interact. It may be thought of as a case study of a technology-led revolution featuring radical changes of business and individual behaviours. The major players and components of the emerging 5G communications network ecosystem are introduced and illustrate an emerging case study on a global scale.

2.2 Telecommunications Ecosystem

The major components of the mobile telecommunications ecosystem are shown in Figure 2.1, where the dynamics between components are determined by 'market forces', i.e. the balance between supply and demand. The ordering of the components of Figure 2.1 is currently arbitrary as the inter-relationship is multi-faceted and complex. Supply and demand of the radio spectrum, digital services, people, technology and energy are key to the dynamics; and spurned by competition amongst network operators and equipment suppliers.

The most basic commodity of the mobile telecoms ecosystem is the radio spectrum (Figure 2.2), where demand currently outstrips supply and thus requires regulation and

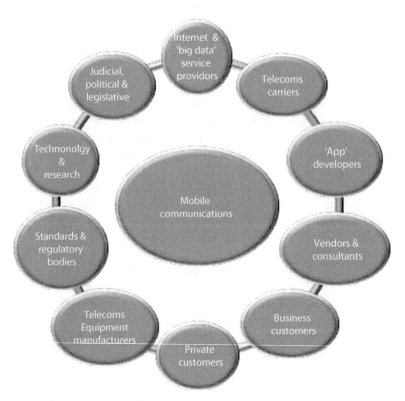

Figure 2.1 Mobile telecommunications ecosystem.

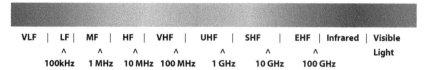

Figure 2.2   Radio Spectrum.

legislation to temper behaviour. It also drives demand for new means of improving efficient and effective use of the radio spectrum, which includes possible changes to the regulatory framework or by technology.

Creating a climate with the right amount of competition at all levels is essential. Not enough competition may result in low service quality, low investment and low levels of innovation. Too much competition may result in market instability resulting from many small market players. Mobility of users and devices brings about the need for service and business compatibility on an international scale, examples of which are spectrum allocations, communications protocols and billing; which require development of international standards.

It is tricky to know the exact starting point for the ecosystem but, without technology, there would be nothing, regardless of the size of market opportunity. So, let's start with **technology & research**. This is where the building blocks are developed that enable mobile telecommunications system to be deployed and operated. The technology element is where the circuits, sub-systems and software are conceived, designed, built and tested. This includes deciding on which approach to adopt, and integration of the different parts. The research element entails determining novel ways of improving the system, including aspects such as antennas, signals and electronic circuits.

Technological breakthrough has an impact on **standards and regulatory bodies** as it may require adjustments to the regulatory framework, such as: tightening, relaxing or changing regulatory requirements. Standards ensure inter-operability between vendor products and countries. Compliance to standards is critical, as a call or data connection between two devices can traverse radio and core networks (CN) run by different operators in different countries.

Regulation, standards and technology impact **equipment manufacturers**. They may need to adapt their product portfolio to accommodate new findings, and to ensure their products adhere to regulatory and standards requirements. Their portfolio may include: smart devices, base-stations, computer centres or any aspect of the mobile telecommunications system.

Private customers are consumers who typically purchase and use mobile services for personal use. They select from the product portfolio on offer. **Business customers** typically purchase bulk services and devices to satisfy the needs of a workforce. They may be able to negotiate preferential terms due to large quantities being requested and may require features to be enabled or disabled. Until the late stages of 4G, both private and business customers consumed voice and data services through the use of (Smart) phones. However, 5G will bring in a sea change, where different vertical industries (like automotive, health and Smart city) will become business customers demanding very specific key performance indicators (KPIs) from the mobile networks to increase the productivity of their sectors. Some aspects of these new business to business (B2B) relationships are covered in Section 2.11. This 'revolution' in the key vertical industries is covered in more detail in Section 12.2.

Although technology and research formed the start of the cycle described here, there is a far more complex interaction between elements, where a change to one impacts a change to several others.

2.3 Regulation and Spectrum

A major aspect of regulation in the context of mobile telecoms is the regulation of the radio spectrum [6]. This topic is covered in detail in Chapter 7 and here we will only provide a brief overview, to show how it fits with the overall ecosystem.

The radio spectrum is the range of frequencies that can support radio transmissions, and ranges from a few kilo-hertz (kHz) to around 200 Giga-hertz (GHz), as indicated in Figure 2.2. An alternative way of quantifying radio spectrum is by wavelength, which is the length one single cycle of the signal stretches over. The longer the wavelength, the lower the frequency. Hence, the radio spectrum spans wavelengths from as long as several kilometres down to wavelengths shorter than one millimetre.

The size of antennas is directly linked to the wavelength, with much larger antennas required for longer wavelengths (low frequencies), and tiny antennas required for short wavelengths (high frequencies) [7]. Frequency (f) and wavelength (λ) are related by the equation shown below, and where c is the speed of light, 3×10^8 m/s:

$$f = \frac{c}{\lambda} \quad \text{Hz} \tag{2.1}$$

Marconi is often credited with turning radio equipment from a laboratory experiment into a deployable system in around 1900 [8]. At that time, radio transmissions used Morse code to communicate at frequencies of around 850 kHz. Equipment was rare and transmissions infrequent. This, coupled with:

- transmissions requiring only a few tens of Hertz of the radio spectrum;
- technological limits at that time restricting transmission power;
- sub-optimal antennas curtailing the energy being launched;

meant that the chance of causing or experiencing interference was almost unheard of.

As more equipment was deployed and wider band voice communications became possible, organisation of the radio spectrum was required to ensure users would continue to communicate with minimum trouble and interference to others. This led to spectrum users being assigned certain operating frequencies, with restrictions on transmission power and antenna locations. As technology advanced further, additional restrictions were applied, such as the need for transmissions to fit within a certain spectrum band without causing undue interference to other spectrum users.

To monitor these demands required advanced measurement equipment such as spectrum analysers, thus stimulating innovation in radio-frequency test equipment [9]. The means of having pre-assigned operating frequencies helped to ensure services experienced minimum levels of interference and required transmitters and receivers to be equipped with filters that could be tuned to the required frequency. The spectrum could be thought of as being 'stove piped'.

More recently, the demand for radio spectrum has become so intense that operating with some interference from other users, albeit at low levels, is the norm. The new objective is, therefore, to develop a means of managing interference and operating in spite of this.

Cognitive radio technology has emerged [10], which can auto-configure the frequency and signal to operate with minimum interference according to the prevailing spectrum profile. This has changed the way frequencies are assigned for some parts of the radio spectrum and has stimulated new transmitter and receiver technologies as well as operating protocols. Furthermore, interference cancellation techniques [11] are emerging which actively cancel self- and third-party interference. This increases radio-spectrum utilisation as it allows transmissions to have closer frequency allocations, and even to co-exist on the same frequency. The new development of the uplink and downlink of a radio transmission co-existing on the same frequency is termed 'Full Duplex'. One implementation of cognitive radio makes use of a database of spectrum occupancy over a geographic area. This enables devices to interrogate the database to help them select a suitable frequency.

The Radio Equipment Directive (RED) is a European Union initiative [12] that requires all radio equipment to comply with strict requirements on filters and signal profiles in order to provide protection from interference and reduce the level of interference caused to other spectrum users. This is a development of the need for signals to fit within a spectrum mask and is a further step towards controlling interference in an over-crowded spectrum. The idea of RED is to require tighter filter requirements such that transmissions can be closer together than before and hence, free up spectrum for additional services.

At the time of writing, much interest is in developing technologies for the millimetre-wave bands, loosely defined as frequencies between 6 GHz and 100 GHz. Due to the very wide bandwidths available in this spectrum, this opens the possibility of delivering huge data rates to high concentrations of users. It is also purging spectrum regulation and coordination activity around the world.

2.3 Standardisation

There are two types of standards, open and closed or proprietary standards. Open standards are managed by bodies such as the European Telecommunications Standards Institute (ETSI) which take input from members before publishing a standard. The membership typically consists of representatives from suppliers and other stakeholders. In contrast, a closed standard is owned and managed by a single company, for example. The objective is for interested parties to adopt this standard and hence become aligned with the products, services and strategy of that company. In both cases, interfaces are specified to enable interoperability across ranges of products and services. An example is mobile phones produced by a range of manufacturers which can communicate effectively with a range of base stations supplied by different manufacturers. Without this, mobile communications would be

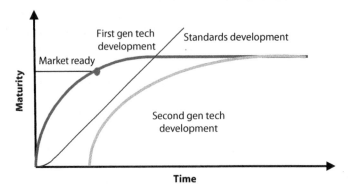

Figure 2.3 Standards the technology maturity time-line.

very much more chaotic. Other examples of standards bodies include ETSI, 3GPP [13] and the IEEE [14] who manage the standard for Wi-Fi networks.

A critical mass of operators, vendors and regulators is needed for a standard to become the industry standard. The suite of 3GPP standards are examples of this, and have driven mobile networks on a global scale, whereas WiMax and Ultra-wideband (UWB) wireless standards failed to gather enough traction for game-changing technologies to form. The evolution of mobile standards, particularly how the 3GPP standards became the foremost in the industry, is covered in Chapter 4.

Figure 2.3 shows that the development of standards lags behind the maturity of first-generation technology development. This is because technology drives standards. Indeed, it shows that first-generation technology may already be introduced into the market before standards are ready. This provides incentive for vendors to be 'first to market'. Indeed, the technology within a generation keeps maturing and a number of new releases within a generation is released before the industry moves onto a whole new generation. There are a number of examples provided in Chapter 4, where later releases of a previous generation have (at least partly) met the KPI requirements for a new generation and these products and services have been marketed with the tag of the new generation.

2.4 Research

Research can be thought of as driving the progress necessary for continuing to address the demand for mobile telecoms services. Areas of focus that have boosted this progress include:

- Improving spectral efficiency, such as: advanced coding and modulation schemes; advanced antenna techniques; network planning; transceiver system design; and protocol schemes [15];
- Improving energy efficiency, such as: light protocol schemes; efficient amplifiers and circuitry; system design considerations [16];
- Suggesting new regulatory frameworks and business models [6].

Rapid improvements have been made over the last two decades in all of the above areas. Coding and modulation is now very close to the fundamental Shannon limit in terms of spectral efficiency [17]. Multiple antenna techniques are now being deployed, which allow transmissions to be highly spectral efficient [18]. Networks have moved to deploying small cells (SC), i.e. very low power base-stations that only cover a small area and enable more users in total to be served than with larger cells [19]. Transceivers incorporate amplifiers that limit interference and waste little energy [20]. Protocol schemes that do not require much spectrum or power to operate have also been developed [21]. Recent research has examined the possibility of powering mobile devices from the energy inherent in the radio signal itself, which is how crystal radio receivers worked in the early twentieth century [22].

Another area of recent research is investigating the possibility of using millimetre wave (mmWave) bands for providing high bandwidth mobile services to a multiplicity of users [23]. This brings about new challenges in terms of transceiver design, antennas, understanding how these signals propagate, network design and performance expectations. Without this research, it will not be possible to address the continued demand for the provision of mobile telecoms. This would result in stagnation and disappointment for private and business customers alike. Importantly, it would affect the economy as mobile connectivity is considered a major catalyst.

As well as technology, research also considers regulatory options, business models and new products and services. Such an example is the recent interest in 'big data', which encompasses data analytics from multiple sources and hence requires storage and connectivity to the required sources of data [24]. Data is considered by many as the 'oil of the modern economy' as it has considerable value for business improvement and development.

Data analytics can help businesses develop optimum strategies based on understanding trends and preferences of users. Thus, the mobile telecoms networks provide linkage with many private and business customers, all indicating their preferences regarding the way they use their devices.

Although research is usually associated with universities, it is also undertaken by industry players trying to attain a commercial advantage through product and service innovation. The difference is discernible by the timescales for the outcomes to have an impact. Industry based research is typically 1–5 years, whereas university-led research is usually much longer term, i.e. more than 5 years. There are also some very effective research programmes that are collaborations between industry and academia. These may involve staff exchange, knowledge exchange or both parties contributing complimentary elements to an experiment. Industry may also directly fund university research of strategic interest and there are several models for doing this.

Very often, one of the research objectives of industry is to generate patents relating to methods and solutions they have produced. These may then form the basis of publications or contributions to form standards. Universities may also choose to patent their findings but, more often, choose to publish them in the open literature to stimulate further R&D activities.

2.5 End Users

This is an area that is rapidly changing within the mobile telecoms ecosystem. Business customers were the first to make use of mobile telecoms technology in the late 1980s, and were typically professionals in the financial sector. Other businesses then began to use mobile telecoms technology to keep in touch whilst on the move. The initial adoption by businesses was probably due to the costs being too high for the public, as well as the advantages of the technology not being immediately obvious for individuals.

This changed as prices came down and initial social acceptance and utility became apparent in the mid-1990s. Nowadays, there are more devices than people on our planet, and the number of devices is still growing. This is driven by M2M and IoT, where the devices are deployed inside other products rather than on a person. One such example is for monitoring and reporting on the performance of cars back to the manufacturer. Another is for reporting and control of central heating systems. This signifies a game changer to the mobile telecoms market in terms of technology, market size, regulation, standards, products etc.

Users' expectations have evolved over the mobile phone generations (from just voice and text to the entire range of applications available today). This can be attributed to the technology 'push' from the carriers and handset makers and the technology 'pull' from users. For technology push or pull to be successful, appropriate technology components must be made available across the entire communications chain. An example was when the provision of data services was pushed through the 3G mobile network ecosystem, yet it did not gain traction until the advent of the iPhone and similar devices. This was because, to be successful, mobile devices needed to have sufficient computing resource, battery life and supporting applications to be available; and for the applications ('apps') to be available, a software development platform was required, as well as a viable 'app' business model and critical mass of app developers.

So, how has the users' Quality of Experience (QoE) changed over the years? Arguably, in some cases, it has reduced due to network overload causing call 'drop-outs' or QoE not being aligned with increased expectations. For example, now we might expect to watch streaming video whilst on the move, but become disappointed when this is not possible. This forces investment in improving the network such that expectations can be met, where the investment may consist of reconfiguring existing network infrastructure or integrating new or more technology to increase capacity or reach in certain areas. This also raises the question of how we can determine QoE for machine-type communications, i.e. IoT devices and services. Philosophically, do machines 'experience' anything and can we define a measure for such a QoE? Perhaps a true 'experience' is only possible by humans and so for this case, the QoE should be referred to as the end human user whom the machine-type network is providing a service to. This may not even be a direct service, but rather a time or cost saving by not having to go and perform a particular task.

2.6 The Role of Operators (Carriers)

The initial role of operators during the early days of mobile networks can be likened to that of wireless operators, i.e. charges made for a connection to another user, whilst operating a single service of voice communications. As mobile network technology evolved from analogue voice to data, this brought about new services, including SMS text messages and text-based email.

This required a completely different network architecture that could deal with user-data services. Without business innovation, operators ran the risk of becoming a data pipe for Over-The-Top (OTT) data services without adding value to the data being transported. The operator can be viewed as the apex of the ecosystem since it is the focus of users, equipment vendors, regulators, research and standards bodies. Thus, it is imperative for network operators to nurture effective relationships between these parties and to form a business strategy that strikes a balance between technology innovation and business stability. In addition, striking the right level of competition between network operators is required to ensure innovation whilst providing commercial stability. There are also strategic decisions to be made, for example, deciding how to manage the roll-out of Wi-Fi such that data can then be very easily accessed through Wi-Fi by mobile devices. Due to cost pressures (cost per bit), consolidation is happening in operator and manufacturer spheres. The different cost pressures faced by the modern operators in the age of network centralisation and reduced income from roaming operations and the role of virtual network operators are presented in detail in Chapter 3.

A central aspect of mobile networks is trust and security of the network and the data it carries as this influences the level of confidence and hence demand. Therefore, ensuring a trusted network and data is essential for guarenteeing customer satisfaction and ultimately continued demand.

2.7 The Role of Vendors/Manufacturers

Equipment manufacturers and vendors are charged with supplying network technology to mobile network operators. They are highly active in technology research and development, as well as in influencing Standards' Bodies and are proponents of technology innovation with the primary goal being more efficient and higher capacity networks. This behaviour is, in part, being driven by the 'data deluge' that came about with the smartphone era. The global reach of GSM and 3GPP standards has allowed equipment vendors to attain a global reach and provide economies of scale. Furthermore, the current focus on Software Defined Network (SDN) technology and network virtualisation will drive down hardware costs and impact traditional vendors. This is because SDN technology enables network routers to be controlled by software that is re-configurable according to a need. Thus, there is a reduced reliance on hardware solutions or the need to change hardware according to prevailing network requirements such as outage or upgrade. At least in theory, SDN should also enable a wide array of vendors (as commercial off-the-shelf (COTS) equipment makers) to compete in this market and hence drive down prices. Although standardisation is aimed at full interoperability and vendor competition at every part of the network, certain network layers have remained vendor proprietary features and have, in effect, resulted in 'vendor lock-in'. This aspect and how it will likely change in the SDN/5G era is covered in Chapter 4.

2.8 The Role of Standard Bodies and Regulators

Standards' Bodies are tasked with determining and providing a standard that all equipment should adhere to. This is often done by means of reaching a consensus of members, where members typically represent the body of manufacturers, network operators and users. Members provide candidate-used cases and technologies which are then used to specify the standard. Regulators, on the other hand, control competition and market access. In the mobile ecosystem, this is largely controlled by means of managing spectrum licenses. As spectrum becomes available, the regulators auction it to one or more winning bidders or, in some cases, provide it for free. For both cases, there are strict restrictions on its use that are placed to limit interference to other users. This is usually by limiting transmit power, determining band-frequency limits and defining the type of signal that may be used. Regulators may also push through Government policy such as stipulating the percentage of the population who need to be covered by a network as a condition of spectrum license grant. Operators are trying out schemes such as 'national roaming' in rural areas to meet these policy related targets at manageable costs.

2.9 Telecoms Ecosystem Dynamics and Behaviour

Figure 2.4 depicts the major components of the mobile telecoms ecosystem, with supply and demand shown in the centre, and strategy and innovation forming a wrapper around the entire ecosystem. Figure 2.4 differs from Figure 2.1 in that Figure 2.1 shows entities of the

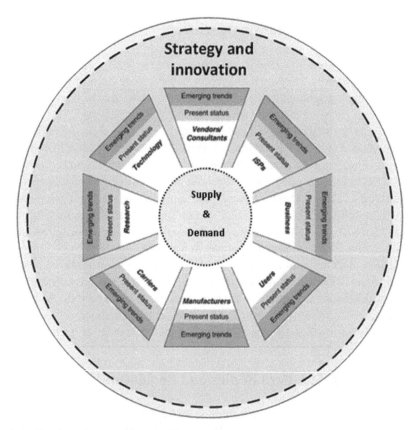

Figure 2.4 Supply and demand in the telecoms ecosystem.

ecosystem so that interactions between them can be described; whereas Figure 2.4 shows the major components of the ecosystem, indicating the split between supply and demand. Figure 2.4 highlights the pivotal role that supply and demand has on the industry. The 'supply' side of the ecosystem is shown at the top and the 'demand' side at the bottom. There are two additional rings included in the diagram that indicate 'present status' and 'emerging trends'. It is the emerging trends that stimulate change to the present status-quo and this may happen from any of the major contributors of the ecosystem. Example changes include:

- New market entrants such as new services providers, equipment providers, applications etc.;
- Changes to pricing of services, technology etc.;
- Opening of new markets such as M2M communications;
- Emergence of a new standard and related intellectual property (IP), i.e. patents pertinent to the standard;
- Technology and service obsolescence;
- Emergence of new technology improving spectrum efficiency, reducing energy consumption etc.;
- Stimulus from strategy and innovation work.

Strategy and innovation concerns determine future direction. This is best determined across the complete ecosystem, where decisions are made in the best interest of all stakeholders, including users. The role of innovation is to stimulate and undertake strategic research and to inform strategic players of emerging trends, technologies and business models.

Four examples are given below, which describe the ecosystem dynamics stemming from: changes to spectrum availability (Box Example 2.1); research and development (Box Example 2.2); service demand (Box Example 2.3); and data demand (Box Example 2.4).

Example 2.1 Spectrum availability

Such an example may be a spectrum regulator making spectrum available to mobile operators that would impact the supply of spectrum. The spectrum would be available at a cost and this is determined by the demand for that spectrum.

The demand is determined by how attractive the spectrum is perceived to be, which depends on aspects such as:

- How wide the band is;
- How much transmit power is allowed;
- Geographic reach of the new spectrum, i.e. is it limited to certain areas, or national, or even international;
- The behaviour of radio signals at the desired frequency, i.e. how well it penetrates buildings, vehicles etc. and the size and directionality of the required antennas (can they be conveniently integrated into a consumer device?);
- Availability and performance of technology to access the spectrum.

Example 2.2   Research and development

This example may start in the R&D community whereby a new way of improving the spectral efficiency of radio transmissions is discovered and developed. This would have a similar impact on the ecosystem as the release of spectrum (Example 2.1), but would also stimulate the need to: protect the invention (i.e. potentially filing a patent); update relevant standards; develop, integrate and deploy technology into devices. Examples of such discoveries include: high-order modulation schemes; advanced-coding schemes; use of multiple antennas (MIMO); emergence of orthogonal waveforms; interference-cancellation techniques etc.

Example 2.3   Service Demand

This example illustrates how service demand affects the ecosystem. This may be stimulated by continued population growth or new applications such as that of M2M communications, whereby there is a demand for more devices and hence more demand on the radio spectrum. A previous example of such a service demand was the surge in demand for mobile data with the emergence of the iPhone and other Smartphones. This, in effect, accelerated development of 4G technologies such as carrier aggregation, where multiple frequency bands could be used to serve a single user in order to provide the peak data rates demanded by some Smartphone applications.

It may be the emergence of a new application that requires considerable spectrum to support it, such as the recent emergence of virtual reality on mobile devices which is predicted to become an important factor across a wide range of businesses and personal applications. For example, it may be used by an engineer to determine from detailed plans what has gone wrong with a product without having to dismantle it, or to help understand what a new house would look like before it is built. Another example is interactive gaming involving connected virtual reality (VR) users. This opens the possibility of a new market and market entrants to develop and sell these emerging applications.

Example 2.4   Data

Over-the-Top (OTT) players in the telecoms ecosystem are having a heavy impact on the traditional revenue streams of mobile telecoms operators. Examples of OTT players include: streaming video content suppliers, video networking companies and social networking companies.

They are driving demand for data, and hence forcing investment in telecoms technologies that provide additional bandwidth to support these applications. Consequently, mobile telecoms network providers are having to spend more on investment to keep up with a massively growing number of users, who are becoming more and more data hungry.

Notice of a forthcoming auction of such spectrum would stimulate studies into the value of the spectrum, R&D into new technologies to access the spectrum, upgrading manufacturing and test equipment to account for new devices, marketing and sales to consumers etc.

A positive consequence may be that consumers find greater productivity because of improved services, and even a reduction in costs due to a greater supply of spectrum or through economies of scale of device sales.

During the year 2000, the UKs Radiocommunications Agency (now called Ofcom) raised £22.5 billion from the auction of five licences for radio spectrum in relation to delivering mobile services based on the 3G mobile telephony standard. The auction saw 13 bidders drive the price of each licence up to approximately £5 billion each. The value of each licence was partly determined by the allocated spectrum that each licence had, but also by competition and other factors mentioned above.

One avenue being explored, as an additional revenue stream, is monetising data obtained from users. An example is usage trends that may be sold back to data suppliers so they can make informed decisions of content selection and business strategy. Of course, this involves a tricky balancing act between providing useful data and preserving the privacy of individual users. A detailed look at this topic is provided in the recent presentation by GSMA on this topic [25].

2.10   5G Ecosystem

The 5G ecosystem consists of players from different sectors. Figure 2.5 identifies the major players in the 5G ecosystem. These can all be mapped onto those depicted in Figure 2.1, where Figure 2.1 shows a more generic ecosystem.

Figure 2.5   Major partners in the 5G Eco-system.

2.10.1 Datacentres

Data centres store and process data and have communication interfaces with other networked data centres and nodes. The primary challenge for data centres in a 5G-communications network is that it is required to handle and process a massive amount of data much faster than existing technologies but with greater reliability. Data centres are expected to play a major role in the 5G ecosystem. With 5G designed to deploy centralised and software-defined architectures (discussed in Section 5.9), the role of data centres will be far greater than what we see today.

2.10.2 RF Chip and Component Manufacturers

Radio-frequency (RF) components such as antennas, power amplifiers, filters etc. in both handheld devices and base stations will be required to enable implementation and function of new 5G specifications. The radio features of 5G will dictate that these components should operate in far higher frequencies and much wider bandwidths than in 4G. These new challenges will invariably push initial prices (which include the significant R&D costs) up. But the component manufacturers will need to be able to meet the optimal supply-demand balance whilst keeping long-term costs down to ensure profits. Another challenge faced by the manufacturers is to meet the high energy efficiency levels. There are now ambitious targets to reduce the 'energy per bit' consumed by 5G from the historical levels in 4G and before.

2.10.3 Telecom Operators (Carriers)

Telecom operators or carriers refer to telecommunication service providers. In 5G, the operators are making promises to offer low latency, high data rate and better coverage from their network to consumers. 5G technology will allow operators to accommodate significantly more traffic and offer advanced features and services for consumers. The 'consumers' will include those vertical industries where the new 5G capabilities can provide value addition. Thus, 5G operators will need to develop new business relationships with these vertical industries to drive new 5G enabled services in an efficient and mutually profitable manner.

2.10.4 Infrastructure Service Providers

Infrastructure service providers are tasked with providing telecom infrastructure to telecom operators, such as fixed network back-haul connectivity. The service typically includes: setup, operation, and ongoing maintenance. Nowadays, telecom operators are seen to be sharing the infrastructure to lower their increasing investments. In 5G, either microcells with shorter distances between cells need to be installed; or upgrading the existing system will be needed. In doing so, infrastructure providers are expected to play a pivotal role.

5G technology brings a new possibility termed 'network slicing'. This allows portions of network resources to be dedicated to certain services and independent from other services. Such an example could be for serving a factory with connectivity from other critical organisations in the supply chain. An end-to-end service agreement could be put in place and dedicated network resources allocated to providing the service, in terms of security, latency and bandwidth. A range of industries could benefit from slicing, such as transport, education, sports etc. Since the 'slice' is end-to-end, it would pervade the full range of technologies and networks that make up the connectivity.

2.10.5 Gaming

The gaming industry can benefit from mass deployment of 5G technology. Gaming, based on VR, such as 'Pokémon Go' demands high data rate and low latency, which are incumbent features of 5G networks. 5G promises to offer a platform for the gaming industry to evolve, and, in exchange, for the gaming industry to support the growth of the 5G eco-system. Gaming can be seen as the main 'consumer' application that will benefit from the low latencies (down to 1 ms) that 5G promises, where the user experience can be even better than that offered by fixed broadband.

To play heavy data-driven online games, gamers need uninterrupted network access, however, they have no control over the lag they often experience. In the era of Augmented Reality (AR) and VR, it is possible to create virtual artefacts and characters. The online gaming industry is developing high bandwidth driven VR and precise geo-location AR-based games, however, for them to provide the best quality experience to gamers, they need a high bandwidth and extremely low latency network. The cost of having such network infrastructure and enabling devices is expensive. Therefore, the online gaming industry needs to balance experience against cost. Development in 5G technology and plans for the mass deployment of 5G networks could significantly impact this balancing act and may motivate the online gaming industry to develop best-quality interactive VR games that provide the ultimate experience to gamers.

2.10.6 Over-The-Top (OTT)

In telecommunication, OTT players refer to content providers who offer a wide array of digital contents to consumers such as video-on-demand, broadcast and unicast audio and video content over the internet etc. OTT players are direct competitors to telecom operators. For example, the Voice over IP (VoIP)-based calling Apps have taken a large portion of the voice income from the operators in the early years of 4G. Although the technologies in 5G will allow OTTs to provide advanced features and better contents to their consumers, it is possible for the profits of OTT players to be affected by the competition of the telecom operators. This is because 5G will allow them to design the necessary architecture and build infrastructure that provides similar services to their consumers as OTT services.

2.10.7 Low-Cost Processing Unit Manufacturer

Baseband processing units need to be redesigned and manufactured to support 5G features, both in handheld and base-station devices. This includes support for massive MIMO, beamforming and Distributed Antenna Systems (DAS), therefore, the processing of the signals will become more complex compared to existing technologies. However, on the one hand, the manufacturer of the processing units requires the manufacture of a significant quantity of the processing units, on the other hand, it needs to keep the costs down to ensure their profits.

A recent development in the 5G sphere is the push towards open RAN, as driven by the O-RAN Alliance [26] and Open Air Interface (OAI) [27]. The emphasis here is to make the radio access network (RAN) much more inter-operable with components potentially from different manufacturers. If this drive gains traction with the operators it could see even smaller manufacturers, specialising in a few specific components of the RAN equipment chain, entering the 5G vendor market. Through this specialisation and increased market competition, the operators could benefit by having lower unit costs.

2.10.8 Investors

Investing in the telecommunication sector could turn out as a very profitable investment for private investors but there are also risks associated with it, such as changes to spectrum license terms and conditions. However, 5G promises special features and growth of the telecommunications sector that has never been seen before. This has attracted investment from both private and public sectors. Typically, a new mobile generation brings opportunities to disrupt the operator market with its innovative technologies. In Japan, the e-commerce giant Rakuten has launched a Greenfield 5G network as a new entrant, leveraging heavily on Open RAN for radio access and NFV/SDN technologies for the core [28]. Even with the expected expansion of 5G services to support various vertical industries, the scale of some of these collaborations (like V2X connectivity in automotive sector) will require massive investments and they open up huge opportunities to potential investors.

2.10.9 Potential Disruptions in the 5G EcoSystem

Every new generation will see some shake-up of the established ecosystems, with new technological and business concepts allowing opportunities for disruption. In 5G, both the breadth of the sectors that can be influenced by 5G and the level of influence some maturing technologies can have on 5G opens up numerous opportunities for disruptive concepts to enter the ecosystem. Some of the key disruptive business drivers for 5G communications networks are described in the remainder of this case study.

2.10.9.1 Data Centre/cloud Access

A data centre, also referred to as 'cloud access', is a pivotal part of the 5G ecosystem that processes, stores and delivers data. By 2022, the number of IoT devices is projected to reach 1.5 billion, with around 29 billion devices connected overall [2]. This means that data centres will have to handle almost inconceivable amounts of data. This is particularly so given that 5G aims to offer much greater data capacity compared to current 4G LTE communications networks. This would result in the need for much greater data processing power and data caching at data centres. Real-time access to data centres will be required for safety critical applications such as autonomous connected vehicles, remote medical surgeries etc.

Data centres use standardised architectures and designs; owners of the centres will need to either redesign or upgrade their existing architecture to accommodate the features to allow 5G services. It is possible that 5G communication networks could leverage these new 5G features in the data centres. Also, the use of virtual machines in the data centres for running different software and processes can enable the features of Network Function Virtualisation (NFV) [29]. In the NFV method, to provide a service, the network provider can use different virtual machines to create multiple virtual networks instead of running from dedicated hardware such as routers and firewalls. Also, NFV can support and simplify a wide area network (WAN) function. A WAN is used to connect enterprise networks over large geographic distance which may require a connection through branch offices and data centres. However, Software Defined Wide Area Networks (SD-WANs), utilises cloud networks to deliver communication services instead of using expensive proprietary hardware and fixed circuits.

In a 5G system, base stations and data centres are envisaged to be located in closer proximity to the network radio towers than they currently are. Hence, it is inevitable that many data centres around the country will be required to support this future technology. This would mean, potentially significant investment will be required by telecom network operators. This will cause major CAPEX (capital expenditure) and OPEX (operating expense) concerns prior to rolling-out 5G. Network operators, to solve this momentous issue, are gradually moving from traditional Radio Access Network (RAN) to the Cloud Radio Access Network (C-RAN) architecture concept. In contrast with RAN, C-RAN deconstructs the base stations to separate Remote Radio Units (RRU) and Baseband Unit (BBU) and relocates the baseband units to a centralised data or baseband processing station [30]. C-RAN can collaborate with many base stations and centrally manage the network access. This will, on one hand, remove the requirement of additional hardware and, on the other hand, will reduce overall footprint and enable large-scale deployment of the base stations. We will cover the technical aspects of SDN/NFV and C-RAN in Section 5.9 and the business related aspects in Section 12.1 in more detail.

C-RAN with Heterogeneous Network (HetNet) architecture will provide 5G services anywhere, anytime, to the desired user's equipment. HetNets can be described as a network of different networks [31]. For example, in a 5G system, a HetNet may encompass Wi-Fi, ultra-small cells, and Distributed Antenna Systems, in addition to Macro cells. Researchers have been working on bringing innovative and effective architectures to bridge the gap between HetNet and C-RAN. In the C-RAN HetNet scenario, it combines cloud computing with HetNet and is referred to as Heterogeneous Cloud Radio Access Network (H-CRAN) [32]. H-CRAN allows co-ordination between microcell, macrocell and SC layers, helps to achieve high spectral efficiency and therefore simplify the radio resources management in complex operating environments. A combination of HetNet and C-RAN technology is thought to be the most appropriate architecture for 5G technology.

This is possible because the Software Defined Networking (SDN) technologies are seamlessly integrated into the H-CRAN [32]. SDN decouples control planes from the data planes and allows software to be redesigned independently from the hardware. In H-CRAN, the remote radio heads (hardware) and base-band processing (software) are separated, therefore SDN can be integrated easily with the H-CRAN and simplify the network access. It is based on cloud-computing technologies; therefore, a cloud/data centre can be used for data storage, data processing and for supporting signal processing functions, such as for suppressing co-channel interference.

A data centre, designed to support 5G technology, can also support the needs of a wide range of other digital industries along with the telecoms industry, such as: VR gaming, simulation driven software, massive M2M type communications etc. Thus, the providers of data centres and cloud-computing platforms will become integral partners of the 5G mobile ecosystem.

2.10.9.2 Artificial Intelligence and Machine Learning

Artificial Intelligence (AI) and Machine Learning (ML) are emerging as key innovations of this decade and they have the potential to influence many areas of our lives. In 5G, network management and optimisation based on AI and ML can provide significant performance gains over the traditional tools like SON (Self Organised Networks). AI and ML techniques depend on having large amounts of data to train models and then apply them real time on performance issues. The availability of massive amounts of data in 5G systems to feed these modules could be seen as a natural enabler for such techniques. 5G promises to venture into many new vertical areas with distinctly different KPIs to be optimised for each vertical. There are no historical mathematical models available for optimising features in applications like V2X, massive IoT and Industry 4.0. So AI/ML-based learning and modelling may be the best option in these situations. An early example of using ML to optimise the mobility load balancing (balance the network load amongst neighbouring cells, considering the user mobility) is provided in [33].

2.10.9.3 Satellite Communications

Satellite communications could be seen more as a complementing technology for 5G rather than a disruptor. There had been attempts to include a satellite component into mobile communications as far back as in 3G [34]. Previously, satellite technologies were discounted by the telecom operators due to three main reasons; cost, latency and availability. The cost of specialised devices for communicating with satellites was driven by the very high sensitivity radio units and the battery power needed to combat the weak signals resulting from very high propagation distances. Although for the geo-stationary satellites at very high orbits this is still an issue, there is now a huge interest in deploying satellites at low earth orbits (LEO) and they do have some major advantages over the terrestrial system. For example, satellites can now provide very wide area coverage at reasonable cost. In this instance, satellite could become an alternative approach for operators even to cover massive sensor networks spread widely over remote areas.

Telecom operators need to remain profitable, but they also need to provide outstanding customer experience. The next generation networks are forecast to need to support 1000 times more data traffic. To relieve congestion from data throughput technologies and to meet the demand of their users, telecom operators are always in search of other mobile backhaul technologies. With this in mind, satellite technology could provide 5G mobile backhaul over satellite communications.

There is now a concerted effort in 3GPP 5G-NR to develop a satellite (more broadly known as non-terrestrial) component, where the same mass market 5G devices can utilise. A more detailed section on satellite communications and the wider non-terrestrial communications (NTN) can be found in Section 11.4.

2.11 Summary

We have shown there are many entities that make up the mobile telecoms ecosystem and that have complex and dynamic interaction amongst them. At present, there is intense activity within the ecosystem in preparation for '5G' mobile connectivity. There are some exciting technologies in the R&D domain, significant developments in terms of regulation and standards, emerging applications and business models. Some of the emerging technological enablers for 5G could pave the way for disruptive entrants to this ecosystem and we have also looked at a few such significant developments.

Although we suspect the basic principles described in this chapter will still hold in ten years' time, we can be confident that the landscape will change significantly as '5G' takes hold and we begin to wonder what might follow.

References

[1] *GSMA (2019). The state of mobile internet connectivity 2019.* https://www.gsma.com/mobilefordevelopment/wp-content/uploads/2019/07/GSMA-State-of-Mobile-Internet-Connectivity-Report-2019.pdf (accessed 14 June 2021).

[2] IoT Agenda (2020). *Internet of things.* https://internetofthingsagenda.techtarget.com/definition/Internet-of-Things-IoT (accessed 14 June 2021).

[3] Ericsson (2019). *Mobility reports.* https://www.ericsson.com/en/mobility-report/reports/november-2019 (accessed 14 June 2021).

[4] NASA (2013). *Your device has more computing power.* https://www.nasa.gov/mission_pages/voyager/multimedia/vgrmemory.html#.Xn4B7mDgrIU (accessed 14 June 2021).

[5] Acharya, S. and Budka, K.C. (2011). Telecommunications in vertical markets: Challenges and opportunities. *Bell Lab Tech J* 16(3): 1–4.

[6] Decker, C. (2015). *Modern Economic Regulation.* Cambridge: Cambridge University Press.

[7] Allen, B. and Ghavami, M. (2006). *Adaptive Array Systems: Fundamentals and Applications.* Hoboken, NJ: Wiley.

[8] The Nobel Prize (2021). *Guglielmo Marconi.* https://www.nobelprize.org/prizes/physics/1909/marconi/biographical (accessed 14 June 2021).

[9] Deery, J. (2007). The "real" history of real-time spectrum analyzers. *Sound Vib* 41(1): 54–59.

[10] Mitola, J. (2009). Cognitive radio architecture evolution. *Proc IEEE* 97(4): 626–641.

[11] Widrow, B., Glover, J.R., McCool, J.M. et al. (1975). Adaptive noise cancelling: Principles and applications. *Proc IEEE* 63(12): 1692–1716.

[12] European Commission (2014). *Radio Equipment Directive.* https://ec.europa.eu/growth/sectors/electrical-engineering/red-directive_en (accessed 14 June 2021).

[13] 3GPP (2021). https://www.3gpp.org (accessed 14 June 2021).

[14] Wikipedia (2021). *IEEE Standards Association.* https://en.wikipedia.org/wiki/IEEE_Standards_Association (accessed 14 June 2021).

[15] Zhang, Z., Long, K., Vasilakos, A.V. et al. (2016). Full-Duplex wireless communications: Challenges, solutions, and future research directions. *Proc IEEE* 104(7): 1369–1409.

[16] Zi, R., Ge, X., Thompson, J. et al. (2016). Energy efficiency optimization of 5g radio-frequency chain systems. *IEEE J Sel Areas Commun* 34(4): 758–771.

[17] Costello, D.J. and Forney, G.D. (2007). Channel coding: The road to channel capacity. *Proc IEEE* 95(6): 1150–1177.

[18] Yang, S. and Hanzo, L. (2015). Fifty years of MIMO detection: The road to large-scale MIMOs. *IEEE Commun Surv Tutor* 17(4): 1941–1988.

[19] Hwang, I., Song, B., and Soliman, S.S. (2013). A holistic view on hyper-dense heterogeneous and small cell networks. *IEEE Commun Mag* 51(6): 20–27.

[20] McCune, E. (2005). High-efficiency, multi-mode, multi-band terminal power amplifiers. *IEEE Microw Mag* 6(1): 44–55.

[21] Mao, B., Fadlullah, Z., Tang, F. et al. (2017). Routing or computing? The paradigm shift towards intelligent computer network packet transmission based on deep learning. *IEEE Trans Comput* 66(11): 1946–1960.

[22] Kim, S., Vyas, R., Bito, J. et al. (2014). Ambient RF energy-harvesting technologies for self-sustainable standalone wireless sensor platforms. *Proc IEEE* 102(11): 1649–1666.

[23] He, D., Ai, B., Briso-Rodriguez, C. et al. (2019). Train-to-infrastructure channel modeling and simulation in MmWave band. *EEE Commun Mag* 57(9): 4–49.

[24] Cukier, K. and Mayer-Schönberger, V. (2014). Big data: A revolution that will transform how we live, work, and think. Boston, MA: Houghton Mifflin Harcour.

[25] GSMA (2018). *Beyond Authentication: Monetising Identity Services*. https://www.gsma.com/identity/wp-content/uploads/2018/10/GSMA-Sham-Careeem-presentation-Turkey-final.pdf (accessed 14 June 2021).

[26] O-RAN (2020). *Operator Defined Open and Intelligent Radio Access Networks*. https://www.o-ran.org (accessed 14 June 2021).

[27] Open Air Interface (2020). *The fastest growing community and software assets in 5G wireless*. https://www.openairinterface.org (accessed 14 June 2021).

[28] Rakuten (2020). *Rakuten mobile lanches 5G service with new plan, same monthly fee: Rakuten UN-LIMIT V*. https://global.rakuten.com/corp/news/press/2020/0930_02.html (accessed 14 June 2021).

[29] Liyanage, M., Gurtov, A., and Ylianttila, M. (2015). *Software Defined Mobile Networks (SDMN): Beyond LTE Network Architecture*. Chichester, UK: Wiley.

[30] Fan, C., Zhang, Y.J., and Yuan, X. (2016). Advances and challenges toward a scalable cloud radio access network. *IEEE Commun Mag* 54(6): 29–35.

[31] Yaghoubi, F., Mahloo, M., Wosinska, L. et al. (2018). A techno-economic framework for 5G transport networks. *IEEE Wirel Commun* 25(5): 56–63.

[32] Meerja, K.A., Shami, A., and Refaey, A. (2015). Hailing cloud empowered radio access networks. *IEEE Wirel Commun* 22(1): 122–129.

[33] Roshdy, A., Gaber, A., Hantera, F. et al. (2018). Mobility load balancing using machine learning with case study in live network. *International Conference on Innovative Trends in Computer Engineering (ITCE),* Aswan, Egypt (19–21 February 2018). IEEE.

[34] ETSI (2021). Satellite for UMTS/IMT-2000. https://www.etsi.org/technologies/satellite/satellite-umts-imt-2000 (accessed 14 June 2021).

3

The Business of a Mobile Operator

3.1 Business Challenges Faced by Operators

The financial pressure on 'end of teens' mobile operators is immense. From their inception in the late 1980s and early 1990s when a comfortable 'cost-plus' business model applied and subscription numbers only ever increased, the encroachment of competition, complementary technologies and regulation have forced operators to consolidate, close or sell outright.

Figure 3.1 shows a graph of Mobile Average Revenue Per User (ARPU) per UK Subscriber for the years 2007 to 2016, the last year for which reliable data was available at the time of writing, showing Pre-pay, Post-pay and blended (i.e. Pre- and Post-paid weighted per number of each type of sub) in both raw UK Pounds and adjusted to the value of the pound in each year relative to 2016 [1, 2].

Faced with dwindling ARPU, operators have adopted a variety of strategies; initially expanding into each other's subscriber base to the extent that each user now has, on average, more than one SIM – to avail of different deals particularly for data consumption – a consumption-led strategy has given way to complementarity. Young households, in particular are opting entirely to dispense with a landline and fixed-broadband connection, relying instead on 4G- or 5G-based Mobile Broadband for data requirements. Total mobile connections comprising personal and 'Internet of Things' (IoT) SIMs and E-SIMS are forecast at 25 billion by 2025 [3], or nearly three per person alive based on world population trends.

5G promises to usher in an era of greatly expanded use-cases for mobile connections. However, until that new market sector evolves, likely out of existing 4G user-base plus previously 'off-grid' IoT connections, increasing the bottom-line means stripping out operational costs. The highest cost for any operator is staff salaries; after this, the next item below the top-line is site rental fees (OPEX is considered here, not CAPEX in new site fit-out costs). Rationalised site fees strategies will be considered in Section 3.2, the next few sections consider other third-party costs operators can optimise.

3.1.1 Third-Party Costs

Mobile operator costs can broadly be broken down into two classes, Capital Expenditure (CapEx), associated with acquisition of tangible assets that depreciate from the moment of

The Technology and Business of Mobile Communications: An Introduction, First Edition.
Mythri Hunukumbure, Justin P. Coon, Ben Allen, and Tony Vernon.
© 2022 John Wiley & Sons Ltd. Published 2022 by John Wiley & Sons Ltd.

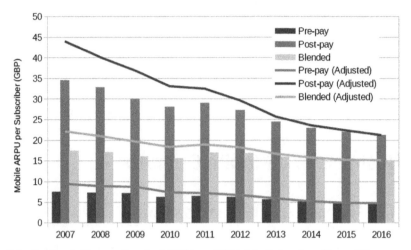

Figure 3.1 Decline in Year-on-Year Real ARPU per UK Mobile Subscriber. (*Source* [1]).

purchase and Operational Expenditure (OpEx) which is intangible and, as such, does not directly have an asset value.

The main cost for any operator is staff salaries which, as amounts that the operator, in effect, pays itself, is not a direct third-party cost – the staff are, by identity, strongly associated with the operator itself. All other financial outgoings have at least some associated third-party aspects. Whilst OpEx is often incorrectly viewed purely as an ongoing cost of running the network, there are OpEx elements involved at the establishment and setup of the fledgling operator [4]; this up-front cost can be broken down into Initial Planning and First-Time Installation expenditure.

Falling into the Initial Planning sub-category are all costs associated with initial feasibility studies, evaluations of modification of the operator's own network and any external network that has to be rented when transmission capacity is insufficient or of the wrong type, consultancy fees to explore scenarios to adopt new technologies and the effect of Value Added Servers and any required new Service Platforms (e.g. new platforms required to assign QoS parameters to IPv6 data flows in 4G/5G). A further large element of the 'roadmap' studies is, of course, infrastructure vendor selection and all the associated incidental costs e.g. travel, accommodation, subsistence etc.

Confusion can sometimes arise around First Time Installation costs, which other commentators sometimes lump together with CapEx. In this sub-category are modifications to buildings and existing plant to accommodate new network equipment and elements such as NodeB and RNC housings, wiring work and cooling plant and, considering base station sites themselves, the support brackets, metalwork, cable-trays, and wiring associated with antenna installation. The antennas themselves are assets that depreciate, the supporting structures are not assets as they are unique to any given installation; in fact, this latter OpEx class would probably have a negative asset value if viewed as such – the costs of removal, that any canny site owner would write into the initial contract, will far exceed any residual scrap value of the material itself.

3.1.2 Radio Access Network Costs

The most visible cost to users associated with a mobile operator (apart, of course, from the handset or UE that he/she is holding) are the numerous base stations (NodeBs and eNodeBs in 3G/4G respectively) that each operator deploys to meet the coverage requirements of their licences in each territory where they have a presence.

Figure 3.2(a) illustrates the 3G Radio Access Network (RAN). This consists of the visible NodeB radio sites and their associated transmission equipment Uu interface antennas as well as the usually invisible Radio Network Controller (RNC) sites principally handling Radio Resource Management (RRM). RNCs co-ordinate 'soft' handover between NodeBs under their hierarchy and with 'drift' RNCs where handover is required to a border NodeB outside the set controlled by the RNC.

External paths for voice and data sessions are entirely separate, with voice calls being routed out to the ISDN via the Mobile Switching Centre and associated Gateway MSC(s), and data being routed to external Packet Data Networks (PDNs, as they were originally defined by 3GPP), via Serving and Gateway GPRS Support Nodes (SGSN/GGSN). Note also that the SGSN must handle the Mobility Management (MM) for data sessions, leading to the frequent optimisation conundrum of the UE being in one Location Area (LA) for voice calls and a separate Routing Area (RA) for data sessions.

The 4G RAN depicted in Figure 3.2 (b) is slightly simpler as the RNC function is absorbed into the eNodeB, so that the RAN is entirely represented by the set of macro, micro and pico eNodeBs; the handover path that in 3G is routed via the RNC is implemented over the inter-site X2 links. There are no higher nodes in the RAN than eNodeBs; within the core network (CN) to which the eNodeBs interconnect via S1 links, the collection of Mobility Management Entities (MME), Serving Gateways (S-GWs) and one or more Packet Gateways are known as the Evolved Packet Core (Figure 3.2b).

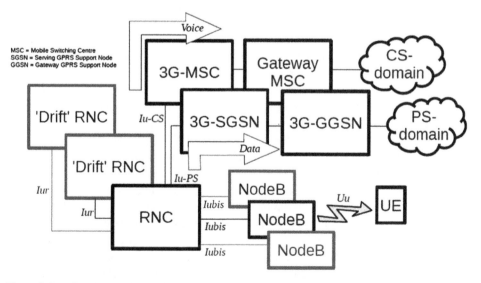

Figure 3.2a Simplified 3G UMTS Terrestrial Radio Access Network Schematic.

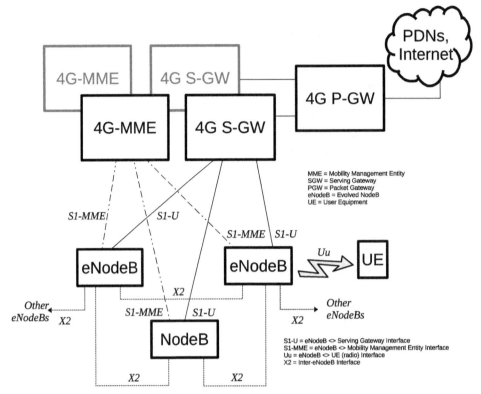

Figure 3.2b Simplified 4G Evolved UTRAN plus Evolved Packet Core Schematic.

Direct cost comparisons between 3G UTRAN (UMTS Terrestrial Radio Access Network) and 4G E-UTRAN (Evolved UTRAN) plus its serving Evolved Packet Core (EPC) are complicated by the fact that data mobility, which in 3G is provided within the unified SGSN, is split in the 4G case between the Mobility Management Entity (4G-MME) which provides Core-to-UE data link maintenance, and the Serving Gateway (4G S-GW) which provides the external routing function. There are other architectural differences, such as the separation of baseband and radio-frequency (RF) processing with widespread use of Remote Radio Heads/Units (RRH/ RRU) mounted on the masthead to reduce cable losses at the higher frequencies used by modern 3G and practically all 4G roll-outs.

A further complication is that network infrastructure deals between operators and system vendors are always 'commercial in confidence', so that operators negotiating new or extended network agreements cannot benefit from vendors' previous discounts; as a result, unit costs for eNodeBs, SGSNs, S-GWs and other network elements are never published.

Some 'guesstimates' can nevertheless be made to cost the E-UTRAN, taking the price of a single-sector antenna installation and RRU as a single unit of cost, the cost of a radio site's Baseband Processing Unit (BBU) can be estimated as approximately 5 units. Inter-site transmission links (X2 connections in 4G), of which we assume an average of three per site, are estimated at 0.5 units, and adding another two units for the fixed fit-out cost (cabinets, cable trays, mast/s, electrical connection etc.) gives a total cost per site of 11.5 units.

Looking at a 'bare metal' EPC, i.e. functions directly implemented on hardware, a 4G MME might cost 10–20 units, a 4G S-GW might cost between 20 and 40 units depending on network

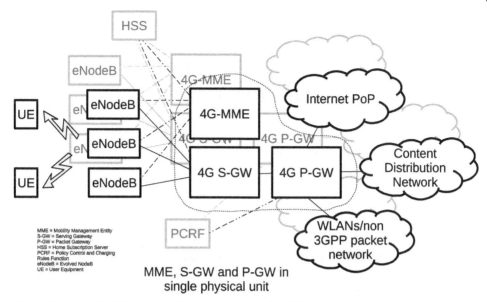

Figure 3.2c Practical Topology and Implementation of E-UTRAN and EPC.

capacity and again depending on capacity, required connections and inter-working functions with other packet networks, a 4G Packet Gateway (4G P-GW) providing in/outbound connectivity for one or more 4G S-GWs might cost between 10 and 20 units. Taking averages of the cost ranges for a basic Evolved Packet System of 200 sites (EPS, the union of the E-UTRAN network and the EPC that serves it) would thus cost 2300 + 30 + 15 = 2345 units.

For simplicity, non-E-UTRAN network elements belonging to the Evolved Packet Core (EPC) such as the Home Subscription Server (HSS) and Policy Control and Charging Rules Function (PCRF) are not costed here, as there is usually only one per operator servicing all EPS (here E-UTRAN + 4G S-GW + 4G P-GW) sub-units. Also, here it is assumed that a 4G P-GW is required for each 4G S-GW whereas, in theory, only one is required for all 4G S-GWs in the operator's EPS. In practice, the EPS will be constructed topographically around 'Points of Presence' belonging to external Internet providers and Content Distribution Networks, Figure 3.2(c); this is for reliability and scalability reasons and to increase resilience by reducing dependency on a single, very high bandwidth connection.

One of the reasons for, and benefits of, splitting the MME and S-GW functions is so that both may scale independently of one another, the MME with total number of UEs and S-GW with the aggregate total downlink plus uplink traffic. In a scenario where an EPS is supported by a single MME/S-GW, then each may be upgraded independently. However, practical topography is driven by the availability of very high bandwidth links at several Internet Service Provider Points of Presence and, unless the mobile operator itself takes charge of Tier 1/2 peering, it makes sense to connect to providers of lit fibre at their geographically distributed PoPs.

In response, equipment vendors have developed hardware platforms for the LTE EPC implementing the combined functions; Figure 3.3 shows an example of such functional integration, the Cisco ASR5500 Multimedia Core Platform, which can implement 4G MME, S-GW and P-GW, in addition to 2G/3G SGSN and GGSN for optimum session continuity between 4G and other network layers such as 2G or 3G, in a single cabinet with dimensions 933 × 438 x 699 mm [5]

Figure 3.3   The Cisco ASR5500 Multimedia Core Platform implements MME, S-GW and P-GW on 4G and SGSN/GGSN on 2G/3G in a unit measuring 933 x 438 x 699mm. (*Source* [5]).

Whereas EPC functions have been consolidated, the modern tendency has been to split apart the E-UTRAN entities. Here, the previously monolithic eNodeB architecture, consisting of antennas, cabling, RF + baseband processing and microwave or fibre transmission links, has been reshuffled for convenience. Towards the end of last century, research into digitising and sending radio signals over fibre was driven by various smart antenna proposals that required multi-sensor antennas and therefore up to one dozen coaxial cable connections per sector between rooftop antennas and ground-level base station [6, 7].

For a variety of CapEx-related reasons, smart antennas in that form were not rolled out but the 'Radio over Fibre' work found new application in the 3G UMTS and 4G LTE era as Radio Remote Units (or 'Heads' depending on vendor) for single-antenna-per-sector site implementations. Figure 3.4 shows a simplified example of a typical roll-out scenario, where three RRUs reduce the insertion loss at a typical 4G operating frequency of 2 GHz from typically 2.5 dB to essentially less than 0.5 dB insertion loss of the RF 'tails' at the mast apex.

Also depicted is an example of one vendor's RRU implementation, in a unit 600 mm x 320 mm x 145 mm weighing 29 kg, which is readily mountable on most mast structures. The ZTE ZXSDR R8884 Radio Remote Unit, one per sector, is capable of 4x4 MIMO with LTE carrier aggregation across 1800 and 2600 either intra- or inter-band and transmit

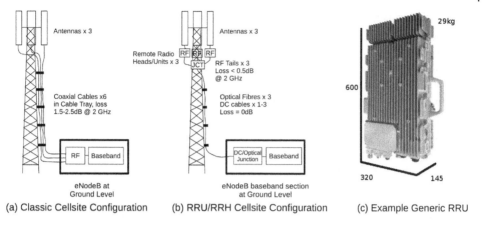

Figure 3.4 Simplified Base Station Topology with Radio Remote Units. (*Source*: ZTE).

power up to 4 x 40 W, all in a unit with volume less than 28l (Q4 2018). More modern implementations have implemented multiple bands and greater cross-band carrier aggregation capability, with higher output power in smaller, lighter volumes.

Now, consider that once the RF and Baseband units are dissociated, and the received signals have been converted to digital form (and the transmit signals from digital), there is no great imperative to have the baseband function physically located at the base site. In fact, the early 'RF over fibre' work was done with the specific aim of permitting large microcell deployments, where a central baseband processor could co-ordinate and control dozens or even hundreds of RRUs (from whence the term originates).

Calculating the costs of a 200-site 'distributed' EPS, each site now costs $3 + 2 + 0.5 = 5.5$ units, assuming the same two cost units for each site fit-out and 0.5 units for the microwave or fibre link, of which only one is now either needed or relevant (neighbouring sites having no geographically co-located baseband function) and for simplicity's sake, assuming that the now-virtualised baseband function costs 0.5 units if implemented as separate instances on a High Performance Computing (HPC) platform.

Into this 0.5 unit per virtualised baseband function estimate is also absorbed the cost of the 4G MME, 4G P-GW and 4G S-GW, as these are also assumed to be virtualised entities running on the same HPC platform. In this scenario, total EPS cost is $200 \times (5.5 + 0.5) = 1200$, slightly over 50% of the cost of an equivalent EPS implemented in hardware but with much improved reliability through not having 200 hardware baseband deployments over the geographic coverage area of the EPS.

It is, therefore, not difficult to see why Network Function Virtualisation (NFV) is such a hot topic amongst operator CEOs and CTOs.

3.1.3 Transmission Costs

At the dawn of digital Personal Communications Networks in the early 1990s, one of the innovations introduced to the market was self-provided transmission links between base stations and the CN, i.e. the Base Station Subsystem (BSS). Mobile operators could thereby avoid the expense of commissioning expensive third-party leased line links between 2G BTS sites and the regional Base Station Controller site and could also self-provide high-capacity links between elements of the 2G CN. In 2008, it was estimated that micro-wave linked 5% of BTSs in North America (where wired T1 links are cheap and readily available), in contrast to Asia at 40% and Europe at nearly two-thirds of all BTS connections [8].

BSS interconnections were typically short- to medium-range microwave point-to-point links in the 18–38 GHz range, with various frequencies used in different world markets depending on spectrum availability. In the UK, the two most frequently used bands in urban and suburban areas were 23 GHz and 38 GHz, the latter for short-range links in the 5–8 km range due to rain fade limitations. Starting from a low bandwidth point in the 2G era, where links would typically only support up to four 2 Mbit/s E1s in 2 + 2 configuration using 4-state, 8-state or 16-QUAM in 3.5 MHz channels, usage of such links in the all-IP 4G/5G era has expanded far beyond 300 Mbit/s with the use of modern forward error correction coding and very high order modulation constellations such as 512-QUAM [9].

Figure 3.5 Illustrates the current frequency plan for these two bands as an example [10], the band plans for other frequency ranges commonly used by mobile operators for

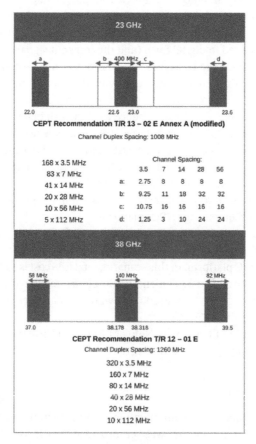

Figure 3.5 UK Frequency and Channel Plan for 23 and 38 GHz Fixed Links. (*Source* [10]).

microwave P2P links such as 13 GHz, 15 GHz, 18 GHz and 26 GHz also being depicted in the source referenced.

Traditionally, mobile operators were given 'self co-ordination' rights to the higher microwave bands which, at the time of issue and being brought into service by the PCNs, were not in use. As such, having paid for a transmission frequency allocation, and a registration fee per point-to-point link brought into service, the emphasis was on re-using this frequency, to the fullest extent possible within a geographic region, to keep overall transmission OPEX down.

Where a large number of links are sharing the same frequency to reduce licence costs, the key factor in planning signal quality for the links is the 'radiation pattern envelope' of the antennas involved, that greatly simplifies calculation of the self-interference between the co-frequency links in order to ascertain whether the signal-to-noise ratio (SNR) of the final demodulated link is sufficient [11]. Taking a generic 38 GHz antenna as an example, Figure 3.6 shows the plan view (not to scale) of a link that is being interfered with by antennas from two neighbouring links, the received power of one of which falls on a side lobe response peak and the other falls on a side lobe response null.

In practical deployments, the lower side lobes of the antenna response pattern are sensitive to the mast or support pole environment on which they are mounted. Rather than

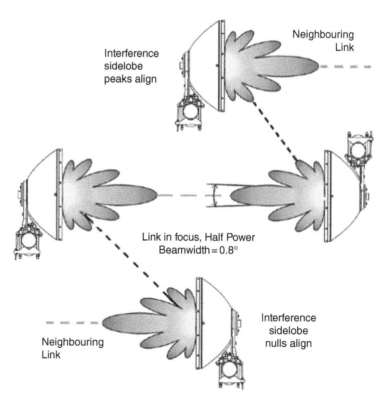

Figure 3.6 Interference in Co-frequency Microwave Links. (*Source:* RFS instruction sheet NMT-30-1001-00e.pdf).

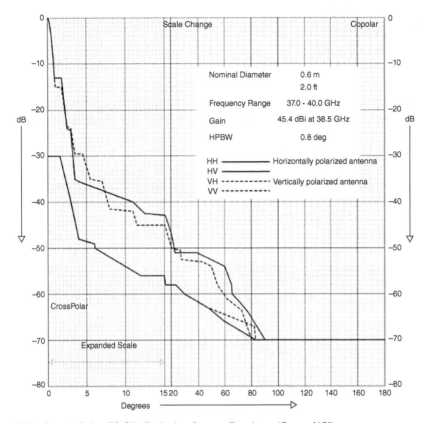

Figure 3.7 Generic 0.6 m 38 GHz Radiation Pattern Envelope. (*Source* [12]).

calculate an exact interference power based on the true antenna polar response pattern, a 'worst case' interference is calculated from the highest side lobe response at each antenna 'look' angle and compiled into a 'radiation pattern envelope' (RPE). Figure 3.7 shows the RPE for the same generic 0.6 m diameter antenna; note that in a dual-polarised antenna, true calculation of interference power suffered requires a figure, not only for the Horizontal to Horizontal (HH) and Vertical to Vertical (VV) co-polarisation responses, but also the Horizontal to Vertical (HV) and Vertical to Horizontal (VH) cross-polarisation responses.

Although an approximation, the use of RPEs is highly effective in practice because they assume a 'worst case' scenario of mutual side lobe interference. This author's personal experience of co-frequency planning was in Stuttgart, Germany where an entire city network was planned on a single 38 GHz frequency using a combination of 0.3, 0.6 and 1.2 m diameter antennas – except for a single link that had to use a second frequency. However, as the GSM1800 PCN in question self-managed its P2P link frequency allocation, the single use of f2, where f1 had been used everywhere else, was more a matter of denting the pride of the microwave planning team, rather than a financial impact to the network.

Due to the 'commercial in confidence' nature of vendor-operator purchases, hard information on the average cost of links relative to the average total cost of the BTS or eNodeB

Table 3.1 5-Year Total Cost of Ownership for 23 GHz links in US and UK Markets (source [13]).

5 Year Costs (Euro in 2008)	Single Carrier (<400 MB/s)	Dual Carrier (<800Mb/s)	Quad Carrier (<1600Mb/s)
Total Cost of Ownership – US Market	28 714	36 279	66 880
Cost per Mb per second – US Market	72	46	41
Total Cost of Ownership – UK Market	37 352	48 361	86 252
Cost per Mb per second – UK Market	93	60	54

itself is relatively scarce. However, in 2008, the microwave equipment manufacturer BridgeWave published a white paper on Total Cost of Ownership (TCO) of microwave circuits [12]. Costs in the study are not amenable to adjustment for inflation, as inflation figures are only readily available for Consumer and Retail Price Indices [14]; it is possible, however, to convert to equivalent Euro pricing valid for 2008, Table 3.1.

It can readily be seen that, far from being an afterthought in the 'bottom-line' cost of designing a BTS or an eNodeB, the transmission link is upwards of 20% of the cost of a site costing between EUR100k and 200k to construct in 2008. This is the difference between the extra traffic and, therefore, marginal revenue brought into the network resulting in the new being site viable or not, especially for new sites nearer the edge of the network where, having immediately to launch 4G or 5G with transmission capacity of up to 400Mb/s all-IP may be prohibitive, even for self-provided links.

3.1.4 Physical Locations

In the mid-1990s, as 2G technology was being rolled out by numerous operators to replace analogue systems, site average cost was an important driver in P/L calculations; over the course of time, with the introduction in the early year of the new millennium of 3G technology then 4G approximately 10 years after, the concept of an 'average site' has decreased in relevance. Not every 2G base station site will have 3G, and not every 2G/3G site will be fitted with 4G technology. In addition, multi-operator configurations on the same building or mast led to sites essentially being 'custom builds'.

The concept of average site cost lives on, however, in calculations of marginal profitability of new sites; if an average site cost for 2G/3G or 2G/3G/4G (and, shortly, 5G) can be derived, then a very good estimate can be made of whether an area not yet being served with the higher data-rate technologies may be revenue positive if a new site is constructed there. These calculations, driven mainly by the operator's knowledge of its ARPU over each type of coverage area were, in the past, the principal determinants of whether a new site would be constructed.

The arrival of 5G on the scene is disruptive of this established, incremental model in at least two ways. Firstly, in operators' rural and suburban areas, where established 2G/3G/4G site technology has traditionally been coverage driven, capacity starts to run out with the push of applications onto 5G frequencies. Use of the new, sub-700 MHz frequencies for 5G mitigates the situation somewhat but there is no real alternative to network densification in these areas; the existing macro sites, even with the advanced Massive MIMO Active Antennas coming for 5G, offer insufficient bits per second per MHz per km^2 to meet the capacity challenge, especially considering that autonomous vehicle traffic will be a major factor in such localities.

In the urban and dense urban areas, cell radii are already down to 200 m average, so further macro densification here is largely impractical. Instead, it is envisaged that the brand new 5G frequencies released around 3.5 GHz will be brought into action from a new generation of both available macro sites, and new-build small-cell (SC) sites. Both these motivating factors are set to increase the cost of new mobile sites, expressed as Total Cost of Ownership (TCO). One study [14] puts the increase in TCO as high as 300%, based on an estimated 50% data year-on-year data growth to 2024 and beyond.

A sizeable fraction, perhaps as much as half, of this additional TCO will lie in the urban and dense urban small cell network; here, although the individual cost of each site will be quite low, i.e. only a fraction of the approximately EUR150k for a brand new macro site, the large number of SC locations required will lead to high total cost.

3.1.5 Power Costs for Multiple Technologies

Although slightly dated, one source (2011, [15]) forecast an annual worldwide power consumption in 2030 by telecommunications of 600 TWh, of which half or approximately 300 TWh will be consumed by mobile, the strongest sector. A medium-sized nuclear power station with reasonable availability outputs between 10 and 15 TWh per annum, leading to the somewhat startling calculation of between 20 and 30 atomic generators being dedicated worldwide solely to mobile telecoms.

The environmental impact of this generation follows naturally, as does the impact on a network operator's bottom line, which is estimated at up to 15% of a network operator's Opex [16]. Around one fifth of that 15% slice is attributed to core network and transmission energy costs, leaving around four-fifths on base station power for BTSs, NodeBs, eNodeBs and gNodeBs, with the same source claiming that due to the high cooling and power system burden of a typical base site implementation, only around 36% of a total power budget of typically 12 kW is used in 'shifting bits', the remainder being expended on cooling and power systems. The situation is summarised in Figure 3.8.

In 2G and 3G networks, possible energy efficiencies in base stations that could mitigate this load were limited by each site's requirement to transmit a palette of broadcast channels, for initial synchronisation, login and camp-on, and mobile terminated signalling. 4G improved the power efficiency somewhat but was still limited by LTE base stations'

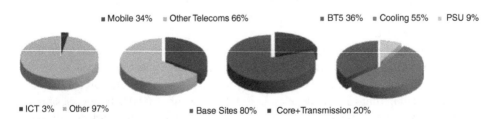

Figure 3.8 Split of Worldwide, Telecoms, Network and Base Site Power Consumption. (*Source* [16]).

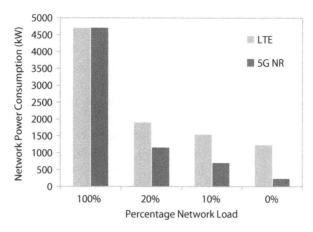

Figure 3.9 Comparison of 4G LTE and 5G NR Network Power Consumption with Low Traffic Load (kW).

requirement to broadcast symbols four times every millisecond, limiting scope for powering down the transmit chain. Small cells and some macro sites could, however, now in 4G be dynamically switched off when not carrying appreciable traffic, e.g. during the hours of darkness, leaving coverage cells to carry the remaining traffic.

5G New Radio greatly improves the situation, with all of macro, micro and pico-sites able to power down when not in use, in addition to other hardware efficiencies when operating below maximum output power. Figure 3.9 shows the great efficiencies that will be possible for network loadings less than 100% from Release 17 onwards, based on a small network of 5000 macro, and 20,000 small cell sites, like-for-like LTE and 5G NR.

3.2 MVNOs – Mobile Virtual Network Operators

Hitherto in this chapter, we have considered only 'real' network operators, bearing CAPEX costs for real Radio Access Network and Transmission sites, and OPEX for energy and other running costs and, of course, staff and contractor costs. A natural consequence of this is that, with only three or four mobile network operators in a nation or territory, there was little competitive difference between the coverage and service offerings of different networks, unless built-in at the licence award stage (e.g. an increased grant of spectrum in return for a higher percentage country landmass before a certain future date).

In the late 1980s and 1990s, some countries including the UK, attempted to introduce a layer of competition between first-generation (1G) mobile network operators and subscribers, both business and consumer. The 'Service Provider' model saw the mobile operators selling 'bulk airtime' (data usage pre-millennium was tiny) to intermediaries who would then retail smaller packages of airtime minutes per week or, more commonly, per month, to end users.

The typical 'value add' was that the Service or Airtime provider would perform the billing and customer-care functions, potentially offering organisations a choice of aggregated billing over their mobile device estate and, for consumers, a choice of enhanced customer care over the basic service provided by the underlying operator itself, potentially at slightly greater monthly cost. Airtime Providers could, in theory, offer a greater range of more capable mobile terminals than the network operator itself, offering these to customers for lower up-front cost in return for a higher final cost recovered over up to three years via the monthly subscription.

Ultimately, this method of reaching the market was very limited in the differentiation that could be obtained; also, the lack of a separate module that could carry the mobile subscriber's personalisation meant that mobile device hardware, and the firmware/software loaded to enable the device to access the network both had to be provided by the network itself. Standalone Airtime providers did not last long into the second-generation era, when the first version of the 2G Subscriber Identity Module (SIM) decoupled the device hardware from the actual subscription that enabled the hardware to access an operator's network.

From this point, networks could get on with their specialisation of tuning and optimising the device hardware and firmware for best interworking with the Radio Access Network, whilst others in theory could package and sell the individual subscriptions. However, due to legislation passed in the mid-1990s, in many countries the first action of new 2G mobile network operators was to acquire nearly all their third-party airtime providers and start selling subscriptions directly to consumers and business users.

The history of various territories' legislative efforts to re-introduce competition in mobile services that had essentially vanished along with the airtime providers is somewhat muddied due to parallel efforts from individual states e.g. the Scandinavian countries, and the framework in which they operated (the European Union). However, arguably – with considerable hindsight – the guidelines that most defined the vision for the future of the mobile network ecosystem that we have now in the early 2020s was 95/62/EC [17], the European directive on Open Network Provision.

Somewhat forgotten one-quarter of a century on is the fact that the Scandinavian countries were at the forefront of virtual operator services, having had a form of international roaming from the outset by virtue of their common 1G analogue Nordic Mobile Telephone System, operating at 450 MHz (long-range coverage) and 900 MHz (short-range coverage). This analogue international roaming incentivised operators who, with a single handset, might be able to offer 'first-party' MNO service in their own Scandinavian country, whilst offering a 'second-party' MVNO service via interconnecting MNOs in neighbouring countries.

Thus it was, that the first putative MVNO to 'blaze the trail', Sense in Norway, laid much of the legal ground for future MVNOs in other countries [18]. However, delays due to their wish to use their own SIM cards in the 2G MVNO pan-Scandinavia service meant that UK MVNO Virgin Mobile, then operating as Virgin Mobile UK on the physical operator One-2-One, was the first to launch service in November 1999.

3.2.1 Economics of an MVNO

In this section of the current chapter, the unique CAPEX and OPEX of virtual operators will be scrutinised. In their 2007 paper [19], Smura Kiiski and Hämmäinen performed a thorough analysis for MVNOs from the early 2000s perspective of voice services. Here, that treatment is updated to add data services to the MVNO profit/loss equation; the original analysis having been exclusively based on voice interconnection profit/loss.

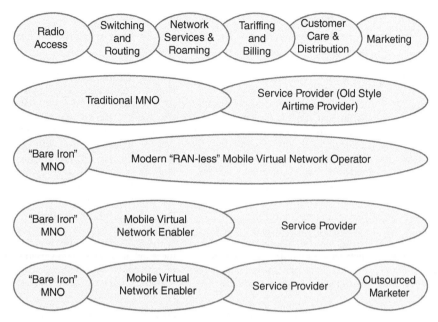

Figure 3.10 Value Chain Possibilities from MNO through to Service Provider.

In deciding which template in the value chain each mobile operating entity 'fits', the authors identify the value chain possibilities shown in Figure 3.10, slightly reworked here for the digital age where data 'screen minutes' now dominates over voice'. A fourth template is added at the bottom of the figure, where in the Web2.0 era 'Marketing as a Service' can be optimised and bought via Social and New Media, and traditional advertising channels were decreasing in importance in the early 2020s. The entities in Figure 3.10 are described in Table 3.2.

In [19], Smura, Kiiski and Hämmäinen develop the ideas of Ulset in [20], which propose that mobile operations ranging from MVNOs through SPs on a vertical axis, can be resolved along a horizontal axis ranging from Cost Leader on the left, to Service Differentiator on the right, Figure 3.10. Reference [19] provides multiple examples from the era of publication of the original paper, of placing MVNOs and SPs in existence at the time of writing of the paper (2005) along the cost leader through service differentiator continuum; today (2020) these examples are either irrelevant, out of business or have moved to a different service model in the intervening 15 years. Figure 3.11 in this current chapter therefore gives indicative examples of services offered by MVNOs and SPs in today's markets.

Hence, for the purposes of comparing where the CAPEX and OPEX falls with Profit/Loss, the four relative extrema were scrutinised, with examples of each scenario's 'selling points':

- Service Provider/Cost Leader (minimum voice/data support e.g. international minutes).
- Service Provider/Service Differentiator (value added services e.g. enhanced roaming).
- MVNO/Cost Leader (unlimited national minutes and data, no specialist content access).
- MVNO/Service Differentiator (unlimited everything plus Content Distribution access).

Table 3.2 Description of logical entities in MVNO value chain.

Entity	Description	CAPEX Status
MNO – Mobile Network Operator	Legal entity that holds the mobile network spectrum licence in a territory. Also owns and operates the Radio Access Network of base stations. In all cases, will also operate a core network to service internal mobile subscribers and customers of Service Providers that do not operate their own core network elements.	Owns spectrum licence, Radio Access Network (or is shareholder in any joint RAN venture) and one 5G-MSC/VLR, HSS, MME and S-GW. P-GW can be virtualised but will also, in nearly all cases, be owned by MNO. Will also own and operate the SMS/MMS servers and any IN platform (despite these being easily virtualised) due to Lawful Interception Gateway considerations.
SP – Service Provider	Very similar to the 1G and early 2G airtime providers, sells subscriptions, billing services and, in most instances, discounted/bundled User Equipment to customers. Each MNO will have their own internal SP.	Owns the 'relationship with the customer'; buys services wholesale from MNO or MVNE and repackages these for subscribers with varying requirements. Owns billing and customer-care system, which may itself be virtualised/outsourced.
MVNO – Mobile Virtual Network Operator	Same as an SP, with the addition of an independent core network (5G-MSC/VLR HSS, MME and S-GW for 4G/5G MNVOs); can make independent interconnection and, in effect, independent roaming agreements with other MNOs/MVNOs.	Owns core network and edge elements required for interconnection with other MNOs/MVNOs. Integrates service, billing and customer-care platforms as either virtualised or 'real' resources.
MVNE	Entity that interfaces between MNOs and their daughter MVNOs/SPs; can operate outsourced activities of these, always on virtualised platforms for 'low cost of entry'.	CAPEX commitments vary depending on relationship with MVNO/SP agreement. Unlikely to own other than hardware platform on which network services are virtualised.

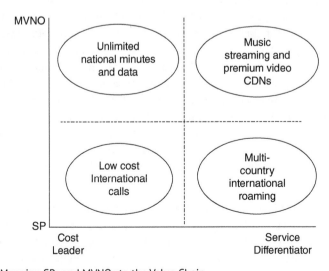

Figure 3.11 Mapping SPs and MVNOs to the Value Chain.

None of the four scenarios is intended accurately to recreate the cost/service combination of any MVNO or SP, much in the same way that no pupil in any class is likely to have exactly the average class height; rather, the notion is to create four generic scenarios against which CAPEX and OPEX 'what-if's' can be run, to see where each lie along the line to viability.

3.2.2   Modelling MVNOs and SPs

Smura, Kiiski and Hämmäinen [19] used a spreadsheet-based model developed initially in the IST-TONIC ('TecnO-ecoNomICs of IP optimised networks and services') research project [21], and further refined in the CELTIC-ECOSYS ('techno-ECOnomics of integrated communication SYStems and services') project [22] covering the time period up until 2005, i.e. the dawn of IP and Internet services via the first 3G operators.

TONIC occupied itself with examinations of two scenarios using the TERA spreadsheet analysis tool running TONIC's 'Titan' model; these were classified as the 'average large European country', and 'smaller, Nordic-type countries'. CELTIC-ECOSYS followed this up with a generalisation from small/large country operators to the four generic scenarios outlined in 3.2.1 above, and enhanced in particular the OPEX model for MVNOs and, as they called them, Service Operators.

Overall, the breakdown of the techno-economic model resembled that shown in Figure 3.12; since the IST-TONIC and CELTIC-ECOSYS projects in the early years of the millennium, data usage has grown nearly completely to supplant voice minutes consumed by the

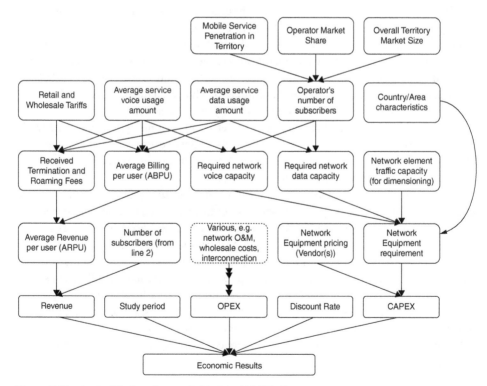

Figure 3.12   Logical Techno-Economic Model of MVNO/SP.

average user – to the extent that so-called Millennials or Generation-Z'ers would be quite surprised and alarmed that anyone might make a voice call to them!

Hence, the model (Figure 3.12) has had to be re-worked to place a greater emphasis on data usage; the logical entities on each line receive the obvious and largely self-explanatory inputs from the entities on each line above except for Subscriber Numbers from line two being repeated on line four for the sake of clarity.

A point of note is that, although the spreadsheet model could have incorporated feed-back loops, for the sake of simplicity recursive calculations these have not been implemented; given the fairly broad assumptions that had to be made elsewhere in the model in terms of other inputs e.g. interconnection costs, vendor equipment pricing etc., the choice has been made to implement a 'stateless' model without feedback.

One consequence of this is that Churn, or the propensity of any individual subscriber to opt out of their existing contract with mobile operator A, and switch to operator B/C/D etc. at the end of the contract term, is not directly modelled. Hence, the logical 'number of sub-scribers' that arises as a function of Mobile Service Penetration, Operator Market Share and Overall Territory Market Size is not necessarily the same identity of subscribers from accounting period to accounting period for the operator under scrutiny, but rather seeks to calculate a general level of 'attractiveness' of that operator (given its contract/Pay As You Go tariffs, UE offers, bundled content offers etc.).

The original 2005 study of [19] forecast ABPU and ARPU levels for 2006 to 2010; this was rescaled to 2012 to 2016, to align with the figures of the Year-on-Year ARPU graph shown in Figure 3.1 at the beginning of this chapter. In the 2005 study, incomes from retail ser-vices had been broken down across a mix of four different service classes, Calling-oriented, Messaging-oriented, Data-oriented and Content-oriented.

These four classes were still, by and large, valid for the 2012 to 2016 timeframe, except by this point in time the effective 'value' charge for mobile voice calling had fallen to zero, no post-paid tariffs from any major operator charging for these any longer. However, MVNOs still receive income from fixed-to-mobile termination charges, a major difference from the Service Provider model where revenue consists only of retail income.

Mobile package charges do not exactly follow Retail or Consumer Prices index either, so scaling up ABPU by 10% across the board results in Service Provider/Cost Leader and MVNO/Cost Leader ABPU staying level at EUR26.40 and for the two differentiator sce-narios increasing linearly from EUR28.6 to EUR33.

Of course, this approximation is somewhat sweeping, as during the initial study period from 2006 to 2010 Content Aggregators such as Netflix and Amazon Prime, and streaming services such as Spotify Premium and Deezer either did not exist or had only just started trading; now it is commonplace for high-end post-paid subscriptions to offer a bundled package to these and other premium content streaming, such as Sky Sports in the UK.

Thus, based on an estimated percentage uptake of these Content-Bundling subscrip-tions, we can now look at an ABPU for the differentiator scenario, at least EUR5 greater at each end of the period – at the beginning of the period there was not a great uptake of these services. Taking this altogether, the differentiator scenario is adjusted to increase in ABPU to EUR38 in 2016, the initial level (before content aggregator offers) remaining at EUR28.60.

Table 3.3   Rescaled and extrapolated assumptions to fit the age of mobile data.

Interconnection Tariff	2012	2016
Percentage mobile-to-fixed from all mobile-originated calls	17%	10%
Percentage fixed-to-mobile from all mobile-terminated calls	17%	15%
Percentage of retail price, fixed-to-mobile termination fee (revenue)	65%	50%
Percentage of retail price, mobile-to-fixed termination fee (OPEX)	20%	20%
Percentage of retail price, incoming mobile-to-mobile termination fee (revenue)	65%	50%
Percentage of retail price, outgoing mobile-to-mobile termination fee (OPEX)	65%	50%

Translating top-line ABPU to (near to) bottom-line ARPU takes a number of factors into account, including traversing a table of assumptions around OPEX; the reader is invited to peruse these in [19], however, it is illustrative to present the changed assumptions around interconnection charges, which will be key to profitability for the MVNO model. Rescaled and extrapolated from the 2006–2010 assumptions, the 2012–2016 assumptions are shown in Table 3.3.

In addition to the service revenues that subscribers pay in all four scenarios under test, both MVNO models also generate revenues and incur costs from voice and data interconnection with the outside world. If a call is being made from mobile operator A to mobile operator B, then A would have to pay a termination fee to B, for B to deliver to the called party. In a situation where incoming and outgoing call distribution and share follow each operator's market share in any given territory, then in a steady state where all operators' termination fees have converged to roughly the same then net inter-operator interconnection fees could be expected to cancel out over a sufficiently long term.

The reality is, however, that there is a great disparity between mobile-to-fixed, and fixed-to-mobile delivery charges, biased in the direction of the latter. Even the advent of IP-based Voice over LTE (VoLTE) services does not seem to have affected this equation much, where MNOs and their daughter MVNOs are substantial net beneficiaries of fixed-line operators' termination fees. In theory, international roaming fees – which are still substantial even at the time of writing this in 2020 – could be an extra source of revenue for MVNOs, however, these by and large rely on the roaming agreements put in place by the parent MNO, to whom the roaming benefit ultimately accrues – there is little point in removing hassle by virtualising your mobile operation, only to have to bring it into the 'real world' by having to manage roaming agreements with all other possible worldwide operator roaming partners.

Putting this all together results in the ARPU projections for the MVNO/Cost Leader and MVNO/Service Differentiator shown in Figure 3.13, where the (Calling + Messaging) and (Data + Content) revenues are the same for the Service Provider models, only the termination fees collected being absent.

The results (bypassing a host of assumptions that can be inspected in [19]) are shown in Figure 3.14; time-to-payback is seen to be greater than 5 years for both service provider scenarios, whereas time to payback for both MVNO scenarios are seen to turn cash positive in 2.5 and 3.1 years for the Cost Leader and Service Differentiator scenarios respectively.

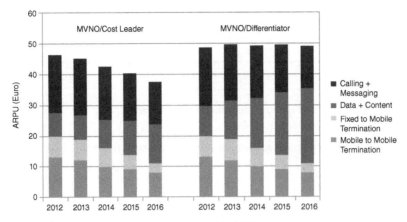

Figure 3.13   MVNO Subscriber ARPU both types 2012–2016.

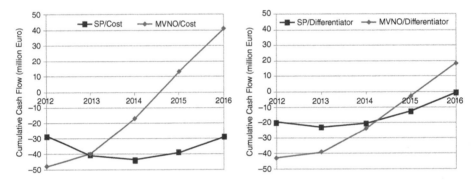

Figure 3.14   Payback Period for Cost Leader and Service Differentiator SP/MVNO Scenarios. (*Source* [19]).

This perhaps helps to explain why Mobile Network Operators prefer and give earlier access to the unprofitable, service provider model and hamper the Virtual Network Operator business model. MNO strategies include, but are not limited to, preventing access to the latest technology layers for an extended period of time, such as happened with ASDA Mobile which was denied access to 4G by host operator EE until 2017, five years after LTE was first launched in the UK.

Latter-day MNO disruption strategies now include preventing MVNO/SP access to features such as VoLTE or even (via SIM identity) the 5G layer in its entirety, whether or not the MVNO/SP is providing its customers with 5G-capable UEs.

Some of the CAPEX and OPEX assumptions in terms of network 'machinery' made in [19] also need to be re-examined. Smura, Kiiski and Hämmäinen had assumed that any MNO would own and operate the RAN elements including SGSNs and one or more GGSNs, with daughter MVNOs owning one or more 2G/3G/4G MSC-Servers, an OMC function, the Service Platform, any walled garden CNs and, of course, the HLR/Home Subscription Server(s). Service Providers would only own the Billing and Customer Care systems, plus any UE discount/loyalty scheme that presumably would be operated with the blessing of the parent MNO. This was the 2G and early 3G 'scene', in the intervening decade-and-a-half things have moved on radically, particularly in the data realm.

Now, in the 2020s environment of 5G beginning to displace 4G, continuing miniaturisation and power efficiencies have made it feasible for MVNOs to own 4G- and 5G-Serving-GW and Packet-GW entities (see Figure 3.2(c)). Thus, MVNOs can, if desired, make carriage and interconnection deals literally from the edge of the RAN all the way to the Internet, providing their own fibre routes.

One disadvantage of so doing would be that the MVNO will be responsible for ensuring continued interoperation of the S-GW and P-GW with the parent MNO and other MVNOs owning such network elements, which must have the same geographic extent as the MNO's own network machinery. Another is that any MVNO encroaching on parts of the RAN and core that would ordinarily be owned and managed by the operator, would be responsible for the security of IPSec flows between virtualised and/or geographically disparate elements, thus getting back involved in exactly the kind of complexity from which the MVNO/SP model abstracts them.

Hence, the future of the now very extensive MVNO/SP landscape may be less about shaving off bottom line costs for subscribers and more about fostering brand loyalty and ability to profile loyal 'brand followers'.

3.3 Operator Business around International Roaming

International roaming enables you to use your mobile device seamlessly in another country and pay any additional fee at the end of your tour to your home operator. This is a significant source of income for many mobile operators; in many 'touristic' countries can be as high as 3% to 4% of their total revenue. The commercial aspects of roaming can be explained by the diagram in Figure 3.15 below, taken from [23].

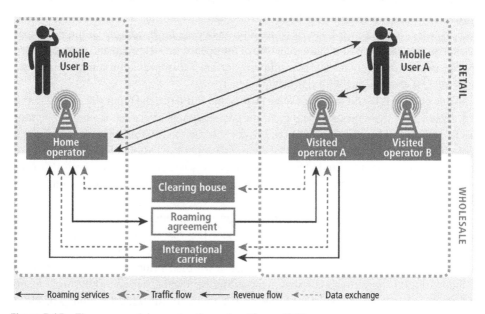

Figure 3.15 The commercial aspects of roaming. (*Source* [24]).

The roaming platform is underpinned by operator roaming agreements, which detail what services are provided at what price, signed between operator A of one country and operator B in another country. Given that there are over 1000 mobile operators in 220 countries and territories in the world, it is no trivial matter for an operator to negotiate and manage hundreds or even thousands of roaming agreements.

The GSMA stepped into this domain at the time of GPRS data roaming introduction, with their GRPS roaming exchange (GRX). GRX are a set of private exchange nodes that roaming operators could connect to so the roaming traffic can be routed without the need for additional data security measures. The associated set of technical and commercial agreement templates greatly simplified the process of entering into a roaming agreement for many operators.

As shown in Figure 3.15, the commercial terms between the two operators is agreed on wholesale traffic terms i.e. without respect to each individual voice-data transaction down the line. Then, each time a data or voice service is used by a roaming user A in a visitor network, the visited operator sends a transferred account procedure (TAP) file to the home operator through a clearing house. The home operator bills the roaming user A based on the usage amounts in the TAP file at the indicated prices to the roaming user A at the end of his/her visit, where this may be very different for disparate subscriber groups e.g. monthly contract versus Pay as you Go subscribers. The usage from multiple roaming users is accumulated and the payments to the visited operator and the transport network provider(s) happen at the wholesale amounts of traffic at the agreed prices, at set time intervals.

3.3.1   The EU Roaming Regulation 'Roam like at Home'

The European Economic Area (EEA) is intended to operate as a single market, with the freedom of movement for goods, services and labour within these European countries without additional regulation or taxes. So, the levy of roaming fees between European countries has long been a bone of contention between the European Union (EU) and European MNOs. Given that more than 65% of Europeans travel abroad and most southern European countries rely on a mass influx of northern Europeans for summer holidays, roaming fees was an issue with significant impact.

At this point, it is useful to look at some factors that influence the pricing of mobile roaming. Technically, there needs to be regular message exchanges between the home operator CN to exchange subscriber information, privacy settings etc., so this represents additional signalling. However, the high pricing for roaming is mainly a result of lack of competition in this roaming market. An individual subscriber is very unlikely to change home MNO or even take up a specific roaming package before a trip abroad, with a view of minimising roaming charges.

Mobile operators used to put forward the point that roaming charges are a part of the total price package they offer and users can choose low roaming charge packages (at the expense of higher home charges) if they wish to. In reality, many people travel abroad only once or twice a year and roaming tariffs are far down their priority list when choosing a mobile tariff. It is only after a trip that many subscribers will experience 'bill shock' when roaming charges become apparent. A side note is that with the wide availability of Wi-Fi hotspots and the use of OTT (over-the-top) VoIP applications, many tech-savvy roamers were able to circumvent these roaming charges to some degree, at least for outgoing Mobile originated calls.

Within this context, the EU and a number of national regulators introduced price caps, to mitigate 'bill shock'. These included retail unit price caps on voice, data and text roaming charges introduced in 2012 and further consolidated in 2014. In parallel, price caps were also introduced for the wholesale roaming traffic. National regulators followed up by introducing maximum charge limits that can be made per trip and a requirement for the operator to alert the roaming subscriber to any approaching limits or even to cap roaming charges after a certain expenditure has been incurred.

The EU was working towards total abolition of roaming charges within EEA, so that the roaming minutes and data would be tallied within the home subscription package. Their argument was that roaming charges were at the retail price levels set and did not reflect the actual costs of providing this service, incurred at the bilaterally agreed wholesale rate.

MNOs also had legitimate concerns. Within Europe, there are significant differences in the per capita GDP levels, especially a division between Western and Eastern Europe. Abolition of roaming charges might be used as a loophole whereby a mobile subscription at a lower tariff from a lower GDP country could be used long term in a higher GDP country of residence. The cost of service provision in such cases might be higher than the up-front billing to the subscriber. Taking these points into consideration, the EU developed the RLAH (Roam Like At Home) regulations, that abolished roaming charges and came into effect in mid-2017. There is an accumulated 4 month maximum time limit per year during which a subscriber can benefit from RLAH. Also, MNOs are able to levy additional fees if it can be shown that the actual costs of providing a specific roaming service is higher than in the home network.

The EUs expectation was that RLAH would actually encourage roaming subscribers to use more of their voice and data allowances whilst travelling. A report published 1 year after RLAH came into effect [24] provides data to support this, as shown in Figure 3.16 below. However, a number of operators did increase their roaming charges to countries outside of the EU, in an attempt to offset their reductions in roaming revenue. This report [24] also finds that whilst the RLAH regulations have effectively abolished roaming charges for the retail market, there is an emerging wholesale roaming customer segment, who is yet

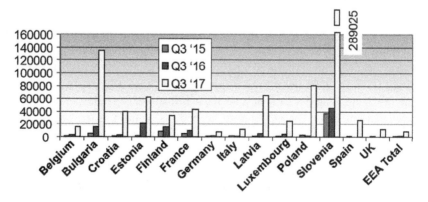

Figure 3.16 Relative data roaming volumes before and since the introduction of RLAH. (*Source* [24]).

to benefit. These include roaming IoT sensor fleets fitted to motor vehicles, sea-craft and aircraft, the connections to which are served by specialised MVNOs. This wholesale roaming customer segment may be the next area where the EU regulations may expand.

With the UK leaving the European common market at the end of 2020, UK mobile operators are no longer bound by the RLAH regulations. Also, the trade deal signed between the EU and UK does not cover this segment. Although UK operators have indicated that they do not intend to re-introduce roaming charges to the EU in the near term, this may change in the future [25].

3.3.2 Covid-19 Impact on Roaming Revenues

As with many other economic sectors, Covid-19 has had a significant impact on the roaming revenues of mobile operators. Worldwide tourism and business travel has come to a standstill and so have roaming-driven revenues. The GSMA intelligence [26] estimates that mobile operator revenues are 4% to 8% lower in 2020 in developed countries due to the Covid-19 impact. A large part of this will be contributed to by the loss of roaming revenue.

However, there has been an increased demand for fibre and other telecom services due to the need for home schooling and home working and it is expected that many of the diversified mobile operators would be able to bounce back strongly from these losses. During the Covid-19-related quarantines, many accounts emerged of stranded in foreign countries, for whom a mobile phone became the only connection with friends and family back home. In many cases, MNOs have stepped in to provide compassionate terms for longer usage, waiving existing retail roaming tariffs.

3.4 The Likely Operator Business Models in 5G

Stagnating operator revenues and the falling ARPU has been a major concern for many of the world's operators for a number of years (see Figure 3.1). 5G presents an opportunity to reverse this trend. Usually, the introduction of a new technology generation generates much hype and excitement, that translates to increased uptake of the services in the near term; 5G potentially offers a much wider and lasting business impact.

The need to diversify their service offering is understood and implemented to a varying degree by many major telecom operators, at least in the connectivity market. In the recent 4G era, many operators have ventured into home Internet and fibre broadband connectivity offerings. 5G in the millimetre wave (mmWave) bands (6–100 GHz frequencies) offer the possibility of enabling fixed wireless services to rival the Giga bits per second (Gbps) data connectivity of fibre. The US operator Verizon is pioneering the use of 5G Fixed Wireless Access (FWA) in the 28 GHz spectrum and has plans to expand this to 30 million US households over the next 5 to 7 years [28].

Diversification can also target other areas like Content Provision and E-commerce. Mobile operators have had some success with Paid-for TV content provision, especially when bundled together with connectivity services. 'Quad Play' service provision, where the mobile, home landline, broadband and pay TV ssubscriptions are linked has permitted many mobile operators (like BT in UK) to offer packages via which entertainment can be viewed on multiple devices as convenient to the user, rather than being restricted to a traditional single device.

Approaching from the other direction, traditional content providers have also ventured into providing mobile connectivity encroaching on the Quad-Play market. Paid-for TV provider Sky in the UK is also a sizable MVNO providing Quad Play bundles as attractive service packages. E-commerce and digital content provider Rakuten has now deployed a 4G and 5G mobile network, becoming the fourth operator in Japan. With the AR-VR and holographic content consumption becoming possible as 5G eMBB services, MVNOs may get a chance to establish in this very competitive digital content market segment.

The biggest opportunity in new markets for 5G mobile operators will come from the service provision to vertical industries. 5G is seen as an enabling technology to satisfy the connectivity needs of many verticals, including automotive, Smart city, Industry 4.0 and mission critical communications. How 5G is developed to handle the diverse KPIs these industries require, will be discussed in Section 5.8. Also, new developments in 5G network architecture, the centralisation of the RAN and virtualisation of the CN (technical details covered in Section 5.9) enables multiple verticals to be supported through the same physical network via network slicing, as briefly explained below.

Revisiting Figure 3.2(C) as a schematic showing the working of a mobile operator, from the end UE on the left, rightward through BTS/NodeB/eNodeB/gNodeB and RAN elements, CN elements then finally outbound connections from the Core to Packet Data Networks on the right, the 5G version of this introduces many changes to how the core layers work and how the 3G/4G logical functions metamorphose into their equivalent 5G versions.

Chief amongst these is the advent of aggregation layers at both RAN and CN levels. The RAN manifestation of this is separation of the previously monolithic gNodeB base station hardware elements into a Centralised Unit (CU) essentially corresponding to the off-mast Baseband Unit previously described in this chapter, and Distributed Units (DU), which are analogous to on- or off-mast Remote Radio Units as also previously described. Where in 4G, a BBU and its RRUs were associated with a single physical site, in 5G, the vastly-more-capable CU can control many DUs and can co-ordinate the interworking of macro and small cells in a wide area.

Exactly which optimum rollout strategies will shape how DUs and CUs are placed in future 5G networks has yet to be determined, and they will not be further considered here save to say that the high bandwidth optical connection between these now supports a full 3GPP-defined interface; the F1 interface. Thus, whereas in 3G and 4G, a single vendor would supply an eNodeB's Baseband Unit and its peripheral Remote Radio Units, now in 5G, it would theoretically be possible to mix DUs from Vendor A with CUs from Vendor B (although it would be a brave and/or over-optimistic operator who would so do).

Much more interesting is the introduction of the aggregation at the CN level. Fitch and Sutton [28] cost and calculate the transit latency of a 12,000 DU network, supported by 10 User Plane Function (UPF) Core elements (corresponding to the data transport elements of Packet Gateway Servers in 3G/4G) and 106 'Aggregation' elements. These 106 new 5G entities are Data Centres, located much closer to the DU locations than the Core UPFs. The number and distribution of aggregation elements, which are virtualised entities and therefore capable of being run by either an MVNO or an SP, depends on the particular use case for which that sub-part of the network is being optimised.

Within any real 5G operator, there could exist multiple virtual 5G network operators, running different numbers and geographic distributions of aggregation elements between the real operator's RAN and Core, essentially creating a '3D' look to the 2D, left-to-right network schematic model. These multiple virtual 5G optimisations are referred to as network 'slices'. A 3D network schematic, with the inclusion of two network slices is depicted in Figure 3.17 below.

To understand the use case, consider the example of a major vehicle manufacturer such as Ford or Volkswagen, who wants to establish a virtual operator in support of the 'Internet of Things' micro-network in its vehicles. These communicate internally and with other vehicles, possibly autonomously steered, via the 5G 'V2X' channel. Th V2X requires Ultra Low Latency, hence the car manufacturer's optimisation would be to bring its aggregation datacentres geographically very close to the DUs and CUs supporting 5G modems in its vehicles, thus minimising circuit delay in the V2X channel and maximising reliability and safety of autonomous vehicle movement.

Conversely, consider the example of a Netflix-like content provider. Here, the aggregation points would be oriented towards security of content, perhaps motivating a centralised location, failover between serving network elements (also motivating a centralised location) and, depending on how the content-provider's on-UE application works, considerations of data buffering may outweigh and reduce the priority of low latency communications. Yet, a third virtual operator, say Hotpoint establishing an IoT slice in support of its in-home products, would emphasise reliability of low-power, low-bandwidth communications, requiring optimisations in position and number of both the core aggregation units and also perhaps a virtualised presence on the DUs themselves.

5G will bring new opportunities for the mobile operators to work closely with these vertical industries and develop B2B and B2B2C solutions. We will look in more detail at the technical and business requirements and challenges in taking 5G to these main vertical industries in Chapter 12.

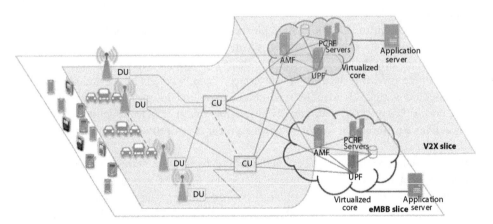

Figure 3.17 Creation of two network slices from the same physical 5G network.

3.5 Conclusion

In this chapter, we have analysed various mobile operator business elements from the subscribers' Average Revenue Per User (ARPU) and the historical development of this, through the various Capital Expenditure elements in Radio Access and CNs, and in the transmission, elements connecting these either over radio itself or superfast fibre networks. Middle-of-the-balance-sheet elements of Operational Expenditure have also been considered, with simplified examples of site rental and power being offered as examples of day-to-day running costs that cannot be capitalised to the value of the network and can be honed and improved over time by operators thus improving the bottom line.

Examples where a business might want to choose a route to market bypassing CAPEX and OPEX complications have also been presented; a very large ecosystem of virtual operators now exists in all countries with developed consumer cellular markets, both of the MVNO-type which own some real hardware and virtualised CN elements and who leave running of the RAN to the host operator, and of the Service-Provider type, who delegate nearly all RAN and CN responsibilities to the host and may own as little as a billing system.

Thus, at this extreme Service Provider end of the spectrum, a cost-leader or service-differentiator Service Provider is much more a vehicle for fostering loyalty with a larger umbrella brand (thus removing some purchasing complexity in terms of purchasing choice from the consumer) than it is about the original motivation for establishing a virtual operator, cost savings. Headline voice minutes and data megabytes are cheap nowadays in the 2020s anyway, with the main operators being forced to compete on price with the virtual operators they host.

We also looked at the Mobile Roaming market and how these roaming prices are determined. Special attention was given to the European regulatory attempts in curtailing and finally abolishing roaming charges within EU. Some recent challenges with the onset of Covid-19 travel restrictions have also been covered.

There is much more 'mileage' to come from the 2020s version of the virtualised operator model with the advent of 5G; whereas the original motivation was cheap(er) voice minutes in the 1990s, through affordable data bundles in the 2000s to 'all you can eat' pricing models for both voice and data in the 2010s, now as rollout of 5G architectures starts there is again an opportunity for genuine service differentiation.

At time of writing (2020 Q4) there are, as yet no examples of optimised 5G MVNOs to which to look as examples but, in due course, it will be possible to update this current work with real-world findings reflecting evolved, optimised 5G business practice.

References

[1] Statista (2020). *Monthly ARPU through mobile subscribers in the United Kingdom (UK) from 2007 to 2016 (in GBP), by pay type.* https://www.statista.com/statistics/273471/monthly-arpu-through-mobile-subscribers-in-the-united-kingdom-uk (accessed 16 June 2021).

[2] Office for National Statistics (2021). *Inflation and price indices.* https://www.ons.gov.uk/economy/inflationandpriceindices (accessed 16 June 2021).

[3] https://www.gsmaintelligence.com/research/2018/02/the-mobile-economy-2018/660

[4] Verbrugge, S., Pasqualini, S., Westphal, F.-J. et al. (2005). Modeling operational expenditures for telecom operators. *Conference on Optical Network Design and Modelling*, Milan, Italy (7–9 February 2005). IEEE.

[5] CISCO (2008). *Cisco ASR 5500*. https://www.cisco.com/c/en/us/products/wireless/asr-5500/index.html (accessed 16 June 2021).

[6] Al-Rahweshidy, H. and Komaki, S. (eds.) (2002). *Radio over Fiber Technologies for Mobile Communications Networks*. Norwood, MA: Artech House.

[7] Al-Rahweshidy, H., Galbraith, D., and Prasad, R. (1998). Performance of radio on fibre for microcellular GSM networks. *48th Vehicular Technology Conference*, Ottawa, Canada (21 May 1998) pp 392–6. IEEE.

[8] BridgeWave Communication (2008). *The economics of gigabit 4g mobile backhaul: How 'wireless fiber' 80 GHz links provide an economical alternative to operator-owned fiber*. https://bridgewave.com/wp-content/uploads/2017/06/WP_4G_Mobile_Backhaul.pdf (accessed 16 June 2021).

[9] Ofcom (2018). *OfW 446: Technical frequency assignment criteria for fixed point-to-point radio services with digital modulation*. https://www.ofcom.org.uk/__data/assets/pdf_file/0017/92204/ofw446.pdf (accessed 16 June 2021).

[10] Ofcom (2018). *UK frequency allocations for fixed (point-to-point) wireless services and scanning telemetry*. https://www.ofcom.org.uk/__data/assets/pdf_file/0016/92203/OfW48.pdf (accessed 16 June 2021).

[11] Lehpamer, H. (2010). *Microwave Transmission Networks: Planning, Design, and Deployment*, 2e. Chicago, IL: McGraw-Hill.

[12] BridgeWave Communication (2008). *The cost and performance benefits of 80GHz links compared to short-haul 18-38 GHz licensed frequency band products*. https://bridgewave.com/wp-content/uploads/2017/06/WP_Lower_Licensed_Frequency_Comparison.pdf (accessed 16 June 2021).

[13] *EUR-USD and EUR-GBP historical data at* http://sdw.ecb.int.

[14] Grijpink, H., Ménard, A., Sigurdsson, H. et al. (2018). *The road to 5G: The inevitable growth of infrastructure cost*. https://www.mckinsey.com/industries/technology-media-and-telecommunications/our-insights/the-road-to-5g-the-inevitable-growth-of-infrastructure-cost# (accessed 16 June 2021).

[15] Humar, I., Ge, X., Xiang, L. et al. (2011). Rethinking energy efficiency models of cellular networks with embodied energy. *IEEE Netw* 25(2): 40–49.

[16] GSA (2016). *5G Network Energy Efficiency*. https://onestore.nokia.com/asset/200876 (accessed 16 June 2021).

[17] Publications Office of the European Union (1995). *Directive 95/62/EC of the European parliament and of the Council of 13 December 1995 on the application of open network provision (ONP) to voice telephony*. https://op.europa.eu/s/skSn (accessed 16 June 2021).

[18] Telenor Group (1999). *Sense Comunications buys GSM network capacity from Telia/Telenor Mobile*. https://web.archive.org/web/20190402015411/https://www.telenor.com/media/press-release/sense-communications-buys-gsm-network-capacity-from-teliatelenor-mobile (accessed 16 June 2021).

[19] Smura, T., Kiiski, A., and Hämmäinen, H. (October 2007). Virtual operators in the mobile industry: A techno-economic analysis. *Netnomics* 8(1): 25–48. https://web.archive.org/web/20200319204223/http://www.123seminarsonly.com/Seminar-Reports/034/52242110-Economic.pdf

[20] Ulset, S. (2002). Mobile virtual network operators: A strategic transaction cost analysis of preliminary experiences. *Telecomm Policy* 26(9–10): 537–549.

[21] Tonic (2009). *Tonic project*. https://web.archive.org/web/20090606155944/http://www-nrc.nokia.com/tonic (accessed 16 June 2021).

[22] Ecosy (2011). *Ecosy project*. https://web.archive.org/web/20110807133631/http://www.optcomm.di.uoa.gr/ecosys (accessed 16 June 2021).

[23] GSMA (2013). *International Roaming explained*. https://www.gsma.com/latinamerica/wp-content/uploads/2012/08/GSMA-Mobile-roaming-web-English.pdf (accessed 16 June 2021).

[24] European Parliament Think Tank (2018). *Roaming: One Year After Implementation*. https://www.europarl.europa.eu/thinktank/en/document.html?reference=IPOL_IDA(2018)626090 (accessed 16 June 2021).

[25] Ofcom (2021). *Using your mobile abroad*. https://www.ofcom.org.uk/phones-telecoms-and-internet/advice-for-consumers/costs-and-billing/using-your-mobile-abroad (accessed 16 June 2021).

[26] GSMA (2020). *Intelligence Shares 'Global Mobile Trends 2021'*. https://www.gsma.com/newsroom/press-release/gsma-intelligence-shares-global-mobile-trends-2021 (accessed 16 June 2021).

[27] Light reading (2020). *Verizon CEO views 5G Home as 'transformative business'*. https://www.lightreading.com/5g/verizon-ceo-views-5g-home-as-transformative-business-/d/d-id/763936 (accessed 16 June 2021).

[28] Fitch, M. and Sutton, A. (2018). Developing a 5G network architecture. *IEEE 5G Summit*, Glasgow.

4

Why Standards Matter

Telecommunications have become a global infrastructure because of their adherence to Standards. From the outset, the ability of one end point – originally a telephone, but now any range of different devices – to communicate with another, has had a dependency on those end points both adopting principles enshrined in the documentation of Standards Bodies. Where differences in protocols have occurred, great care has been taken to define the interworking between the differing definitions, and also to limit the number of permitted options.

Fixed-line telecommunications were bifurcated by the differences between 'a-law' companding algorithm devised in Europe, and 'μ-law' companding created in North America and Japan and, as a consequence, differences in transit protocols. This historic split generated a need to interwork between a-law and μ-law for any telephone call between countries using differing principles. Looking at this split retrospectively, it is clear that using either one method or the other globally would have been much more sensible, but the split is so ingrained in network implementation that it is impossible to reverse. The fact that there are only two options should be seen as a victory, however, telecoms could be in the same state as electricity supply, with their mix of voltages and socket form factors.

The geographic split in fixed-line telecoms was at least initially reproduced at the advent of mobile telecoms. Technological research naturally leads to many parallel streams of development and consequentially can create differences of opinion. Whilst superiority of technical solution can often be the determining factor in deciding which strand of research becomes 'the standard', there is also scope for other influences, not least of which are politics, commercial weight, or a desire for rapid progress forcing compromises to be made.

This chapter outlines a brief history of mobile telecoms Standardisation activities, reflecting on the bodies that have averted the potential for a highly segmented industry, the reasons why unification around a common global standard ultimately has come about, the tensions that have existed along that path and the importance of Standards in the future, when even the process of Standardisation is being disrupted by technology, and more recently, a global pandemic.

The Technology and Business of Mobile Communications: An Introduction, First Edition.
Mythri Hunukumbure, Justin P. Coon, Ben Allen, and Tony Vernon.
© 2022 John Wiley & Sons Ltd. Published 2022 by John Wiley & Sons Ltd.

4.1 The Creation of a New 'G'

At the advent of mobile telecoms, the idea that multiple generations of technology would exist must have been at best, a very distant thought. However, when the first mobile phones reached commercial launch, there was already work taking place to improve on the quality and security of the service that had only just begun. This put in place a cycle that has perpetuated from initial analogue mobile connections, which we can now think of as first generation, or 1G, through digitisation, the addition of data, the consolidation of voice and data onto a single system and now to the expansion of mobile connectivity to be about much more than just making voice calls and sending text messages.

The idea that a new 'G' is created every ten years is a widely-used adage. This self-perpetuation of the industry is driven, in part, out of commercial necessity – vendors must always be thinking about what the next thing to sell may be, and operators have a need to expand upon their product offering, particularly when the consumer market reaches saturation – but also because of the ingrained innovation of the industry as a whole. So, whilst a new G may appear every ten years, there is considerable work done prior to the launch of a commercial service. The period of time between initial conceptualisation and full commercial adoption and scale is far longer than ten years, with each new generation's development process overlapping the previous generations. Let us consider that process in detail.

4.1.1 Research

Research into innovations to improve the communications network continues regardless of where in the cycle of the definition of a new 'G' the industry happens to be. In particular, academia are persistent in their pursuit of both methods for better utilisation of existing limited or expensive resource (Spectrum, computer resource, backhaul bandwidth, electricity and more) or in developing new paradigms to shift the implementation of networks or to open new possibilities (softwarisation, functional fragmentation, open interface definition, and so on).

Alongside this, commercial participants in the industry are also innovating, although often with a focus that is, at least in part, influenced by the Standardisation process that has developed around the definition of each generation, as well as based on their internal product development cycles.

The first aspects of a new Generation to be identified are the service targets that will be used to separate this Generation from the previous one. How these targets are set, and by whom, has shifted from one generation to the next and, at some point in the overall process, usually, well in advance of the 'Standardisation' of requirements, the industry forms a shared opinion on what needs to be achieved by a new G and why – this will be discussed for each generation later in the chapter. These requirements focus on flaws in the previous generation or address fundamental shifts in the usage patterns of consumers. In every case, however, what is defined provides focus for research efforts by all parties.

As the industry became increasingly fixated on new Generations, so a geopolitical aspect to research into the technologies to be included in each new G also developed. Research activities have become increasingly sponsored by regional groups, where consortia of academia, equipment vendors and network operators are funded to advance their chosen areas of development. This has further led to the idea that each new Generation has some form of race associated with it. Regional research efforts are considered to be competing for dominance or superiority, and intense focus is placed on being the first region to demonstrate technology that meets the requirements laid down for that Generation. This idea of there being a race associated with a new generation perpetuates into commercial implementation by vendors and deployment by operators.

Regardless of research source, all efforts are funnelled into the Standardisation process.

4.1.2 Standardisation

Standardisation of mobile telecoms is fundamentally important to the overall ecosystem that drives the industry, and the adoption of the specifications that the Standards' process defines sits behind the commercialisation of the innovation driven out of the research work that was discussed in the last section. As mentioned, requirements for a new generation of mobile telecoms are set loosely to allow for consolidation and focus of research work, but these requirements are formalised as the first step of a standardisation process.

In fact, the process for standardisation is so vigorously adhered to that it has become a standard itself. During the standardisation of ISDN, ITU-T defined a three-stage process for Standardisation which is now documented in ITU-T Recommendation I.130 [1] – specify requirements (Stage 1), develop a logical or functional model or architecture (Stage 2), and define protocols (Stage 3). This process is now widely used to retain order and to ensure features included in Standards are justified, and the overall body of work is complete and coherent. If a Stage 2 feature or function can be shown to advance the definition within a specification towards meeting a Stage 1 requirement, then it can be included (and conversely, if there is no requirement to base a Stage 2 feature upon, it can be rejected and not included). Similarly, Stage 3 protocol definitions meet the protocol functional requirements set for reference points or interfaces in Stage 2 architectures.

This process makes the documentation that results from Standards relatively easy to follow for the uninvolved reader. By reading a Stage 1 specification and following references into Stage 2 and Stage 3 documentation, the reader can identify the basis, principles and details of any feature included within a specification set. Applications of this process vary from one Standards forum to another, but broadly, good specifications adhere to these steps.

Most Telecoms Standards Definition Organisations are organised around Working Groups and Plenary Groups. Working Groups are where subject matter experts from all stakeholders discuss fine details of the Standards that fit within the scope of that group. The working groups often fit into families around broader themes, and each of those themes then has a Plenary Group that aims to ratify the work of the working groups, and maintain the alignment of the work completed by the working groups. A variety of models exist for how this can be organised. Generally, working groups hold regular face-to-face meetings, with increasing levels of work continuing between those meetings via e-mail lists and conference calls. Plenary groups then hold periodic meetings where only the most controversial subjects are discussed in detail, and where new versions of specifications are approved, which can then be published.

There are negative implications to this approach when applied to a global industry however. First, as we will see later in the chapter, where an industry is structured around a single dominant standard, the definition of that standard is a slow process. This is unsurprising as all of the work from the Research phase of the creation of a new Generation could potentially be relevant to the Standard that will define it. That work needs to be discussed, evaluated, refined and documented in a form that is agreed by all parties involved in writing the Standard. The operators, equipment manufacturers and academic institutions that were involved in Research, provide their conclusions to SDOs and a single implementation is chosen. This inevitably means that there are winners and losers in the discussion of every facet of a new specification set.

As a consequence, discussion around specification can become intense and heated. Where the standardisation of a feature is running in parallel to Research, often a sense of collaborative work sets in and consequentially, agreement on a final solution or proposal is sometimes less controversial, but when two competing proposals for the solution of a particularly important feature are placed on the table, reaching a conclusion where one is included in the final specification and the other is rejected it can lead to the need for voting which, in turn, has offline lobbying, bargaining, political wrangling and significant meeting time being consumed. It is often noted that the real work in a Standards forum does not take place in the meeting itself, but in the corridors outside during coffee breaks, and over dinners.

3GPP has been the primary global Standards Definition Organisation (SDO) for mobile telecoms since the beginning of Standards' work on 3G, and has defined the dominant global Standards for mobile telecoms for 3G, 4G and 5G, by adoption of the three-stage approach outlined in ITU-T Recommendation I.130. 3GPP is structured around three Technical Specification Groups (TSG), each of which serves as the Plenary group of multiple Working Groups (WG). These are currently organised as illustrated in Table 4.1, reproduced from 3GPP's webpage [2].

The TSG Service and System Aspects (TSG-SA) is responsible for defining the system architecture and service capabilities of mobile networks, and so creates Stage 1 and Stage 2 Standards, and some of the Stage 3 aspects of end-to-end architectural concepts such as Security and the use of Codecs. TSG Radio Access Network (TSG-RAN) defines all protocol (Stage 3) aspects of radio-access networks, whilst TSG Core Network and Terminal (TSG-CT) defines the protocol interaction between user devices and core network (CN) elements.

Work in 3GPP is structured to promote progressive development of new concepts through the introduction of new ideas as discussion papers. If sufficient interest is found within a working group for an idea, it is proposed as a Study Item which is then allocated time to be worked on within meetings, in order to gather all perspectives on a topic. Once a Study Item has progressed to a good level of agreement on the main principles and concepts in a topic, a Work Item is proposed, which then allows for the inclusion of the concept within a Technical Standard (TS).

Participation in 3GPP groups is open to any company that is a member of one of the 3GPP Organizational Partners, these being the seven regional SDOs that form the partnership of 3GPP. This leads to two main blocks of participants – vendors and mobile network operators. Often, it falls to the Mobile Network Operator community to set direction on contentious issues, as ultimately, operators buy what vendors produce, and so, if the operator community can express a strong preference on a specific topic that will often be

Table 4.1 The current structure for technical specification groups in 3GPP (source [2]).

Project Coordination Group (PCG)

TSG RAN	TSG SA	TSG CT
Radio Access Network	Service and System Aspects	Core Network and Terminals
RAN WG1	SA WG1	CT WG1
Radio Layer 1 spec	Services	mm/CC/SM (lu)
RAN WG2	SA WG2	CT WG3
Radio Layer 2 spec	Architecture	Interworking with external
Radio Layer 3 RR spec		networks
RAN WG3	SA WG3	CT WG4
lub spec, lur spec, lu spec	Security	MAP/GTP/BCH/SS
UTRAN O&M requirements	SA3-LI	
	SA3 subgroup on Lawful Interception	
RAN WG4	SA WG4	CT WG6
Radio Performance	CODECs	Smart Card Application
Protocol aspects		Aspects
RAN WG5	SA WG5	
Mobile Terminal	Telecom Management	
Conformance Testing		
RAN WG6	SA WG6	
Legacy RAN radio and protocol	Mission-critical applications	
RAN AH1RAN AH1		
RAN ad hoc group on ITU-R		

sufficient to conclude any debate. Equally, proposals that do not gather any operator support can become sidelined in the meeting agenda. This dynamic has started to change during the definition of 5G however, with increased participation from representatives from outside industries who will be the customers of operators, creating a new bloc with increasing influence on requirements and service definitions.

3GPP works on the basis of consensus decision making, but if a compromise position cannot be negotiated, voting takes place where a 71% share of a vote will force a result, and if that cannot be achieved, a second vote with a straight majority can be used, although this is a very rare occurrence.

Further information on 3GPP working practises can be found via the 3GPP webpage [2]. More details of 3GPP's creation and evolution will be provided later in the chapter.

4.1.3 Commercialisation

Once a Standard for a generation has been defined, there is usually a clamour for the new innovation to be launched as a commercial service, but before that can happen fully, both network equipment and end-user devices need to be commercially available. Development cycles have become shorter for network equipment as the shift towards software-based implementation has gained pace, but there is still a need, after any Standard is finalised, for network equipment to be purchased, deployed and tested prior to any live service being launched. Also, there may well be a need for spectrum to be acquired which can, in itself, be a lengthy process, particularly at lower frequency ranges.

Often, the first networks to launch a commercial service do so to claim some reputational kudos as 'winning the race', or to indicate national or operator technological superiority. This can mean the network implementation costs are high since the equipment being bought for a new G is usually at the bleeding edge of a development cycle. Early adopting network operators carry a high cost for the development work of their chosen suppliers, and this is passed in turn to the early adopting customers who need to purchase premium handsets at pay premium rates, to have the latest phones with the newest connectivity advances on board.

Over time however, the cost of equipment, devices and subscriptions all reduce. The premiums paid by those operators who are first to launch reduce as equipment and device component ecosystems reach scale, and development costs are amortised across higher numbers of units.

Commonly adopted standards underpin this commercial ecosystem from a number of perspectives. First, architecturally, the closer you are to the end-user device, the greater importance there is on scale factors to reduce costs. Whilst there is a wider range of device form factors and manufacturers, the number of manufacturers of key components is relatively small. This means that device-component manufacturers reach very high scales of unit production, to generate high revenues at comparatively low margins. When a new generation of technology is finalised, the first chipsets that support that technology carry a high premium, meaning that they can only be incorporated into the very highest value devices. But, as consumer adoption grows, with more networks deploying services, and cost points for subscriptions reducing, so the scale of component manufacture increases, at the same time as the maturity of the technology naturally drives price points down.

As discussed previously, a standard that is shared as widely as possible drives Interoperability. Inter-operability is important for promoting competition in network and device components which, in turn, serves to reduce network Capital Expenditure (CapEx), since differentiation on functional capability is naturally limited when all network vendors have to support common standards' functions. Whilst functional differentiation of network products remains possible, it is usually via proprietary extension of protocols. As a result, vendors are obliged to compete on price and on operational factors such as energy efficiency, reliability and support capability, all of which feed in to the operator business model as Operational Expenditure (OpEx). Operators want to implement a network that is as cheap as possible to acquire and operate, to allow them flexibility to either maximise their margins by charging a premium to customers for a new service, or to undercut their competitors at the expense of some margin, in an attempt to increase their subscriber base, and hence derive more overall profit through greater scale of connections provided.

Inter-operability elsewhere in the network is also essential for revenues associated with interconnection and roaming. Few customers would adopt the services of a network where they could only contact other customers of the same network. The expectation of customers is that they can reach any contact and any content, without constraint exerted explicitly or implicitly by the operator of whom they are a customer. As noted previously, this is a key factor for telecoms generally, but in mobile telecoms an additional complexity is the support of roaming. Roaming, the ability for a device to be taken out of its home network and still receive service, even when it is in a different country, is driven by the global adoption of a common radio interface (to allow the device to attach to a radio network of a different operator) and common implementation of interfaces that support authentication, charging and services that pass between the 'home network' – the network the customer uses when they are connecting in their home country or location, and who the customer pays for service – and the 'visited network' – the network the device is attempting to attach to. Revenues from roaming are an important part of the business model for some operators and, in extreme cases, can even be the driving force behind an entire business model. Roaming is only possible due to widespread adoption of common standards.

4.1.4 Continued Innovation

The finalisation of a new generation is often associated with the completion of a specific 'Release' of 3GPP standards. 3GPP Standards are organised around Releases, with each Release being a self-contained and complete description of an implementation of a specific set of features. 3GPP releases support two key principles for continuous development of technology – backwards and forwards compatibility.

Backwards compatibility is a simple principle that can be summed up as 'if you add something new, it can't create a problem with anything that is pre-existing'. As a result, any new innovation which is built on top of a previously defined feature or function, cannot break the operation of the previous versions of that function. This principle is important as it prevents one vendor or operator forcing all others to upgrade systems based on a new piece of functionality. This usually means that new functionality that builds on top of a pre-existing feature has to be 'optional' in its operation and protocol impacts – it can be used by two interacting functions, but if one end of an interface does not support the new feature, the other end that does support it cannot rely on it being available.

Forwards compatibility is the pre-emptive decision in protocol and feature design to allow for extensions to be added in the future. This takes the form of extensibility methods in protocols so that when new features, as described above, are identified, they can be included without breaking anything that already exists. This means that forwards compatibility through protocol extensibility enables future features to be backwards compatible.

All of this results in 3GPP releases being made up of the previous release's body of definitions, plus backwards compatible extensions to existing features, as well as the addition of new features.

Between one new generation and the next, a number of 3GPP releases can be finalised. The first 3GPP Release defining a technology that could reasonably be identified as '4G' was 3GPP Release 8. This was the first Release that included definition of Evolved Packet System (EPS), made up of E-UTRA (more commonly known as Long-Term Evolution (LTE) radio) and Evolved Packet Core (EPC). 3GPP Releases 8 to 14 extended the definition of EPS with new features and functions, but without ever being considered to contain such a fundamental shift as to be considered a new 'generation', despite order of magnitude improvements in theoretical maximum throughput per connection. All of the improvements and refinements that took place to EPS during Releases 9 to 14 are enabled as a consequence of the inclusion of forwards compatibility protocol designs, and have been implemented to ensure continued backwards compatibility. Indeed, 2G and 3G standards are also perpetuated into future Releases and any changes that may impact those standards are reflected back into the current definition of all previous Releases. 3GPP's body of work is thus a definition of every feature implemented in every Release for every previously defined specification from 2G to 5G. How this came to pass is discussed later in the chapter.

It should also be noted that the extensibility of protocols using optional fields and features can be used for the inclusion of non-standardised features. This is often a source of some controversy, since it has been known for vendors to add a non-standard, proprietary feature into their implementation of a protocol which adds service value to an implementation, but is only available in that vendor's implementation. By not making such extensions available in Standards, the result is a vendor-specific interface implementation, which can then be leveraged for commercial value.

Vendors that undertake this type of extension see two potential advantages of doing so. First, when bidding for business with operators, the vendor in question can propose additional functionality over and above that described in the Standard, through support of the additional proprietary features within their implementation. If this appeals to an operator, the operator may select a vendor as a supplier of equipment on both ends of the interface that includes the proprietary extension. This could happen at initial build or as an up-sell of additional functionality once the vendor has been selected as an equipment supplier.

Once an interface supporting additional proprietary features is included in a network, it is likely that an operator would become dependent upon such an extension as a feature for the operation of the network or as part of a commercial service. This then makes it very difficult to subsequently change the vendor, since the operator may have revenue or operational efficiency associated with the proprietary feature that would be difficult to lose if an equivalent proprietary extension were not supported by a new equipment supplier. This creates what is often referred to as 'vendor lock-in'.

Often areas prone to vendor lock-in can be identified in Standards. Definitions of protocols or network functions that are vague or seem to be open-ended are sometimes kept that way to make vendor lock-in easier. Particularly problematic are systems such as network management and charging. Here, the tight coupling between network equipment and the management systems that monitor and control the network result in an obligation to purchase the management system of the vendor that is also supplying the network equipment. It has only been in recent times that inter-operability between management systems from multiple vendors has begun to occur. Even then, in the early stages of implementing a single management view of the network, this was done by the addition of an expensive integration layer to hide the multiple proprietary management implementations of multiple vendors in a single network.

It can be seen then that whilst Standards promote ecosystem and economy of scale that allow for the creation of a global market in telecoms equipment and mobile phones, in other areas, it is the exploitation of Standards' structures for non-standard implementation that allows markets to exist.

4.1.5   Intellectual Property as a Metric and Political Currency

Throughout the process, from research to standardisation and commercialisation, innovation is a continuous trend. During Research, innovation around the formation of new technologies and ideas to meet new requirements takes place across multiple areas and from a diverse group of parties. Under standardisation, there is a degree of compromise and consolidation that, in turn, results in innovative solutions to combine the paths of research into a solution that draw on the many different facets of the research stage. Commercialisation invariably identifies shortcomings in the standardised solution as it is turned into a robust, scalable and inter-operable implementation – the demand for inter-operability and commercial success means that innovation is still required to resolve the issues that are identified in ways that the entire industry can agree on.

At every point, there is potential for companies who propose the solution to a problem to have included Intellectual Property (IP) in that proposal. IP, usually in the form of patents, is created by companies and institutions to protect innovations they have created that could become a part of a Standards definition, with particular value placed on 'Standards Essential Patents' or SEPs. SEPs are patents that must be used in order to be compliant with a standardised feature, and hence can be licensed by the patent holder, at a price which is levied against other companies implementing the feature covered. The potential for a company to exploit their SEP portfolio has led to many companies investing huge amounts of time and effort into patent filing. When a standard is finalised, often an evaluation of who holds most SEPs is completed as a way of determining which companies are demonstrating greatest levels of innovation, which companies have been most successful in their standards strategy, but also which company has the best negotiating position for trading their Intellectual Property Rights (IPR) with others in the industry to minimise their licensing burden.

4.2   Shifting Political Power and the Making of an Ecosystem

When the first mobile phone call was made in 1973, it not only signalled a new service emerging into the telecommunications industry, it also triggered a round of innovation that has fundamentally impacted the way the world communicates, the ability for people to access untold levels of information, and a radical shift in the overall productivity of business.

By the end of 2019, GSMAi estimated there were over 8 billion cellular connections worldwide, with more than 5.2 billion individual subscribers [3]. It is safe to say that none of those connections are using the same technology as the first call in 1973, but since the creation of 2G during the 1980s, the standardisation principle of backwards compatibility has meant that 2G is still maintained in many networks, and supported on the vast majority of mobile devices capable of making phone calls.

In the intervening years between 1973 and now, during the process of standardising how networks and end-user devices work, there have been shifts in the mobile telecoms industry in technological principles and areas of innovation. These have, in turn, caused changes in geopolitical power and in the fortunes of the companies that make up the industry. As we will see below, this has been felt within, and influenced by, the development of standards that have underpinned the industry's progression.

4.2.1 2G GSM – Europe Leads

GSM can trace its origins back to 1975, when European ministers began to discuss the need for allocation of a specific spectrum band to be set aside for the deployment of a shared, digital telecommunications technology. The band to be used was later identified to be at 900 MHz, and the requirement for the technology to be digital was because, whilst the first mobile phone call was an extraordinary achievement, the use of analogue technology meant that calls could easily be eavesdropped upon.

By the time detailed work began in earnest in Europe to define what the technology would be, the USA and Japan had begun their own work on similar projects. European governments could see the need for Europe to be present in this technological step, and agreed to work collaboratively to pool the resources of academia and industry in the definition of a single standard, which all mobile telecoms network operators in Europe would deploy. The process for writing the Standard was to take place in a new group – Groupe Spécial Mobile, or GSM – established by the European Conference of Postal and Telecommunications Administrations (CEPT).

The relative complexity of Europe geographically, when compared to the USA and Japan, meant that specific principles for a European technology were needed, which would, in in the longer term, propel the technology to a far wider market. The key principle was a common spectrum band, managed and licensed in a coordinated fashion. This was essential to allow networks that were operated across national boundaries by different operators, to co-exist without interfering with each other. This, in turn, meant that a hierarchical addressing structure was also needed, to identify the country a mobile network was hosted within, the specific network itself (since it was anticipated that each country would have more than one network) and an identifier of the individual subscription. Networks would advertise their identity and only customers that recognised the network identifier would be allowed to attach to that network.

In 1984, the French and German governments agreed to collaborate to develop a technical specification that would then be used by their state-owned mobile network operators. Italy joined this agreement the following year and the UK joined in 1986. This agreement pitted different national interests and companies against each other to have their proposals agreed for inclusion in the new standard that was to be developed. Aside from the differing technological approaches, one of the major issues was IPR, with different approaches being taken by state-owned companies, who tended to provide free access to their innovations, versus companies that were from an enterprise background, who viewed IP as a leverageable asset and hence enforced licensing of their IPR. The gap between the two sides of the debate was wide and was the beginning of IPR battles within the industry that continue to this day.

Despite this, 1987 proved to be a momentous year. Set against the backdrop of national governments that, in many cases, still owned the national fixed and wireless operators, and so were defending national interest, challenger operators were beginning to find their feet in an embryonic market, vendors also had some national interest but, in other cases, were seeking to gain a foothold into a market that they could see being lucrative if they could embed their technological advancements, and any number of academics, making progress towards a point where all parties would agree to a common technical definition in GSM must have seemed like a ridiculous ambition.

As it turned out, the agreement between France, Germany, Italy and the UK proved to be the catalyst. In February 1987, the first GSM specification was finalised [4]. It was by no means final, nor perfect but it provided sufficient momentum for the four governments to back the technology through the signing of the Bonn Minister's Declaration in May and then progress to the proposal of the 'GSM Memorandum of Understanding', or GSM MoU [5] which was signed in September.

The GSM MoU put in place commitments from operators across 13 European countries, to build networks to the same spec and within the same allocated 900 MHz spectrum band. Whilst the combined markets were not to the scale of the USA, and the technology was in some ways not as advanced as that being worked upon in Japan, the public policy approach of the MoU was fundamental to its success. The MoU was open to others to join and, because of that, it also attracted attention outside Europe.

The GSM Standards group transitioned to operate under ETSI, the European Telecommunications Standards Institute in 1989, which afforded greater openness and membership for companies who wished to contribute to the evolving standard. With broader participation came a tightening of the definition of GSM and, in 1990, GSM Phase 1 specifications were published. This offered the first fully implementable definition of the technology, and triggered product development to push towards service launches.

The first GSM networks launched across Europe in 1991, signified by a first GSM call made by the Finnish prime minister. The launch was coordinated across multiple networks, using equipment from different vendors and, with a broad range of handset manufacturers developing, early GSM mobile phones.

GSM soon became the dominant 2G technology in the world, commanding around 80% of market share. The US CDMA-1 and Japanese PDC standards also gained footholds in other markets, but nowhere near the scale or spread of GSM. In 1993, Telstra became the first non-European operator to sign up to the GSM MoU, which triggered an influx of others, so that by 1994, over 100 operators from around the world had signed. This rapid scaling of the participation led to the formal registration of the GSM MoU Association in Switzerland – the GSMA.

The GSMA's role was the perfect complement to ETSI. ETSI did not have the capability, nor the inclination to take on the burden of more administrative tasks around the management of an increasingly globalised ecosystem. There was also a need to separate some of the more commercial aspects of networks interworking and interconnection from the technical specifications of intra-network technology. The GSMA was able to perform these roles, without being seen as 'European' since the ecosystem that was developing was clearly global, and without being tied into the complex definition of the radio interface. The GSMA was also able to help construct a commercial ecosystem.

Under the agreement in the GSM MoU, Roaming had already been conceived – the idea that a device could move out of the network coverage of the operator that it would normally be served by, and attach to a different network, either in the same country or in a different one. The GSM specifications defined this possibility but did not include any guidance on the underlying commercial or technical principles to enable this to take place. GSMA provided a forum for operators to discuss these matters, walking a tightrope between commercially sensitive agreements between pairs of operators and the shared benefits of having a common set of principles under which the agreements required could be reached. The net result of the combination of ETSI GSM Standards and the GSMA, as the organisation that maintained the principles of the GSM MoU, was a meteoric expansion of adoption.

This was further accelerated by the addition of SMS messaging – a feature that was originally defined to allow short messages to be sent to control machines, but which has proved so popular as a person-to-person method of messaging – and with the expansion of service from just voice calls, to support fax communications and data in 1992. Between 1994 and 1995 GSM connections moved from 1 million to 10 million, and to 50 million by 1996. This demand for connections meant that additional spectrum bands were needed, driving expansion into the 1800 MHz and 1900 MHz band, allowing networks to increase capacity, The use of 1900 MHz allowed GSM to be adopted in the USA as a competitor to CDMA-1, by operators ranging in scale from nationwide coverage to localised small-scale businesses.

Innovation in technology continued as a consequence, with chipset manufacturers producing chips that supported all three GSM bands, meaning that mobile handsets and devices could connect to networks outside their home operator, even if the visited network supported a different frequency band to the customer's home network. With a handset ecosystem driving consumer choice, operators reaching subscriber scale that was not foreseen and, in turn, able to amortise fixed costs across a large customer base, prices for service began to decline and the service became accessible to a much larger population, thus driving renewed demand. An ecosystem perpetuated by a standard, and unified by economy of scale and Roaming, had been created.

The only question that remained was, what next? The wider trend in the industry to extend beyond early attempts to connect to the internet and to instead provide a broadband service seemed to be the answer. Whilst Circuit Switched data had been added into the GSM service suite, this was limited to the bandwidth supported by the voice codec of GSM – 9.6 kbps. Despite this being a relatively good data rate for the time, it was clear this would not suffice for long, so ETSI made various attempts to expand this rate, the most successful of which was General Packet Radio Service, or GPRS. GPRS initially supported data rates of up to 53.6 kbps and was later extended to 114 kbps. The service-enabled devices to attach to the Internet and, from there, to company networks allowing users access to e-mail. The addition of higher speeds of data proved to be successful, but was also causing controversy within ETSI behind the scenes. There was a danger of a rift, because work had already begun to define 3G.

The addition of data service had the consequence of also requiring that service be available during roaming. GSMA had empowered voice and text roaming, but required individual technical implementation details to be agreed between every pair of operators. For an operator community of several hundred operators, each would need to maintain potentially different technical implementations and commercial agreements with every other operator. Inadvertently, a large administrative overhead had been placed on the mobile operator community. Each roaming agreement had to be negotiated on commercial and technical terms, such as agreement on which long-distance transit networks to use to transport traffic, addressing domains and cost for providing services to roaming subscribers. For data roaming, GSMA decided to take an alternative approach, in the form of the GPRS Roaming Exchange, or GRX.

The GRX was a set of technical agreements supported by GRX carriers, often in the form of transit operators, who would provide a single, consistent GRX product. This simplified the technical implementation of the links between home and visited networks for roaming data traffic. Alongside this, GSMA drafted framework roaming agreements, which amounted to a template contract that could be completed very quickly between any two networks who wished to enter into a roaming contract. The GRX and framework contracts massively lowered the barriers to agreeing data roaming, and as the GSMA membership continued to grow, it was for this ease of access to potential lucrative roaming revenues that many smaller operators signed up to the GSMA MoU and began to pay their membership fees. Today, one of the most significant benefits of GSMA membership for operators is access to RoamFest events, where operators can agree multiple roaming agreements in a number of days, at events hosted by GSMA.

4.2.2 3G UMTS – Universal (Except Not Quite)

In 1987, as the GSM Standard was being finalised, Europe had already begun to consider what 3G might be. The European Union funded 'Research and Development in Advanced Communications Technologies in Europe', or RACE, within which, funding was allocated to develop the Universal Mobile Telecommunications System, UMTS. The ambition of UMTS was to be much broader than just European, which was a highly ambitious goal given that GSM networks had not launched at that point. However, the principles and collaboration around GSM had already indicated the mass appeal of mobile communications, and it was clear that a unified global approach would offer the greatest benefits.

UMTS had, at its core, the need to deliver higher data rates as well as voice calls, and as the work gathered pace to deliver Research findings, this served as a catalyst for the development of GPRS in ETSI. There was a view that, by extending GSM to include higher data rates, there may never be a need for a new Generation and, as the research into the principles and architecture of UMTS progressed, there was a steady leakage of transferrable concepts into GSM, many of which enabled GPRS to achieve the data rates it did.

The one thing that GSM could not easily change however was the RAN multiplexing systems. GSM used Time Division Multiplexing, which constrained the throughput achievable for any individual connection due to its narrowband nature. UMTS proposed the use of two, competing broadband mechanisms. One was Code Division Multiplexing (CDMA), enabling any individual user to use all available resource at any given time, and the basis of the US 2G technology. This, however, proved controversial, since CDMA was considered to be 'not invented here'. The other candidate technology, Advanced Time Division Multiplexing was also proposed, offering an extension to the principles of 2G, and importantly, very much developed in the European R&D community. Both were found to perform equally well.

By this point, it was 1996 and the appreciation of the scale of mobile telecoms had begun to emerge. GSM was approaching 100 million subscribers in markets across the world, and had totally supplanted the 1G services which had been based on North American AMPS technology. Europe was, therefore, in a position of technological strength but needed to also be inclusive of the markets that had adopted its 2G technology as it progressed to 3G. Most significantly, Japan had indicated interest in sharing in the 3G vision of Europe, and some American companies had also been brought into the wider Research work by the potential use of CDMA. ETSI continued to strengthen the GPRS specifications and tabled proposals for support of higher data rates through the definition of Enhanced Data rates for GSM Evolution, or EDGE. Despite this, those supporting UMTS in Europe, with a preference for TDMA, had formed the UMTS Forum to coordinate their technical advances and to act as a lobbying group.

The ambition had always been to define a global technology, and so it was proposed that a third Generation Partnership Project (3GPP), should be formed. 3GPP was made up of major regional standardisation bodies from around the world – ETSI in Europe, ARIB and TTC for Japan, CCSA from China, TTA in Korea, and ATIS for USA. Whilst ATIS are accredited to ANSI, the over-arching US Standards definition group, so too is TIA, and it was TIA that documented the Standard for cdmaOne. TIA were determined to pursue a North American 3G technological path and ultimately established the alternative group, 3GPP2, which went on to define CDMA-2000 as the 3G evolution of cdmaOne. The ambition to define a single-global standard was lost.

3GPP brokered a deal to secure a single Standard for their preferred 3G solution. Japan was supportive of a Wideband CDMA technology variant, and European companies were happy to compromise to this solution to gain potential access to Asian markets. The US preferred a narrower band CDMA solution and so, initially, two variants were adopted but these later converged to a single W-CDMA-based definition. It was agreed that UMTS would adopt the GPRS CN architecture, thus smoothing the adoption of UMTS for both voice and data services.

There was, however, still a problem. Despite the divergence between 3GPP and 3GPP2, there was a high degree of commonality between the two specifications and considerable commonality in essential patents. Major players in 3GPP2 refused to license their IP for use with UMTS.

As well as all of this, the International Telecommunication Union (ITU), an agency of the United Nations focussed on information and communication technologies, established a process entitled IMT-2000 [6] to ensure global alignment of spectrum usage for approved technologies. With 2G networks having become a global success, it required a body with a full set of international stakeholders to assist in the future coordination of spectrum usage, which included the identification of technologies that could be operated within licensed and regulated spectrum bands. National regulators took ownership on the allocation of spectrum to mobile operators within their respective countries, but the alignment of the bands to be used needed full co-operation of every national regulator on the planet, and only the ITU had this level of participation.

IMT-2000 laid down a set of requirements that candidate technologies should meet in order to be included within the resulting family of Standards. Both W-CDMA/UMTS and CDMA-2000 met these requirements but the dispute over granting licenses to use common

IPR to those attempting to manufacture W-CDMA devices and networks was seen as a political move to hold the industry hostage. The matter became so politicised that ITU issued a statement that they would exclude CDMA-2000 from the IMT-2000 family if a resolution was not found. One month ahead of the IMT-2000 deadline a compromise was agreed, which paved the way for the deadlock to be broken. Whilst disaster was averted on this occasion, moves within the industry began to ensure that a similar situation could not occur again.

The target of UMTS had been to deliver data rates of 2 Mbps, but the first set of Standards, delivered in 1999, had only a theoretical maximum data rate of 384 kbps. This was a significant step forward from the data rate of GPRS and there was also an expectation that the service would be able to get closer to those theoretical maximums than GPRS had been able to. When the first UMTS network was commercially launched by 3 in the UK, the hype associated with 3G had reached fever pitch, not least because of the high costs that had been paid by operators for spectrum bands to operate 3G networks in. It was with no small amount of disappointment that it soon became apparent, that what was delivered by an operational network was sometimes only one-tenth of the promised performance. UMTS connections were only used for data connectivity in extreme situations and it seemed that, whilst mobile phones continued to boom, 3G was really more about increasing voice call capacity by providing access to new spectrum bands.

It was not until much later that the promise of mobile broadband connectivity was met. In a similar way to how the creation of UMTS stimulated innovation that also contributed to enhancing 2G data rates from GPRS to EDGE, as research took place into what might push boundaries further in 4G, it became clear that some of that innovation could also be used to improve 3G performance. Higher-order modulation schemes, Multi-input multi-output (MIMO) techniques to allow a device to establish multiple radio connections in the same frequency band, and Carrier Aggregation allowing use of multiple frequency bands at the same time, lifted the peak download rate of 3G to 3.6 Mbps, then 7.2 Mbps, and then 14.4 Mbps. High Speed Packet Access (HSPA) delivered such high bandwidths that, in some markets, it had a serious impact on fixed-line broadband connectivity rates, particularly for customers on lower incomes. It allowed travelling workers to be connected to their company networks with a bandwidth that enabled something close to 'in-office' connectivity. Demand for HSPA modems was high, and was coupled with availability of USB plug-in variants that self-installed, and were much easier to use than older PCMCIA cards.

HSPA also coincided with the most significant shift in mobile phone usage patterns since 2G was launched – the smart phone. Smartphones, with larger screens, and accompanied by applications that were developed in developer communities, drove a data-demand boom. This initially caught operators by surprise, since the usage patterns of HSPA, whether as a technology for connecting to corporate VPNs, or for the consumption of video and highly connected applications, had a very different nature to previous services. Whereas a 2G or 3G data service consumed on a phone with a small screen might involve downloading a text page and the customer then reading the content whilst the connection was idle, now customers were using the connection for multiple purposes and utilising higher bandwidths for longer periods. Network engineering principles had to be rethought and backhaul capacity had to be significantly upgraded. Some smartphone applications tried to minimise the bandwidth impact of their implementations by consistently attaching and detaching from the network, but this brought a different problem since it generated huge amounts of network signalling and also drained device battery. To get around this issue, Standards for network attach and detach signalling were modified.

Many applications involved the download and consumption of video content. Video had been available over mobile for some time, but had never taken off because of the length of time it took to load a relatively short video clip. With the combination of HSPA and smartphones, video could be downloaded at a data rate faster than it would play out, and viewed on a screen large enough to add value from higher resolution content. Soon video was making up more than 60% of the overall network traffic of operators that had launched HSPA. Demand was outstripping supply, but fortunately, 4G was already in development to meet that demand.

With the increased focus on data as the basis of many services, it was soon noted that Standards were now supporting two parallel network technologies. Voice and SMS were supported over Circuit Switched networks, whilst data services were implemented on Packet Switched (PS) networks. This split was a source of concern, and there was also worry about the growing influence of 'Over-The-Top' services such as MSN and Yahoo! Messenger. The popularity of messaging services delivered via the Internet on fixed-line networks was seen both as a threat and opportunity – a threat to SMS revenues but an opportunity to extend beyond just SMS, and deliver some enhanced real-time interactive services. When services like Skype, delivering voice calls from computers using the Internet as a transport network, became popular, it was clear that mobile networks could see significant revenue erosion unless operators' own services could be improved and expanded. 3GPP began the definition of the IP Multimedia Subsystem (IMS) to enable this during 2000.

IMS was defined using protocols that were specified by the Internet Engineering Task Force (IETF). IETF had very different working practises to 3GPP, and the resulting clash in working culture and general ethos between 3GPP and IETF caused considerable friction. 3GPP had a strict membership format, which restricted the ability to be included in discussion and to have voting rights, to those that were members of the regional partners and that attended meetings. IETF was open to anyone to contribute, regardless of their affiliation and background, and even allowed for individual private citizens to carry as much significance as huge companies. 3GPP worked to a strict 'Release' structure with plenary meetings approving work plans and specifications on a quarterly basis, whereas IETF documents, known as 'Request for Comments' or RFC's, went through extensive approval processes that operated a rolling two-week approval, where any comment could reset the two-week process.

More significantly, the broad principle of the Internet was one of a decentralised, ownerless construct allowing freedom of traffic flow and access to all. Mobile Telecoms was owned by increasingly powerful network operators and built by their chosen vendors. This meant that whilst IETF protocols were intended to provide a construct that anyone could access and implement, 3GPP wanted to use IETF protocols – in particular SIP (Session Initiation Protocol) and Diameter – and extend the protocol definitions to include information for charging end-users for use, or to restrict their access using Policy Control and extensive authentication. These constructs that 3GPP wanted to put in place flew directly in the face of IETF principles.

The result was that 3GPP delegates were persistently pushing the relevant IETF groups in an attempt to have the extensions to SIP and Diameter ratified in IETF RFC's, which could then be referenced in 3GPP specifications. IETF delegates objected to the 3GPP extensions and were able to prevent the completion of 3GPP proposals by using the IETF approval principles to slow specifications down. 3GPP and IETF leadership formed a joint task force to try to overcome the issue, and this allowed for a solution to be brokered, but the result in 3GPP specifications was that many of the extensions to SIP and Diameter that the IMS implementation depended upon, were specified as optional. Whilst those writing the specifications in 3GPP knew that many 'optional' aspects were required for equivalent operation of IMS services compared to CS, implementers would not have the same level of insight, and the 3GPP concern was that inter-operability would be problematic. As it turned out, it would be many years before any commercial IMS services were implemented at scale, as a consequence of decisions taken in the specification of 4G.

When the ITU completed their IMT-2000 assessments, both UMTS and CDMA-2000 were successfully included. Both technologies utilise 'paired' spectrum bands, meaning that separate bands are allocated to upstream and downstream. However, many spectrum licenses also include allocation of 'unpaired' spectrum, where upstream and downstream paths can be dynamically configured within a single band. The technological challenges of minimising interference in such a technology had been considered too complex to make unpaired spectrum useable. However, Chinese operators and manufacturers brought a candidate 3G technology to 3GPP to be considered for standardisation at a point when 4G technology was already well in discussion. This technology was TD-SCDMA.

China's mobile telecoms market was booming with three operators all registering explosive growth in 2G services. When China did issue 3G licenses, it took a controversial step of insisting that each of the three operators deploy a different technology – China Telecom being obliged to use CDMA-2000, China Unicom being allocated UMTS, and China Mobile, the largest of the three based on subscriber numbers, to deploy TD-SCDMA. Chinese vendors and operators brought a close to complete Standard for TD-SCDMA into 3GPP which, whilst following the 3GPP standard process, needed little additional work other than to be simply rubber stamped. Such was the size of the Chinese market, and the growing influence of the Chinese vendor community, that the specification of TD-SCDMA passed into Standards relatively unaltered, and was then ratified as an IMT-2000 technology with little objection. This was indicative of a shift in influence from European-led technical definitions, that would grow stronger as work on 4G progressed.

4.2.3 4G EPS – Avoiding Old Mistakes (and Making New Ones?)

3G had ended up being something of a contradiction within itself. It had been intended to add broadband data to mobile telecoms, but the initial headline data rates of 384 kbps were some way short of the promised 2 Mbps, and had been shown to be theoretical, since many networks were struggling to deliver more than one-tenth of this data rate consistently. However, mobile telecoms were still in a massive growth period, and so, 3G networks and connections had grown rapidly, along with the continued use of 2G. Voice and SMS revenues continued to grow, but so did demand for more connections and higher data rates. HSPA had addressed some of the pent up demand, but the advent of the smartphone had simply moved demand for bandwidth to a new level. Networks were struggling to meet demand as a result of HSPA's success, so the clamour for 4G grew.

4G was tabled as delivering data rates over 100 Mbps and with significantly reduced network latency which would then allow voice calls to be carried over the same packet core as data traffic. To achieve this, 3GPP moved on again from W-CDMA as the radio modulation scheme and selected Orthogonal Frequency Division Multiplexing (OFDM), which allowed a single end user to be allocated the entire frequency band of the service, if it was available. OFDM also paved the way for use of frequency bands of differing sizes – GSM and W-CDMA had to be used in bands of 200 kHz and 10 MHz respectively for upstream and downstream paths, but OFDM allowed for, in theory, any frequency band size to be employed. By now, it was well understood that there was a difference between the theoretical headline rates of a new G, and what the end user would actually receive, but even so, 4G was widely anticipated to be a further significant improvement above what even HSPA was delivering.

The ITU had laid out the timetable and service expectations for 4G through IMT-Advanced [7], the follow up to IMT-2000, which had essentially defined which technologies could be called 3G. 3GPP were working on their candidate technology for IMT-Advanced under the name 'Long-Term Evolution' or LTE. There was much excitement about LTE's potential to be a single standard for the whole industry since many of the major CDMA-2000 operators had indicated their intent to also deploy LTE. Indeed, it was CDMA-2000 operators who were at the forefront of potential commercial launches. The scale of the UMTS and HSPA market was now so large that the global volume production of chips and modems offered significant economies of scale. CDMA-2000 operators who were in markets where UMTS and HSPA were available from their competitors, were seeing customer numbers being eroded as customers switched to a technology with greater handset choice, often at cheaper prices. Added to this, CDMA-2000 customers could not easily roam into markets where all operators had deployed 3GPP technology, so CDMA-2000 operators were also missing out on a revenue stream to which their 3GPP competitors had access.

Having made the decision to swap to a 3GPP-defined 4G technology, the CDMA-2000 operators wanted to launch a commercial LTE service as quickly as possible, with perhaps the most aggressive in their plans being Verizon Wireless. Based in the USA, Verizon were suffering from the moves that AT&T and T-Mobile USA had made to HSPA. CDMA-2000 also had an upgrade path to EV-DO, the CDMA-2000 equivalent to HSPA, but the upgrade cost was required to be seen merely to be keeping pace with their competition on data rate but without the breadth of device choice offered by HSPA. Verizon curtailed their EV-DO investment, switched to LTE and began to market their plans as 4G. This presented AT&T and T-Mobile with something of a problem. They had recently invested in widespread roll-out of HSPA, and could not immediately move on to LTE until that investment had returned some value. So instead, they began to market the HSPA services as '4G-like', '4G-ready' or simply '4G'.

This blurring of the lines of what 4G amounted to, was further confused by a second Standards group developing a candidate for IMT-Advanced. IEEE, the group who amongst many things, defined Wi-Fi, was looking to define a related wide-area technology, called WiMAX. WiMAX had actually existed for many years, originally as a Fixed Wireless Access technology. The development of WiMAX had begun with the formation of the WiMAX Forum in 2001, and then continued with the development of specifications for implementation by IEEE 802.16e. WiMAX was originally delivering services that challenged DSL and Cable Modem, but IEEE and the WiMAX Forum also developed a mobile variant, 802.16m. The perception was that WiMAX would be related to Wi-Fi, and would be able to leverage the huge pre-existing Wi-Fi supplier base, potentially at a cheaper price point than LTE for base stations and device chipsets. Wi-Fi was present on more devices globally than even 3GPP technology, and so WiMAX was seen as a real threat, particularly when one of the early supporters of WiMAX deployment emerged as another US operator, Sprint. It also had considerable support from Intel, and commercial deployment of the Fixed Wireless variant in South Korea had shown that the technology was mature and viable.

At one point in 2008, in the US mobile telecoms market, Verizon were priming the market ready for the launch of LTE as 4G, AT&T were describing HSPA+ (the 21 Mbps variant of HSPA) as '4G-like', T-Mobile had launched HSPA+ as a 4G service, and Sprint were branding their proposed WiMAX launch as 4G. The irony was that none of these technologies would be IMT-Advanced candidates.

In the marketing push to launch anything that could be sold as 4G, the IMT-Advanced process became somewhat lost. In October 2010, the ITU announced the results of their assessments of IMT-Advanced candidates, and identified 3GPP 'LTE-Advanced' (defined in 3GPP Release 10) and IEEE 802.16 m 'WiMAX2' as the technologies that were 'true 4G' [8]. This statement drew much consternation since LTE (Release 8) and WiMAX had already been commercially launched by some operators as 4G-branded technologies. It became clear that the term '4G' and the rigour of the ITU-R IMT-Advanced process had become irreparably decoupled. ITU acknowledged as much when issuing a further press release in December 2010 which included the statement that 'it is recognized that this term, while undefined, may also be applied to the forerunners of these technologies, LTE and WiMax, and to other evolved 3G technologies providing a substantial level of improvement in performance and capabilities with respect to the initial third-generation systems now deployed' [9]. What had received less attention was that 3GPP Release 8 LTE and WiMAX had already been accepted by ITU as IMT-2000 technologies some years before. 3GPP had completed the definition of LTE within Release 8 in 2008, and the first LTE network had been launched, complete with 4G branding in December 2009. In making its statement in 2010, ITU-R had failed to appreciate that the 4G boat had sailed some time earlier.

The industry tangled with the question of what to do with WiMAX? WiMAX was a TDD-based technology only, whilst LTE had a widely supported FDD variant and a TDD variant, defined greatly by the same Chinese contingent that had defined TD-SCDMA. Operators that held paired spectrum allocations would have to deploy LTE as this spectrum implied an FDD-technology deployment, but many operators also had unpaired spectrum holdings that had sat idle, but which now could be leveraged for either TDD-LTE or for WiMAX. Chinese operators and vendors had a clear preference for TDD-LTE, but many other operators were torn, whilst others again were still unconvinced by the feasibility of TDD-based technology in general, despite the growing customer base China Mobile had using their TD-SCDMA network. With demand from end users increasing and availability of spectrum in useable bands being limited, the adage that 'there is no such thing as bad spectrum' meant that, at some point, a TDD technology would gain a footing. The question was, which one?

Whilst WiMAX had powerful backers, most also had a stake in the 3GPP ecosystem. Intel, with its pedigree in Wi-Fi, were the only large company that might benefit significantly from the adoption of WiMAX and TDD-LTE as parallel tracks, but there were also large companies, and most significantly, large network operators, who had thrown their weight behind LTE as a single global standard. These companies did not want to see the industry turn back from a single standard to a market that would potentially be bifurcated, and would result in the same issues with inter-operability and lack of capability to roam on to any other network that had caused them to swap track, re-emerging with the potential rise of WiMAX operators. The scale of the 3GPP ecosystem, and the alignment of the 4G Evolved Packet Core between TDD and FDD LTE radio variants meant that LTE could be seen as a single network standard for roaming and interconnection purposes at least, and chipsets supporting TDD and FDD LTE variants, and the capability to aggregate FDD and TDD carriers in a single user connection, were already in development. This put WiMAX very much on the outside of a technological juggernaut, that also had backwards compatibility, network integration and a smooth evolution path with UMTS and HSPA working in its favour. Operators that had supported WiMAX, began to turn their attention towards TDD-LTE, and WiMAX which had seemed to be a significant competing technology, gradually withered away. Parts of the WiMAX spec were integrated into TDD-LTE but, otherwise, WiMAX was relegated back to be a Fixed Wireless technology, in terms of its commercial deployment base.

Part of the wariness around TDD in general had been because of Chinese companies pushing so strongly for the technology. 3GPP participants were still getting over the IPR arguments that had surrounded 3G, and feared a similar situation occurring in 4G with a new powerful bloc of companies bringing technological principles to the industry that were enabling concepts that, whilst understood, had never been thought practical to be used in wide area mobile networks. 3GPP had adopted 'Fair Reasonable and Non-Discriminatory' (FRAND) principles within its IPR policy in its rules when it was formed but the controversy surrounding 3G had left a bitter taste, and various efforts were being made to offset the impact of essential IPR. FRAND principles are intended to prevent situations of unreasonable licensing costs or conditions from being enforced by IP holders, where that IP is included in a standard. 3GPPs FRAND policy includes the condition that any IPR included in a standard document should be licensed to other companies at a reasonable license cost and without discrimination between parties wishing to implement. This has the secondary effect of making licensing of IPR a competitive market with a cap on 'reasonable' costs for each piece of IPR, which then has the consequence of companies seeking to amass large IPR portfolios for potential inclusion in Standards as the best course of action for gaining licensing fees.

Against this backdrop, patent pools emerged, with multiple parties consolidating their patent portfolios together under favourable terms, but then using their shared patent mass to attempt to negotiate better terms with the large individual companies that held bigger portfolios. For all of this effort, not much changed except that there was not one single dominant holder of IPR – the strengths in IP were evened out a little, but IP remained an important metric for determining who was influential in Standards, and it was clear now that Chinese companies were becoming a significant force.

In defining LTE, one fundamental decision was taken that would force a shift in mobile architecture – the Circuit Switched Domain, which provided Voice and SMS functionality in 2G and 3G was dropped. The rationale for this harked back to the advent of IMS, with the intent that the network would become easier to run by having all applications and services provided over an IP-based packet switched core. There was also a growing concern about voice and SMS revenue erosion because of the growing number of alternative internet-based applications providing similar services. With LTE providing such high bandwidths, services like Skype and Viber were offering voice calls for free, using high-definition codecs that, in good network conditions, were of much higher quality than a legacy mobile circuit switched voice call that was limited to a fixed bandwidth by standards defined 15 years earlier. 3GPP had defined the availability of a wideband codec to improve the quality of circuit switched calls, but adoption in devices and networks was low, and interconnection of a call using such a codec from one network to another was non-existent.

Curiously, having taken the decision to remove circuit switch, 3GPP did not make any concerted effort to replace it, in the belief that IMS would enable a voice-call profile which would be inter-operable. It soon became clear that this belief was not correct. IMS had so many options included within it to appease the IETF when the original standardisation of IMS first took place that inter-operability was far from guaranteed without additional work. A small group of mobile operators identified this issue and set about writing a profile of the IMS specifications to enable the rigorous definition of how voice service would be implemented using IMS for LTE networks.

In parallel, however, others sought to fill the gap left by the removal of Circuit Switched from LTE. One approach, which was proposed in 3GPP was Circuit Switched Fallback. This involved users having their LTE radio stopped when they made or received a telephone call, which would force the handset to attach to a 2G- or 3G-radio connection, where the CS Domain would still be available. The solution was ungraceful, and defeated the object of removing the CS domain from LTE – it would oblige operators to perpetuate the CS Domain for voice calls, and not offer the extensions to the voice service that a higher bandwidth LTE bearer would offer, such as use of wideband, high-definition codecs. In the absence of a willingness in 3GPP to define an alternative solution, CS Fallback standardisation progressed.

A further alternative was proposed, called Voice over LTE using Generic Access, or VoLGA. VoLGA was based on a technology previously standardised in 3GPP called Generic Access Network, or GAN. GAN allowed for the emulation of a CS voice service on other access technologies, primarily targeted at Wi-Fi. The technology encapsulated a CS voice service within the connection of an alternative access and the concept of encapsulation is described later in the book. VoLGA repurposed GAN to encapsulate CS calls within LTE bearers, but again would result in the need to maintain the CS domain CN elements that had been intended to be removed. VoLGA gained some support ahead of CS Fallback since it was at least using the LTE radio connection, but it was still some way from the intended goal.

The need for an IMS-based solution was clear, and the mobile operators working on the IMS profile for this solution needed a recognised industry body to document their proposal to gain wider exposure and support. Without 3GPP as an option, they turned instead to GSMA. GSMA had the advantage of being an operator-led organisation and so could convene support from a wider operator community, which would then lead to vendor support. GSMA began their work defining the Voice over LTE profile of IMS in October of 2010, and announced the publication of the completed specification in February 2011. This was viewed as the long-term solution for voice calls on LTE, but there was still felt to be a need for an interim solution, and with LTE networks having already been launched, CS Fallback was also productised, whilst VoLGA support waned.

Voice over LTE did still have one major issue to address, again within GSMA's remit – how to support roaming. The problem existed because of differences in how voice and data roaming services were implemented. CS voice calls for roaming customers were largely handled in the Visited Network, and could then be routed directly to the terminating network (the network where the person being called was present). This was of fundamental importance to maintaining the quality of the voice call, since a person making a call may be halfway around the world from their home network, but calling a number that was local to their current location. It made economic and technical sense to allow the call to be routed locally within the country, rather than traversing across the globe and back again.

But this principle did not apply in PS networks. In PS, all sessions were routed to the home network before breaking out to the location of the application being used, because of the use of the GRX as the basis for Packet Roaming. Now, the voice service defined in VoLTE was implemented on packet networks, but voice-call roaming economics were based on the roaming principles of Optimal Routing. Discussions in GSMA became entrenched with two opposing camps each having justifiable positions as to which principle should hold, and concluded in a compromise allowing the definition of both a 'Local Breakout' (which has two sub-options) option and a 'Home Routed' option. The consequence of this is that VoLTE Roaming remains greatly unimplemented to this day – as recently as July 2020, GSMA launched the 'VoLTE Implementation Guide' [10] to assist operators in their efforts towards inter-operable and roaming capable services, and maintains a webpage illustrating the current status of roaming agreements in countries [11]. However, the lack of agreements means that CS Domain still remains within networks.

As LTE networks continued to be enhanced with higher-order MIMO, higher modulation, greater levels of carrier aggregation, the data rates being delivered pushed into 100s of Mbps, and latency reduced into 10s of milliseconds. Mobile internet connectivity was achieving close to the performance of fixed-line networks, but inevitably, a push towards a further fifth Generation of technology had already begun.

4.2.4 5G NR – New World Order?

There is a marked difference in the motivations behind 5G compared with each previous generation. 2G, 3G and 4G were all trying to address specific shortcomings of previous generations with a move from analogue to digital, the addition of data connectivity, increasing bandwidths for those data connections and then reducing latency to enable a single infrastructure to support real-time interactive services and data services. Once LTE-Advanced was completed in 3GPP, there was a feeling that there may not need to be another 'G'. 4G had a roadmap of improvements that were expected to take it to 1Gbps and beyond, and latency could be kept to 10s of milliseconds.

5G had been discussed as early as 2012, and by 2014, the hype cycle that sprung up around 4G was already in full swing. Vendors were discussing their R&D activities which, in turn, led industry analysts and journalists to question operators on their plans. The premise for 5G was that there was huge nascent demand for a connectivity technology that would enable enterprise customers to reliably and universally connect any device to a network which would, in turn, allow data to be gathered, behaviour to be observed and controlled, and consequentially offer insight and value for all aspects of business which would then deliver gains in productivity, improved efficiency and consequentially, higher profits.

The problem was that no one really knew what the requirements of 'enterprise customers' would be. In 2015, the NGMN Alliance, a mobile operator-led trade organisation established during the definition of 4G, prepared a white paper [12] focussed on possible use cases and, from these, derived a number of performance metrics, many of which had never been considered previously. These included further extension of bandwidth to 1Gbps, and a further reduction in end-to-end latency to 1 ms, but also included expectations for reliability, network coverage, connection density, minimal performance in any location, equipment power consumption reduction, spectral efficiency, high-speed mobility and bandwidth per unit area. None of these metrics had ever been critical to a new generation before.

GSMA had also produced a paper in 2014 [13], identifying the same expectations of what 5G might be targeted to achieve, but going further to note that some of these goals were not technical challenges. For example, to achieve 100% network coverage, a network operator would need to make the necessary infrastructure investment to put cells in areas where, until that point, there was no business case to make that investment. 100% network coverage, as a goal, would oblige operators to make the Capital Expenditure outlay for the equipment, the site purchase, planning permission, infrastructure costs associated with powering and backhaul links, and civil engineering works to build base stations in remote locations. The operator would also then have to pay the OpeX costs to power and operate these cell sites, including maintenance and spares inventory. Some of these cells may be in places where little revenue would be generated, so the business case for many cells may be negative.

When the ITU-R produced Recommendation M.2083 'IMT Vision – Framework and overall objectives of the future development of IMT for 2020 and beyond' [14], laying out the basis for assessment of candidate technologies for IMT-2020, conspicuous by their absence were expectations for network availability and network coverage. The methods employed by the industry to define the technologies that met the requirements identified by ITU-R are described in the other chapters in this book, although it should be noted that meeting some of the requirements are still dependent upon the specific implementation of networks that operators select, rather than purely being about implementing specifications.

In developing 5G-New Radio (5G-NR) and 5G Core, defined as a Service-Based Architecture (SBA), 3GPP forged ahead into a number of new technological areas. These included the use of higher frequency bands, popularly referred to as 'millimetre wave' (mmWave) bands, and embracing software-based network implementation. This has, in turn, opened paths to a broader Management and Network Orchestration (MANO) model as defined by ETSI ISG NFV, and the concept of 'Network Slicing', whereby multiple logically separated network instances can be run in parallel on common hardware platforms. The breadth of innovation in 5G, combined with an increasing participation as companies from other industries – automotive, manufacturing, healthcare and emergency services to name but a few – has resulted in the scale of the Standardisation groups in 3GPP swelling.

This has, in turn, resulted in huge numbers of contributions being handled in 3GPP meetings, and IPR becoming an increasingly competitive domain. Following their successes with TD-SCDMA and TDD-LTE, Chinese companies have invested hugely in resources to throw behind their Standardisation efforts. Chinese vendors in particular are now the greatest contributors in 3GPP meetings, and also hold very high proportions of 5G Standards Essential Patents. This is a marked contrast to 2G and 3G, and even within 4G, whilst the influence and scale of the Chinese market, and the operators and vendors that serve it, was increasing, 5G has seen a marked increase in their efforts. This has, in turn, caused other companies to step up their involvement to the point where the number of contributions submitted to individual 3GPP working group meetings cause an additional digit to be added to the document numbering scheme, so that a working group could handle up to 100 000 documents in a single year.

3GPP had laid out an aggressive timeline for completion of its specifications, aiming to have Release 15 completed by December 2018. It became clear that this was unlikely to be achievable due to the volume of work that had been included in Release 15, but rather than slip the date by 3 months, an alternative proposal was tabled. Release 15 included two different implementation options, known as 'Non-StandAlone' (NSA) and 'StandAlone' (SA). NSA would use the 5G-NR RAN, but also require a device to attach to an LTE base station, and then have data and signalling route from the 5G base station, through the LTE base station and then to a 4G CN. This model had been proposed to allow for a gradual roll out of 5G service without the need for an entire network upgrade, and it was widely recognised that whilst SA might be the longer term goal for most network operators, NSA would be the basis of the first 5G network launches. NSA and SA were being developed in parallel, and there were (and still are) questions about the compatibility of devices between NSA and SA.

Now, faced with the prospect of missing their target completion date, 3GPP took the decision to split the NSA and SA specification work. NSA would still be completed in December 2018, but SA would be allowed to slip out for six months, to June 2019. In turn, this reinforced the position of vendors with extensive LTE contracts, and was objected to by those vendors who were trying to make headway in a new 5G market. In addition, the work to complete NSA remained onerous, and was only completed with compromises and some features being dropped, but despite this, the completion of Release 15 was still broadcast widely and sparked the drive towards commercial launch of 5G services. 5G mobile services were launched in 2019, based on 3GPP Release 15 specifications using NSA implementations, first in Korea and then rapidly afterwards in the USA and some European markets. To date, there are no 5G SA commercial service launches and, at the time of writing, the ITU IMT-2020 process is still on-going.

During 2020, the first steps have been taken into 5G Roaming, but this is still based on the home routing model that was put in place for GPRS during the 1990s. At some point, this will present a fundamental issue for 5G services that are expected to support 1 ms latency. When the requirement for end-to-end latency reduces to 1 ms, physical distance becomes a fundamental problem. With a roaming model that requires a connection to traverse across what may be a long transit connection to a subscriber's home network, the delay budget for a 1 ms latency service may be exceeded simply because the speed of light is not fast enough.

The industry has tried to address this through the advent of Multi-access Edge Computing, or MEC. MEC was originally called 'Mobile Edge Computing' but the name was changed to allow for wider applicability of the principles and technology employed. GSMAs whitepaper in 2014 [13] identified this problem, noting the need for content ingress for a low-latency service to be much closer to the end-user. The paper also went on to discuss how this would increase the CapEx requirement for such a roll out to maintain reliability through equipment redundancy, and the need for additional power supply to keep equipment turned on. These principles flew in the face of the requirement for 5G networks to have a much lower total cost of ownership (TCO) compared to their predecessors. Indeed, at the same time as MEC would be pushing the content for a service closer to the base station, 3GPP were also defining new separated architectural elements for the RAN, which were intended to enable some components to be moved more towards the centre of the network.

In addition, one of the highest profile use cases demanding low latency connectivity was 'vehicle-to-anything' (V2X) connections. V2X would allow roads to become far safer and to some level 'autonomously controlled', to predict when an accident is imminent, or detect if a pedestrian is about to step in the path of a moving vehicle. All of this would mean very low latency and very high reliability within a highly localised environment. However, no expectation of this environment being operated in a single operator market is included, meaning that every operator would need to interconnect to every other operator to allow a vehicle connected to one network to detect a pedestrian connected to any other, for example. This, in turn, implies a fundamentally different physical interconnect environment between operators.

The combined effect of ultra-reliable low latency services for vehicles that cross international boundaries is a need to implement entirely new constructs for roaming and interconnect. Those constructs impact technical implementation and also commercial frameworks and, as a result, sit firmly in the domain of the GSMA. The disruption that 5G brings still has some way to go.

4.3 Future Standards

This chapter has looked primarily at the way that the Standards process of each generation of mobile technology has evolved and the roles that ETSI, 3GPP, GSMA and ITU have played in taking technical research and turning into a global ecosystem. However, 5G has coincided with some other fundamental changes in how networks are built and maintained that may change both how networks evolve, and hence how Standards are written. This cumulative effect draws into question whether there will ever be a further, explicitly identifiable step change in network principle that could be recognised as a 'generational shift' in the way that previous Gs have been discretely obvious.

There are those who would argue that 5G itself is not really as radical a change as 2G, 3G and 4G. With each of these, the radio network saw a fundamental new technology introduced, with GSM being TDM, 3G being CDMA, and 4G being OFDMA. 5G, whilst introducing sub-carrier spacing, still has OFDMA at its heart, despite other more radical options being available. 5G initial deployments being dependent on simultaneous connection to the 4G RAN means that the feeling of 'evolution' versus 'revolution' is only strengthened.

One place where 5G is different however, is in the extent to which integration of hardware and software has been broken. 'Open' as a software development principle has been embraced, meaning that the 5G CN is now divided into its fundamental components. The Service-based Architecture allows visibility of the network functions as clusters of microservices, and hence network functions themselves are increasingly becoming false constructs that merely allow the industry to recognise the building blocks from previous CN implementations are still there.

But the direction that Open software development takes network build in is fundamentally different. The principles that underpin Open software come from Enterprise networks where network elements can change on a daily or weekly basis, and can be developed to incorporate many additional features independently of other elements and the presence of other vendors. If the mobile network implementation principles continue in the direction that Enterprise networks have moved in, the rate of Standards development in 3GPP will become a major encumbrance, and the idea of a new Generation every 10 years will seem to be a glacial pace of development.

Having said this, some of the more fundamental principles that underpin mobile telecoms, and telecoms in general, will still need to be retained. Reflecting on where this chapter began, inter-operability is so key to consumer choice, roaming and interconnection, that this must remain enshrined in the industry which, in turn, would seem to enforce a role for Standards to defend that principle. The danger of 'Open' networks in that environment is that they result in many 'published proprietary' implementations. Licensing and FRAND will need to be revised to prevent Open actually becoming the battle ground of a renewed drive towards vendor lock in.

With network softwarisation comes the idea of network slicing, whereby any customer could have a separated implementation of a network in software, which could evolve as they wish. This may actually represent the reason why there is never a 6G – what would a new generation of technology amount to it if it could be introduced in a logically separated set of software that occupies the same hardware as every previous generation? At what point would '5.9G' which differs a little from what was implemented the previous week following a software update, be found to push past some arbitrary counter to become 6G?

The link made in the industry between 5G-NR, and the advent of a truly software-based network, is really a fortunate coincidence of timing, but from here on, the rapid changes that a software-based network can implement may outstrip the pace of a Standards process that is already struggling to keep pace. The final component of this sea-change could even be Covid-19.

Telecoms industry standards' meetings have always primarily been held face-to-face. Those involved in ETSI SMG meetings in the 80s often reminisce of revising typed documents with correcting fluid and desperate searches for functioning photocopiers to be able to create enough copies of the revised document to share with the delegates of a meeting to make sure a document was agreed during the meeting. In exactly the same way that mobile phones have changed from simply making calls, to calls and texts, then e-mail, internet, and applications, so the velocity of Standards' meetings has been revolutionised by word processors, then computers, and finally laptops connected to servers dedicated to the meeting group that is in session. Correcting fluid and photocopiers have long since been replaced by change tracking in documents and email distribution at the touch of a button. What Covid-19 has forced Standards groups to give up is the last hangover of the original way that Standards were prepared – the face-to-face meeting.

Covid-19 has seen 3GPP meetings forced entirely online, and the effect on the processes has, to date, been counterproductive. When many hundreds of technical experts are in one room arguing a point, and are also able to have discussions over coffee, lunch and dinner, compromises can be found. A conference call or e-mail discussion is not as fluid a method for discussion and does not foster the personal negotiation that can lead to log jams being broken and deals being brokered. 3GPP has tried to retain the feel of a 'meeting' by still having document deadlines and times for 'meetings'. However, what would have been at a one week face-to-face meeting is taking two weeks to complete online, usually with a restricted agenda. 3GPP are already seeing their timelines for Release 16 and Release 17 slipping. There is also widespread reports of delegate fatigue since, when a meeting is online, it is always in the middle of the night for someone in a global industry. Whilst face-to-face meetings meant dealing with jet lag, it was always daylight when the meeting was happening, and there was always caffeine.

Covid-19 has coincided with a point in time when the principles of how a network is built, scales and evolves is becoming far more dynamic and fluid. This may also be how Standards will change.

References

[1] ITU-T Recommendation I.130. *Method for the characterization of telecommunication services supported by an ISDN and network capabilities of an ISDN.*

[2] 3GPP webpage. Available from https://www.3gpp.org/specifications-groups/specifications-groups

[3] (2020). GSMA Intelligence. *The Mobile Economy 2020*, March.

[4] *Decisions of the CEPT/CCH/GSM Meeting in Madeira, 16-20February 1987 Concerning the Recommended Technical Standard for a Pan European Digital Cellular Radio System.* Available from http://www.gsmhistory.com/wp-content/uploads/2013/01/3.-1st-GSM-Tech-Spec.pdf

[5] *Memorandum of Understanding on The Implementation of a Pan-European 900 MHz Digital Cellular Mobile Telecommunications Service by 1991.* Available from http://www.gsmhistory.com/wp-content/uploads/2013/01/5.-GSM-MoU.pdf

[6] (2003). ITU-R Recommendation M.1645. *Framework and overall objectives of the future development of IMT-2000 and systems beyond IMT-2000.* June.

[7] ITU News Magazine webpage. *Development of IMT-Advanced: The SMaRT approach.* Available from http://www.itu.int/itunews/manager/display.asp?lang=en&year=2008&issue=10&ipage=39&ext=html

[8] ITU Press Release. *ITU Paves Way for Next-Generation 4G Mobile Technologies*. Available from http://www.itu.int/net/pressoffice/press_releases/2010/40.aspx#.X8PxcSw3aUl

[9] ITU Press Release. *ITU World Radiocommunication Seminar Highlights Future Communication Technologies*. Available from http://www.itu.int/net/pressoffice/press_releases/2010/48.aspx#.X8PzOCw3aUk

[10] GSMA. *GSMA VoLTE Implementation Guidelines*. Available from https://www.gsma.com/aboutus/workinggroups/wp-content/uploads/2020/08/VoLTE-Implementation-Guide-July-2020.pdf

[11] GSMA Webpage. *2G/3G sunset and VoLTE Roaming*. Available from https://www.gsma.com/aboutus/workinggroups/key-areas/volte-roaming

[12] NGMN *NGMN 5G White Paper*. (2015). Available from ngmn.org/wp-content/uploads/NGMN_5G_White_Paper_V1_0.pdf

[13] GSMA Intelligence. (2014). *Understanding 5G: Perspectives on Future Technological Advancements in Mobile*.

[14] ITU-R Recommendation M.2083. *IMT Vision – Framework and overall objectives of the future development of IMT for 2020 and beyond*

5

The Mobile Network

In this first deeply technical chapter of the book, we will look at how mobile networks operate. Today's mobile networks are some of the most complex engineering systems ever to be invented. They are capable of supporting millions of users with seamless, ubiquitous coverage and providing diverse communication services (voice, data, video, text, gaming etc.). The mobile phone usability even extends across national boundaries, with international roaming. You may travel hundreds of miles without using your phone, but the network seems to know exactly where you are, when you want to activate a service. Furthermore, today's mobile users interact with many other wireless technologies like Wi-Fi and Bluetooth, yet the network architecture ensures seamless connectivity for all kinds of different technologies. We will look at how this seamless connectivity is provided from an architectural point of view. We will give special emphasis to radio connectivity, as the following technical chapters of this book are focussing mainly on radio connectivity (technically known as the Radio Access Technology). We will also provide accounts of the evolutions of core network (CN) and radio access network (RAN) over the years, through the mobile generations and mainly look at how and why these changes have happened. We will delve into more detailed descriptions for more recent mobile generations and their subsequent releases, as these are likely to be in operation for a good number of years to come. We will try to cover many of the network concepts and features succinctly, at an introductory level. A more detailed account of mobile networks from 2G to 5G can be found in [1].

5.1 Mobile Network Architecture

The mobile network architecture refers to the arrangement and interactions of components that make up the network and the interfaces between them. The ultimate purpose of network architecture is to connect your mobile device to any other device or a server (depending on the service type) as seamlessly and efficiently as possible. As the service types and applications we are using the mobile devices for have evolved over the years, so has the mobile network architecture to best support these applications. We will first look at the basic building blocks and functionalities of the network architecture and then

The Technology and Business of Mobile Communications: An Introduction, First Edition.
Mythri Hunukumbure, Justin P. Coon, Ben Allen, and Tony Vernon.
© 2022 John Wiley & Sons Ltd. Published 2022 by John Wiley & Sons Ltd.

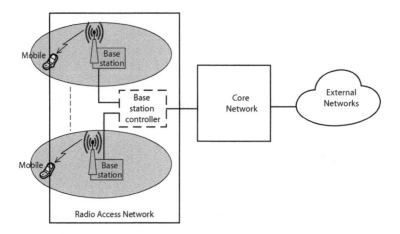

Figure 5.1 The generic architecture of the mobile network.

investigate how and why the architecture has evolved over the mobile generations. Figure 5.1 below depicts a very basic and generic architecture diagram.

Mobile networks, as the name implies, provide the freedom of connectivity while on the move, virtually anywhere. This mobility is supported through the use of radio waves for the 'last mile of' connectivity – technically the 'air interface', between the mobiles and the base station. Although this last mile is a very fluid way for communication, a well-planned, rigid structure of radio cells needs to be in place to orchestrate this. The network of cells, activated by the base stations that provide radio coverage throughout the span of the cell are shown in Figure 5.1. In earlier cellular generations (up to 3G), there is a control node for the base stations – which decides on the attachment of a mobile to a certain base station, the process of switching the attachment (called handover) and the resource allocation to the mobiles for communication. In later 4G standards, this node would disappear, with its functionality absorbed mainly by the base stations. Thus, we have shown this controller block in dashed lines. The interface that connects the base stations to the controller (and to the core in 4G) is known as the backhaul links. The Base stations, BS controllers and the radio interface, as well as the backhauling interface, constitute the RAN. As the name suggests, the overall function of the RAN is to provide and manage the radio connectivity to the mobiles. The main RAN components will be explained in detail in this chapter, with later sections detailing how these components evolved/changed across the mobile generations. The planning of the RAN, to provide the expected radio coverage and capacity (the cumulative data rates to the active mobiles) is covered in Chapter 6.

The function of the CN is to connect the Access Network to the other external networks. The CN carries out the switching and routing functions needed to connect a specific voice or data connection request to the correct paths. One of the key evolutions of the cellular CN is the transition from a circuit-switched network in early 2G to a mixed circuit and packet switching (PS) core in the later 2G (2.5G) and 3G and to a fully PS core in 4G and 5G. We will explain circuit and packet switching in the CN section and these evolutionary changes in each of the generations. Thus, in the earlier 2G, the core would contain switching centres, evolving to IP packet switching gateways in 4G. As noted before, some of the mobility related control functions moved into the CN in 4G. The CN is also responsible for user Authentication, Authorisation and Accounting (AAA) functions when an end-to-end connection is made. Thus, it would contain various data bases to check mobile credentials and also record resource usage.

Each external network that our mobile network connects to should provide the services the user requests. In earlier generations, these services would exclusively be voice calls or text messages to another mobile phone, or voice calls to another landline. Thus, the external networks would be PLMN (Public Land Mobile Networks, including the user's own PLMN) and PSTN (Public Switched Telephone Networks) for landlines. In later generations, mobile data communications became a key feature and thus connections to PDNs (Packet Data Networks) including the internet would become necessary.

5.2 The Radio Access Network (RAN)

As noted above, the function of the RAN is to connect mobiles to the CN through radio connections. This is achieved by the distinct and contiguous span of radio cells, each controlled by a base station, which apparently gives the name for cellular communications. The RAN is basically responsible for the delivery of data or voice calls that come down from the CN intended for a specific mobile (called the 'downlink (DL)') and connecting up a mobile to the CN when it initiates similar activity (called the 'uplink (UL)'). The radio-resource allocation and management for these functions is done by the RAN. The radio interface consists of data channels to carry the user data and control channels to provide and manage this connectivity to the mobile. The radio resource can be differentiated in time, frequency and code domains. Many of the cellular generations (including 2G and 4G), utilised only the frequency and time domains to differentiate allocations to the active users. The code domain was utilised in 3G and we will discuss its application in Section 5.6. The time/frequency arrangement block of the radio resource is generally known as a radio frame. The radio frame will contain a number of 'physical channels', which are essentially decodable data patterns carrying specific data or control information from the base station to the users in the DL and vice versa in the UL. The manner in which the data (or control) 'bits' are incorporated onto an attribute of the radio waves (this can be amplitude, phase or frequency) is called modulation. The full set of functions in the transmission and reception of digital signals is covered in Chapter 8. The wireless or radio connection between the base station and the mobile is also known as the 'air interface'. Multiple users have to share this common radio resource (technically termed as multiple access) and these methods can involve separations in time, frequency or in code domains. The multiple access scheme is a defining feature for each of the cellular generations and we will study these later in this chapter in some detail.

There are a number of distinct procedures that the RAN has to carry out, covering any eventuality expected in the behaviour of mobiles. We will look at these key generic procedures of the RAN in this section. When we look at the individual cellular generations, we will investigate how these functions are executed and why they have evolved in a particular manner through the generations.

5.2.1 Synchronisation

The radio-frequency (RF) generation in the base station and the mobile is done independently by frequency synthesisers (or oscillators). Due to the size, power consumption and cost constraints of the mobile, it is more likely to operate a lower accuracy frequency synthesiser and this will have less stability. Hence, there is the need to synchronise carrier frequencies between the transmitter and receiver. Also, there needs to be timing alignment, to de-modulate the signal and extract out the data bits by accurate sampling. The propagation delay between the base station and mobile will differ according to how far the mobile is from the base station and this will impact timing alignment. Also, multiple reflections of the transmitted signal arriving at the receiver with different delays (what we term as multipath propagation) can cause timing mis-alignment and the Doppler effects (due to relative movement) can cause frequency shifts. Hence, timing and frequency synchronisation between the base station and the mobile is essential to initiate communications.

For the time and frequency synchronisation, known sequences are transmitted from the base station, in pre-arranged (time/frequency) positions of the radio frame. The mobile will carry out correlation (essentially cross multiply and sum-up) operations on the received sequence and the internally generated sequence, so time and frequency alignment can be achieved. We call this process 'auto-correlation'. Additional information such as the cell ID is also carried with these synchronisation signals, which will enable the mobile to initially connect onto a serving cell.

5.2.2 Broadcast Messages

There are a number of system information parts which need to be transmitted all over the cell, for any UE to be able to decode and utilise for accessing the cell. These are contained in broadcast messages from the base station and transmitted in the broadcast channel in the downlink. It is common to categorise this system information in two parts. The first with critical information (like the system bandwidth and system frame number) are transmitted more frequently and are usually contained in a Master Information Block (MIB). The second, with more detailed information, is transmitted in a number of System Information Blocks (SIB), to be acquired by the relevant users once the MIB is accessed.

5.2.3 Paging

Paging is the procedure through which the network indicates to a mobile that there is an incoming message intended for it. If the mobile is in idle (or sleep) state, it needs to 'wake-up' through the paging message. There is no predictable pattern as to when a message intended for a particular mobile would arrive. If the mobile were to listen for paging messages all the time, it would not have a sleep mode, and a lot of power would be wasted. The solution provided for this is through the DRX (Discontinuous Reception) cycle. The DRX cycle ensures that the mobile periodically changes between sleep mode and 'listen' mode. There are various timers associated with the DRX cycles and these control the pattern of 'listen' and 'sleep' modes of the mobile. The paging messages are sent during the time intervals when the mobiles are in 'listen' mode and are included in the downlink control channels. Different mobile generations have optimisations to make paging and DRX more robust and efficient. For example, in LTE and 5G-NR, the DRX is applicable even when the mobile is in connected mode (actively receiving or sending voice/data) to further enhance power usage efficiency.

5.2.4 Random Access

Random access procedure is usually enacted when a mobile station wants to get 'active' (to initiate a call or text, or to connect to the internet etc.). We refer to this as the mobile moving into the 'RRC connected state'. From the base station perspective, the timing when a mobile wants to become active is seen as random, hence the use of the term 'random access'. This can be one of the most complex procedures in RAN, but we will try to provide a simplified explanation here.

Cellular random access is based on the mobile transmitting a known pre-amble to the base station, on the random access channel (RACH). This occurs at permitted time intervals (or slots), within the radio frame. The random access process is typically contention based, i.e. multiple mobiles have to select a pre-amble from a common set known to all the mobiles, which can lead to collisions. When the mobile initiates a connection request, it uses a pre-amble from this known set. There is a chance that another mobile may use the same pre-amble in the same time interval in the same cell and hence a contention may happen at the base station. Then, the base station will execute a contention resolution process, and mobiles may attempt once more.

Although the mobile will acquire synchronisation from the downlink synch signals before attempting random access, it does not know how far it is from the serving base station. Hence, there is an uncertainty about the delay to reach the base station, which is dependent on the path distance it has to cover (called the propagation delay). When designing the pre-amble sequences, the impact of this propagation delay (up to a certain maximum cell range distance) must be considered. To estimate the timing difference in the uplink RACH signal, the base station will multiply the received signal with its own stored version of the signal. This multiplication will yield a peak value only when the two sequences are time aligned. Thus, pre-amble sequence design has to show this very good 'autocorrelation' property as in the synchronisation signals.

5.2.5 Scheduling

Scheduling is the process of allocating radio resources to each of the active users in a cell. This process is conducted by the MAC sub-layer (discussed in Section 5.4) of the base station that controls a particular cell. Multiple active users will simultaneously request access to radio resources and the scheduler is configured to a certain policy to carry out the allocations. The scheduler (particularly in later generations) will consider the quality of the radio channel reported by each of the users, the Quality of Service (QoS) requirements of the users and factors such as whether the data type of a user is delay tolerant or not in making scheduling decisions. If the radio channel quality is better for a certain user more data could be transmitted, contributing to increase the cell throughput. There are three broad policies for making scheduling decisions. The max-SNR policy schedules users with the highest radio channel qualities. Whilst this policy can maximise cell throughput, it lacks in fairness as users with lower radio channel quality are excluded from the process. At the other end, we have the round-robin scheduler, which rotates through all active users requesting resources. This policy gives the most fairness, but the overall cell throughputs can be low. The proportional fair scheduler policy sits inbetween these two extremes and it is widely used in practical cellular networks. One of the fundamental questions in a scheduler design is the trade-off between performance and the amount of control information needed to be transmitted to/from users.

Scheduling algorithms are not standardised and the operator is free to enact any of the above variants (or a combination) to suit their network. What is standardised is the framework and signalling around the scheduling process. The scheduler complexities and interactions have grown over time, to accommodate the complex user requirements in later generations. For example, in 4G LTE, the scheduler interacts very closely with entities managing the hybrid automatic repeat request (HARQ) and link adaptation processes.

5.2.6 Power Control

Power control is needed in the mobile to manage its transmit power level according to the distance it is positioned from the base station (or more accurately the path loss of the uplink signal). If all mobiles transmit at the same power, the nearer (to the base station) mobiles can drown out signals from the further-away mobiles due to interference. Also, base stations need to receive signals with very low and very high signal strengths and thus need to have a high dynamic range. Transmitting too much power than necessary would quickly drain out the limited battery power contained in the mobile. To mitigate these problems, the base station continuously monitors the received signal strengths from the mobiles and sends power control signals back to each active mobile. Rather than dictating the absolute power levels, these periodic control signals inform the mobiles (some of them may be moving) to ramp up or down the transmit power levels by ΔdB amounts. This type of power control is called closed-loop power control, where the mobile responds to feedback commands from the base station. The mobile itself can make measurements of the downlink signal and estimate the power level it needs to transmit back to the base station in the uplink. This is called open-loop power control.

Due the properties of the air interface, some of the mobile generations require very tight and accurate power control – called fast power control, whilst others can manage with slow power control. We will discuss these in detailing the RAN for specific features of the mobile generations.

5.2.7 Handover

When a mobile in the active (or connected) mode moves out to a neighbour cell from the serving cell, the connectivity should be transferred to the neighbour cell. This is the handover process and it is critical to ensure a good QoS in the mobile network. The base station will detect the mobile is moving towards its cell boundary by the periodic signal strength measurements of the serving cell and the neighbour cells the mobile is reporting back. When these signal strengths cross a threshold, the decision is made to enact the handover procedure. In the earlier 2G and 3G, there was a central base station controller in control of multiple base stations and the handover process occured through this. In 4G LTE, the base stations are more autonomous and hence the handover process is different. We will detail these differences in the following sections for each mobile generation.

The handover process is determined purely by the signal strength measurements reported by the mobile, hence this is an example showing the importance of such measurement reporting. The propagation conditions and interference levels from neighbour cells could vary dynamically, as could the reported measurement values. There is also a wide spectrum of mobile makes and there can be slight differences in the accuracy of their measurements. Due to these reasons, the handover point cannot be pre-determined. As a result of the unpredictable nature of fading in the radio channel, the reported signal levels can go slightly up and down, as the mobile moves towards the cell edge. Hence, the serving base station would wait until the target base station signal strength reaches a certain hysteresis value above the serving base station signal strength threshold, to initiate the handover process. This is illustrated in the Figure 5.2 below.

Sometimes, there is a need to handover from one radio access technology (RAT) to another. Thus, all subsequent generations would ensure that connected users can be handed over to the cells operating the RATs of previous generations. Also, this would

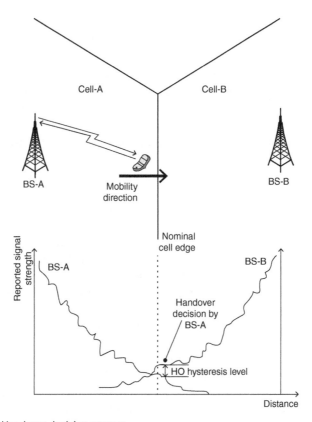

Figure 5.2 The Handover decision process.

mean that the mobile handset should be capable of taking measurements in different frequencies of different RATs and switching over to these RATs when handover is completed.

5.2.8 Link Adaptation

The link adaptation is a feature that came about in 3G and 4G systems. Depending on user location and mobility, the quality of the radio channel can vary significantly. Link adaptation effectively adapts the modulation and coding scheme used at a certain time instant to the user's radio channel quality. The user frequently reports back the channel quality through a CQI (Channel Quality Indicator) as a SINR value (Signal to Interference plus Noise Ratio) and the base station allocates a higher MCS (Modulation and Coding Scheme) if the SINR is higher. Coding implies error correction coding, i.e. extra bits are used to detect and correct errors in the data stream as it is received through the radio channel. In higher MCS, more data bits are packed to a constellation (as you can study in Chapter 8) and thus good-quality radio channels will deliver higher data rates to the user.

5.2.9 HARQ, Error Correction

The hybrid automatic repeat request (HARQ) is a process through which data received in error is detected, repeat transmissions are requested and repeated data packets are combined with the original data packet to generate a (good enough) data sequence. Usually, the HARQ process is carried out on transport blocks, which is the data unit managed by the physical layer (we will discuss this in detail later, in relevance to how LTE operates HARQ). The transport blocks are received by the next layer (MAC) and it checks the correctness of the data received by checking the error control bits (a commonly used method is CRC – Cyclic Redundancy Check). If the transport block is received in error, a re-transmission occurs, but this time with a lower coding rate, to ensure more robustness to error. The MAC layer will combine this new data block with the previous erroneous block(s) to attempt to generate a reasonably accurate data block. If this is successful, the MAC will send an ACK message for the particular data block and the HARQ process for this block will finish. This type of combining (with re-transmitted packets with different code rates) falls under incremental redundancy (IR) combining. Chase combining is another type that can be used in HARQ, where the same code rate and error control bits are used in the re-transmission. It is possible to run concurrent HARQ processes for successive transport blocks, as some may need more HARQ attempts than others. The retention of the erroneous blocks by the MAC and then using them to re-construct an accurate block makes HARQ the most efficient data repetition process.

There is also another repetition process carried out at the RLC layer (detailed in Section 5.4) level, the ARQ (Automatic Repeat Request). In ARQ, the re-transmitted packet is not altered in code rate or any other way, it is simply a re-transmission. The receiver will send an ACK only when the correct packet (or data block) is decoded (at the check bits). The repetition will happen a maximum N times and the ARQ will come out of the loop.

5.2.10   MIMO Techniques

The use of multiple transmit and receive antennas in the physical layer (known as Multiple Input Multiple Output or MIMO) is a key technology enabling higher data rates and will be discussed in detail in Section 9.4. MIMO techniques offer the ability to linearly increase data rates as per the minimum number of antennas at either the transmit or receive ends. This is termed as the spatial multiplexing gains in MIMO and a pre-condition to achieve this full capacity increase is that transmissions between antenna elements have to be fully de-correlated. After about a decade of theoretical and practical developments, MIMO technology was first introduced to 3GPP Release 7 HSPA standards as a 2×2 (two transmit and two receive antenna) system, enabling a factor 2 data rate increase. In practical systems, there are a number of limiting factors that work against the arbitrary increase of antenna numbers. Each transmit antenna needs an independent transmission path with a power amplifier and this particularly impacts the size, cost and power consumption constraints in mobile device. As the antenna spacing in a single mobile device cannot be made large, achieving the full signal de-correlation for higher numbers of antennas is challenging. Another issue in increasing the number of antennas is that for each of the transmit streams, separate pilot bits (discussed below) need to be included for the decoding of the streams and this incurs additional overheads. A variant of MIMO, known as Multi-user MIMO (MU-MIMO), where a number of mobiles can make up the antenna numbers on one side and each mobile can independently receive its data, became adopted in 4G LTE systems. In 5G, a further evolution of MIMO, known as Massive MIMO is being discussed in 5G-NR and we will cover the basics in Section 9.4.5.

The theoretical discussion on MIMO in Section 9.4 will show that both transmit and receive ends of the link need to pre-code the data and post-code the receive vector (respectively) with specific unitary matrices (related to the MIMO radio channel), in order to simplify decoding. As these matrices cannot be computed for all possible channel variations or shared between the two ends, a practical adaptation in cellular standards is to use a codebook. This codebook will have a number of code words which can approximate to the real matrices and only the index of the code-word can be transmitted to the receiver. The code book design was a major part of the MIMO-related development in later standards (3G HSPA onwards) and we will take a brief look at this under 4G LTE.

5.2.11   The Control/data Channels and Reference Signals

As we have discussed the main procedures of the RAN, now we can explain the need for control channels a bit more clearly. Procedures like random access or power control do contribute to make the main task of mobile communications, i.e. carry voice or data messages, feasible or more efficient. The information related to these procedures is carried in special control channels in the uplink and the downlink. The 'channel' here refers to a predefined part of the radio transmission frame used by the specific control bits and is not to be confused with the radio channel, which is the radio propagation path. The control channel information is specific to a user and also conveys messages about which part of the radio resource is allocated for the specific user for their data transmissions. Similarly, the data bits are carried on the data channels of the radio resource. These data channels are termed as 'shared' channels in 3G, 4G and 5G, as parts of which are allocated to specific users on a temporary basis. The overall signal paths for the control and data information through the protocol stack (which we will introduce in Section 5.4) is termed the control plane and data plane. You will see that in later cellular generations, there are significant differences in how protocol stack layers are arranged for the control plane and the data plane.

In addition to the data and control channels, reference signals are also carried in the radio resource. These reference signals carry known bit patterns called pilots, which can be altered by the radio channel for example. Then the receiver can work out what the radio-channel properties (or co-efficients) are, by evaluating the changes to the pre-known pilot bits. Thus, when we investigate the radio-interface structure for different generations, we will look at these three components.

5.3 The Core Network (CN)

The CN connects the RAN to the external networks, which provides the data and voice services the user is seeking. Voice and data service requests that originate at the UEs (at the edge of the network) are accumulated and switched through to the correct path in the CN. Thus, the CN sees the largest aggregation of switching and routing as it supports virtually millions of users in the network. The CN physically consists of very high capacity switches and routers, to direct individual traffic to the correct paths. It will also consist of gateways which can connect the mobile network to the other external networks, providing voice or data services. The CN elements are typically built to achieve speed and handle high capacity, although not necessarily to possess high intelligence, as these are routine tasks.

The CN also provides other functionalities, i.e. commonly known as the AAA. This ensures that only the registered users can utilise the mobile network and they are properly charged for the services they use. The CN is supported by a number of data bases in implementing AAA features. The Home Location Register (HLR – for network registered users) and Visitor Location Register (VLR – for roaming users) are two such example data bases. CNs also feature Operations and Maintenance (O&M) centres or Operations Support Systems (OSS). These monitor alarms and other abnormalities in the network and can do automated servicing or create alerts for manual intervention.

One of the key CN evolutions that happened over the mobile generations is the transition from circuit switching to packet switching. We will explain the differences between these two techniques below.

5.3.1 Circuit Switching and Packet Switching Networks

These are two fundamentally different approaches to routing voice or data through a network. In circuit switching, the end-to-end circuit is first established before voice or data is transferred. Circuit switching is fundamentally not dissimilar to the two cups and a string play-phone, where the physical circuit established is represented by the string. In packet switching, the content is broken up into small packets and transmitted through many possible routes to the destination. These packets contain a header, which consists of the destination address and packet index number amongst other information. At the destination, the packets are put back into the correct order. There are elaborate mechanisms for re-transmission etc., if some of the packets are received in error. The most common standard (or protocol, as we define in the next section) for packet transmission is Internet Protocol (IP) based and this is also widely used in mobile networks. You will see this term IP in a number of discussions in this chapter.

Packet switching is inherently more efficient than circuit switching as the usage of radio and network resources is very high with packet switching. In circuit switching, content will not be transferred if a dedicated end-to-end link cannot be established first. By dividing the content into small packets and utilising multiple possible paths, packet switching maximises the chances of content transfer even in a busy network. Circuit switching has traditionally been used for voice while the packet switching has been used for data. With voice, only limited delays can be tolerated, so a dedicated link ensures smooth continuity of the call. With circuit switching, network resources are reserved for the entire duration of the call, guaranteeing quality. Many of the data services can tolerate delays (like file transfer FTP) and suits the packet switching mechanism. However, the mechanisms of packet switching has developed so much that 'real-time' services like video and voice can also now be effectively transferred though packet switched networks. With development of VoIP (Voice over IP), voice calls are transferred more and more through packet switching. In the 4G LTE standards, there is a provision for packet switched voice transfer (called VoLTE), which can effectively make the whole of the LTE core packet switched. There are numerous calling Apps now which allow video and audio calling through packet switching, using the Wi-Fi or cellular data connectivity. A good account of circuit and packet switching can be found in [2].

5.3.2 Tunnelling and Encapsulation

Another key concept used in cellular packet CNs is tunnelling. Through tunnelling, dedicated pathways for user data can be created between various nodes in the CN and between the radio control nodes and the CN. These tunnels can be likened to veins in the human body carrying blood (similar to data packets) between vital organs. It becomes clear that there is a control signalling part associated with tunnels as well as the purpose of carrying user data packets. The data packets from a mobile reaching the base station will be configured as IP (Internet Protocol) packets, with a unique source and destination headers. Within the base station, another layer of 'packaging' is carried out, where the packets traversing through the CN are routed using the tunnel end-point IDs (TEID). This kind of additional packaging is termed as 'encapsulation'. Tunnelling and encapsulation can bring many benefits, as we will examine a bit later.

The same tunnelling protocol (or methodology) is used in 2G GPRS, 3G and 4G standards and this is known as GPRS Tunnelling Protocol (GTP). There are two variants of GTP, GTP-U to carry user data between the CN nodes and also the base station (as discussed above) and GTP-C to carry control signalling to set-up, manage and later dismantle the GTP-U based data sessions. You will see these GTP-U and GTP-C entities as protocol layers in the subsection below and when we look at the 4G LTE protocol stack in detail.

There are many benefits of using GTP-based tunnelling in the CNs. These benefits are a big reason why GTP has remained unchanged from 2G GPRS to 4G LTE, despite the network architecture having undergone significant changes. Tunnelling allows the abstraction of the physical nodes so user mobility can be smoothly accommodated. Imagine a mobile user travelling at high speed, traversing many base stations and CN nodes, while connected to the same server on the Cloud. Thus, the source and destination addresses of the IP packets will remain the same, but they have to go through different nodes. The GTP will create data (and control) tunnels appropriately between these nodes, with the TEIDs indicating different endpoints of the tunnels as the user moves. This dynamic nature of tunnelling within a single data session enables seamless mobility. Also, with diversification of data services in the later mobile generations (particularly 4G), it became important to provide different QoS provisions for different services. For example, non real-time applications like file transfer can run as a best effort service class, while real-time applications like interactive gaming need a low latency, guaranteed bit rate service class. The tunnelling in the CN can easily be adapted to meet the requirements of these different QoS classes. The encapsulation provided with tunnelling also increases data security and privacy. This is especially useful in roaming situations, when the encapsulated data can travel through third-party networks.

5.4 The Protocol Stack

Another way to look at the mobile network is through a functional description of specific layers, which together make up a protocol stack. In this section, we will explain what a protocol stack is and also broadly look at the protocol stack evolution within mobile communications.

A communication protocol is a pre-defined set of guidelines or rules through which two equipment can communicate. Nowadays, there is the need for equipment operating under different wired or wireless standards (like Ethernet, LTE, WiFi, Bluetooth) to communicate with each other. Also, the communication type can vary widely – voice, text, data transfer, video transfer, web browsing to name a few. One 'brute-force' solution would be to define a specific protocol for each of these specific permutations. This would result in thousands of protocols and the list would grow quite rapidly with new additions. Also, even a minor change in a standard would require the whole protocol to be modified.

A far more manageable solution to the above situation is presented by defining a protocol stack. A protocol stack is made up of specific protocol layers with clearly defined functions. Each layer communicates only with layers above and below. The signalling and information transfers between these layers are also well defined. This stack architecture would enable changes to be made in each layer without affecting the functionality of the others. Also, this solution enables the same layers within the two communicating nodes to add/ peel off similar data formats and extract information where needed. This is called the horizontal communication within protocol layers.

In a simplified example, imagine a courier service between two remote offices of a company. Each office would have a mail man, who would collect letters or parcels internally, package them into one big parcel, address them to the remote office, make a record of the number of parcels delivered and hand this over to the courier. In networking terminology, the mail-man is a 'middle layer', dealing with 'packets' coming from the 'higher layer', processing them and delivering to a 'lower layer' (the courier). The processing in the middle layer involves adding the packets up (known as 'concatenating') to a big parcel with outer packaging with address notification (known as a 'header'). The whole outer parcelling process of the mail man can be viewed as an 'encapsulation'. Information on the number of parcels is also marked for the benefit of the mail man in receiving office, and this is a form of horizontal communication. Also, he could include a form where the other mail man could indicate if the big parcel is received in order and is sent back with the courier on return delivery. This is known as an 'ACK' or 'NACK' (if things are not in order) message in networking terms. After opening the big package, the other mail man would distribute the parcels he receives from the courier inside his office. He uses the names and departments in the 'packets' (another instance of a 'header') to correctly route 'packets' to the higher layer. The contents within the letter or parcels ('packets') are of no concern to the mail-man – the 'middle layer'. Also, the mode of transport used by the courier ('the 'lower layer') is of no relevance to him. He has a specific, well-defined function to perform – as duties to the higher layer, duties to the lower layer and 'horizontal communications' with his peer at the other office. This sounds very robotic and mechanical in a human-related example, but this is exactly how it is intended to be in a communications network.

With this three-layer set-up, specifics within each layer can be changed without affecting the other layer. Imagine the company wanted faster delivery times. If it is viable, they may decide to change the courier company and employ a new courier who uses air transportation. The functions of the middle layer (the mail man) need not change to implement this. Alternatively, if this is a large warehouse or a factory, the company may decide to limit the mail-man's function to interaction with the courier and replace internal delivery with an automated parcel-routing system. The lower layer (courier) functionality or letter or parcel creation in the higher layer need not change to implement this either.

In the wireless communication sphere, there are many examples where we change the 'courier' i.e. the lower layers of the protocol stack, without affecting the higher layers. We can change the connectivity from LTE to Wi-Fi within a web-browsing session with your mobile device. Similarly inside a car, we can direct a mobile voice call to connect to the car audio system through Bluetooth. All these interactions and changes are made possible by the fact that specific functionalities of the end-to-end communication chain are contained within well-defined layers of the protocol stack.

5.4.1 The OSI Model of 7 Layer Protocol Stack

The Open Systems Interconnection (OSI) is a conceptual reference model of a protocol stack, which demonstrates how applications can communicate over a network. The OSI model is the basis for almost all modern communication protocol standards, although it is rarely implemented in its entirety. This is because none of the major standards keep all related functions together as defined in the 7-layer OSI standard. Today, it is used mostly as a reference model, where most vendors of telecommunication equipment today attempt to detail their protocol stack functions in relation to the OSI model.

OSI development started in 1977, with the involvement of major computer and telecommunication companies at the time and with the auspices of International Standardisation Organisation ISO [3]. Originally, it was meant to be a standard architecture with detailed specifications for building computer networks, regardless of the type of equipment in use. Instead, in the early 1990s TCP/IP was widely adapted for computer networking, creating the Internet as we know it today. The role of OSI was relegated to a reference model, yet it is included here to understand the modular functionality of the communication protocols. A comparison of the early development of OSI and the TCP/IP models can be found in [4].

We will begin the description of the OSI protocol layers with the Physical layer. This layer connects the source device to the destination device through the physical medium. The medium can be wired (for example, copper or optical fibre cable) or wireless (radio waves or even visible light rays). The data is transmitted as 'bits' in the physical layer. The bits may take the form of voltage levels or pulses and other modulated signals on the radio or optical wave. The physical layer deals with aspects like definition of bits (voltage levels and durations), connection establishment and release, duplexing method (how bits streams are arranged in both directions – uplink and downlink). The physical layer inserts a lot of control and pilot symbols into the data stream (particularly in wireless communications) to enable accurate decoding at the receiver. These control and pilot information are not transferred to the layers above the Physical layer.

The Data link layer is the single most complex layer in the protocol stack and is often divided into MAC (Medium Access Control) and LLC (Logical Link Control) layers in many of the actual protocol suites. After obtaining the 'bits' from the Physical layer, the data link layer groups them together into 'frames'. In the other direction (from Network layer) the incoming data is packaged into frames with source and destination addresses and sent with error control and flow control functions. This is the part played by the LLC.

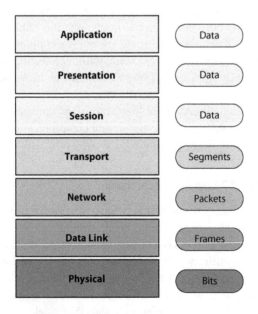

Figure 5.3 The protocol layers as per the OSI model.

The MAC functionality controls access to the common transmission platform (this could be the wired links or wireless links (as discussed above)) for multiple contending data streams. The scheduling function we detailed in Section 5.2.5 is executed in the MAC layer.

The network layer is equivalent to the IP layer in the TCP/IP protocol. This layer deals with routing of the frames (from the Data link layer below) to the correct address. The Network layer examines the source and destination addresses on the frames coming from the layer below. If the destination address (IP address) matches the final destination where the router resides, the Network layer re-assembles the frames into packets and delivers them to the transport layer above. If the address does not match, the network layer pushes down the frames back to the lower layers. Thus, the functionality of a physical router is executed within the Network layer.

The Transport layer has the capability to deliver data across network connections. It receives packets from the Network layer and controls the reliability through flow control, error control and segmentation. TCP (Transmission Control Protocol) and UDP (User Datagram Protocol) are the most common Transport Layer Protocols (TCP). TCP is a connection-oriented protocol, when a TCP data transmission occurs, a session is established first and transmitted packets are numbered and re-ordered at the receiver to control the data flow. UDP is a connectionless protocol meaning that the data is transferred in a 'best effort' manner, without ordering. While TCP can guarantee packet delivery and hence have high reliability, UDP is much faster due to the removal of overheads and constraints. UDP is widely used when speed is of the essence, as in live video streaming and on-line gaming. The four layers discussed so far are termed as the lower layers of the protocol stack.

The Session layer, together with the Presentation and Application layers, make up the upper layers of the OSI protocol stack. The functionality of the session layer is to initiate, manage and terminate communication sessions between two end points. The session layer is also known as the port layer, because each communication session is associated with a port – a number that is indicative of a particular upper layer application. NETBIOS is broadly categorised as a Session layer service, which is used to connect two end points for the transfer of large amounts of data.

The Presentation layer is responsible for format conversions and encryption/decryption of the data coming to/from the Application layer. These format conversions may include conversions to ASCII, Postscript or Binary. This layer performs the simplest function of the OSI protocol stack.

The Application layer is on the top of the protocol stack and provides network services to end-user applications. Typical examples of Application layer protocols are HTTP, FTP and SMTP which support end-user applications of web browsing, file transfer and e-mail transfer, respectively. It is safe to say that whenever you are using a software in your device that operates a networking function, you are dealing with the Application layer.

5.4.2 Protocol Stacks for Mobile Communications

In mobile communications, the protocol stack has evolved significantly from the voice-centric 2G to the data-centric 4G. Also, all of the above OSI protocol layers are not necessarily present. The more complex data link layer and network layer are broken down into several sub-layers in mobile communications.

Another significant point to note is that the information flow through the protocol stack can be categorised as user data and control signalling to control/manage the flow of this user data. These are viewed as two separate planes, the user (or data) plane and the control plane. From the mobile network perspective, the user plane data originates in the Application layer and is routed through the protocol stack layers down to the Physical layer. The control plane signalling can originate at different protocol layer entities (particularly the RRC sub-layer) and again transmitted down to the Physical layer. In earlier mobile generations, the control and user plane information would flow through the same protocol entities. As mobile services diversified with many data applications in later generations, the control and user planes would be optimised differently, with different protocol layers dealing with these planes. We will look at these differences in detail from 4G onwards.

Figure 5.4 below depicts the protocol stack for the 2G GSM and 4G LTE Networks. We present a brief description below, with a more in-depth description on the LTE protocol stack, including the interconnections between each of the physical nodes, being provided later in the 4G network section.

The protocol stacks show the difference between the requirements for voice communications in 2G and data communications in 4G. The 4G network is a packet-switching network, while the 2G network is a connection-oriented circuit-switching network, and these changes are clear in the network layer. In the 2G GSM network, the network layer is the highest in the stack. Within this layer, the connection management sub-layer contains functions for establishment, management and release of service connections (including call control, supplementary services and SMS). The Mobility management sub-layer is responsible for the location updating of a powered on mobile (as it moves) and mobile authentication and security aspects. The RRM sub-layer is responsible for the establishment, maintenance and release of the radio connections. These include paging, handover, power control, radio channel access for the mobiles. The whole network layer in GSM is thus geared towards making and breaking of service and radio connections and the authentication of mobiles. The link layer of GSM is known as LAPDm (Link Adaptation Channel D – mobile) and is a modification of the link layer used in fixed landline communications of ISDN (Integrated Services Digital Network). This LAPDm layer is responsible for the establishment, maintenance and breakdown of links through which data frames are communicated. It also provides error and flow control for these frames. The physical layer provides the radio interface, and in GSM, it characterised by the TDMA/FDMA multiple access. We will discuss this in the 2G network features in Section 5.5.3.

The network layer of 4G LTE is all geared up for packet-based communication. At the top, it has two distinct sub-layers for user plane and control plane. The user plane is all driven by IP connectivity, through routers (known as gateways) of the CN to the external data networks. Above this, a separate Application layer is present (similar to the Application layer functionality in the OSI model). The control plane has a sub-layer known as Non Access Stratum (NAS) and this is the top-most layer from the control plane protocol between the central control unit of the core (MME) and the mobile device. The NAS in general, are protocols that connect the CN and the UE and are transparent to the

Access network (the eNodeBs). The sub-layers below the RRC (Radio Resource Control) sub-layer is common to both user and control planes, but provide quite different functionalities. The data link layer is sub-divided into three sub-layers and is much more complex than in GSM, with functionality to support packet transmissions. We will discuss the LTE protocol set-up in detail in Section 5.7.2. A good reference to 4G protocol stack can be found in [5].

In looking at the generic protocol layer structure, one of the features to emphasise is the manner of information exchange between layers. The information (control and data) passes through the PHY layer and the PHY layer of the other node (UE or eNB) through physical channels. Between the PHY layer and the MAC layer (of the same node), information passes through transport channels. Between the MAC layer and RLC layer, information passes through logical channels. Between the RLC and PDCP layers information passes through radio bearers. On the data plane, user traffic passes between the PDCP and IP layers. On the control plane, the RRC layer passes and requests information from all the layers below it (not shown explicitly in Figure 5.4). With the PHY layer, the RRC passes layer 1 configuration information whilst also requesting measurements from PHY. With MAC layer, RRC exchanges MAC control information. Similarly with RLC and PDCP, RLC control and PCDP control information is exchanged respectively. Generically, the packets transmitted by a protocol layer as output are termed as the PDU (Protocol Data Units) and the packets received by a protocol layer as input are termed as the SDU (Service Data Units). At each sub-layer, specific headers are added to the SDUs and these headers help decode related control and data information in corresponding sub-layers.

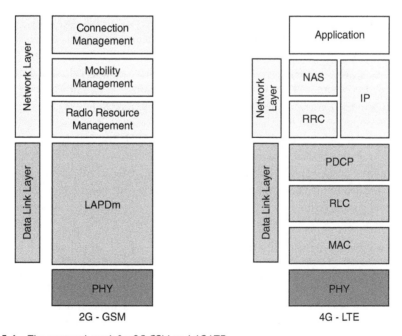

Figure 5.4 The protocol stack for 2G GSM and 4G LTE.

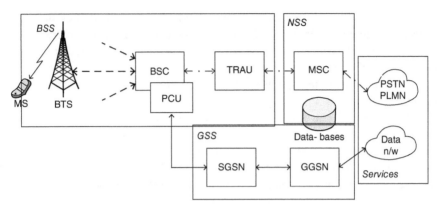

Figure 5.5 The 2G network architecture. (*Source* [6]).

5.5 The 2G Network

We will start investigating how and why mobile networks evolved with a detailed look at 2G. 2G was the first Mobile Generation to truly capture global traction. What started as a specialised service accessible only in a few locations at a very high cost in 1G, grew exponentially to cover a global populous in 2G. 2G was the first to use digital transmissions in the RAN. At first, the 2G control centres and CN were very much similar to the landline networks. Then there were significant architectural changes within 2G itself, with the introduction of packet switching. We will focus on the network architecture, RAN and core evolution of the GSM strand of 2G, which is by far the mainstay of the 2G standards.

As noted in Chapter 1, 2G ushered in the era of digital mobile communications. However at the beginning, 2G inherited circuit switching as the means to route calls and limited data applications. A conventional way to describe the 2G network is as two sub-systems, the BSS (Base-station Sub-System) and the NSS (Network and Switching Sub-System). Later within the generation, packet switching was introduced in the form of GPRS, also termed as 2.5G. This evolution basically added another subsystem, the GSS (GPRS Sub-System) connected to the BSS to enable packet routing to the external packet-data networks. A functional block diagram of the GSM and GPRS networks are shown in Figure 5.5 and the component blocks are detailed in the sub-sections below. A detailed account of the 2G network architecture can be found in [6].

5.5.1 The Network Architecture of 2G

The base station sub-system (BSS) consists of the Mobile station, the Base Station (called the Base Transceiver Station (BTS) in 2G), Base Station Controller (BSC) and the Transcoder/Rate Adaptor Unit (TRAU). The BSS oversees radio transmissions to the mobile, radio resource management and ensures smooth handover between neighbouring BTS. We will look at the functionalities and interfaces of each of the units in the BSS in the RAN functionalities section.

A number of BTSs are connected to a node called the Base Station Controller (BSC). The BSC oversees the radio resource management (RRM) of several BTS under its purview. The RRM involves scheduling suitable resource blocks (i.e. time slots in 2G) to active users with a BTS controlled cell. The BSC also controls handover procedures for users moving between the BTS cells under its watch. As explained before, the handover procedure involves UE measurements and these are reported to the BSC by the serving BTS of the active users.

The TRAU is positioned between the BSC and the NSS. The functionality of the TRAU is to convert the GSM voice codec data rates to the 64kbps data rate used in the Public Switched Telephone Network (PTSN). The GSM can utilise several speech data rates, including the full rate at 13kbps, enhanced full rate at 12.2 kbps and half rate at 5.6 kbps. The TRAU conducts rate conversion between these various codec data rates to 64 kbps.

The BSS remained largely unchanged within the 2G and the architecture evolutions centred around the CN. At the start of 2G (GSM) the NSS performed as a circuit switching core. With the evolution to 2.5G (GPRS) the packet switching core (the GSS) was added to the architecture.

5.5.1.1 Network and Switching Sub-system (NSS) of 2G (GSM)

The NSS of the 2G network can generally be considered as its CN. The overall control of the network and the interfacing to the external networks is provided by the NSS. The NSS for 2G networks consists of the Mobile Switching Centre (MSC) and various registers in the form of databases.

The MSC is the main component of NSS. It provides the switching (or call/data routing) for connection requests originating from MSCs within its own network or coming in from external networks. The MSC's switching function is complemented with additional functions unique to a mobile network. These include registration, authentication, call location and handing over the calls to other MSCs or PSTNs if the connection request requires this. The MSC operates as the core circuit switched exchange of the GSM network.

The Home Location Register (HLR) is one of the data bases maintained in both GSM and GPRS cores. HLR is the reference database for all administrative details of each user in the PLMN and their last known locations. The HLR contains details such as customer ID, customer mobile number and billing details. An entry in the HLR database is made when a user registers with a mobile network. There is essentially only one single master copy of the HLR database. Even when mobiles are not actively communicating, but switched on, the location status is periodically updated in the HLR. This helps to route an incoming call to the correct BTS. The Visitor Location Register (VLR) is another database which contains a copy of the HLR data for a particular area of the network. Thus, VLR is a more local copy of the HLR and is more frequently updated with mobiles' location within the area. The VLR eliminates the need to access the (master copy of) HLR all the time and provides more up-to-date tracking information of users. The VLR can access HLR's of other PLMNs to retrieve certain user details, for example, to enable roaming.

5.5.1.2 GPRS Sub-system (GSS) of 2.5G

As noted in the Introductory chapter, 2G was the era which saw exponential growth in mobile connections across the globe. Alongside this subscriber growth, there was also a significant uptake of mobile data services. These dual growth factors ensured that the CN was considerably evolved from 2G to 2.5G.

During the heydays of GSM, mobile data applications also began to surge in popularity. With its digitised radio access scheme, GSM inherently appealed to narrow band data applications like SMS, MMS, web mail and web browsing, on the radio access. In the fixed data networks, this was the era when benefits of fully IP-based packet switching networks were fully appreciated and implemented, particularly the IP-based internet and intranet (corporate/private) networks. The circuit switched GSM CN posed serious limitations in the connectivity for these IP-based data services and the need to support packet switching became evident. This evolution came to the GSM networks as GPRS (General Packet Radio Service). The key difference in GPRS from GSM is the ability of GPRS to support end-to-end packet switching. Packet switching allowed more optimal use of radio resources, as detailed in the previous Section 5.2.

In the network core, where the GSS is added, data packets are separated at the Packet Control Unit (PCU) and directed through parallel packet switched gateways to external data networks. Two new functional components, the Serving GPRS Support Node (SGSN) and the Gateway GPRS Support Node (GGSN) were added to the GPRS CN. The main functions of the SGSN were to authenticate and register GPRS mobiles, support mobility and charging functions. The GGSN acts as the interface to the external packet data network (PDN) and also collects charging information related to the use of external networks and as a filter for incoming traffic. With end-to-end packet switching, GPRS users were charged for the actual data amount they consume, not the time they would stay connected.

5.5.2 The GSM Frame Structure

The main unit of GSM radio resource allocation is the frame, which contains eight distinct time slots. These slots are allocated to different active users in a cell. The slots allocated to individual users can be used for data or control signalling. A typical data burst in a slot is shown in the example, where there is a specific structure to carry the data, training, synchronisation, tail and guard bits at the end. A sum of 26 data frames (or 51 control frames) make a multi-frame in the time domain. These multi-frames are further concatenated to super frames and hyper frames in the time domain. In the frequency domain, these frame structures span 200 kHz, hence, GSM is considered as a narrowband deployment. Multiples of these narrowband structures can be grouped together in one frequency allocation. In the commonly used frequency division duplex (FDD) mode, the uplink and downlink will have different frequency segments (of a few MHz wide) containing multiples of these narrowband allocations, as shown on the top of Figure 5.6. Essentially, the same frame structure is used in GPRS and EDGE evolutions.

GSM numerology is based on 200 kHz bands each acting as one RF carrier which can occupy up to eight users as there are eight time slots per frame. Each of these time slots last 0.577 ms and is termed as a GSM burst. A data burst is shown in Figure 5.6, but equally there can be control bursts and other types (for example synchronisation bursts – which we will cover in Section 5.5.3.2). A number of these narrowband RF carriers can be created in a single band to perform as the downlink and a similar configuration can be done in an adjacent frequency band for the uplink. The GSM system ensures that a given user will be allocated different time slots in the uplink and downlink, so that it does not have to transmit and receive simultaneously.

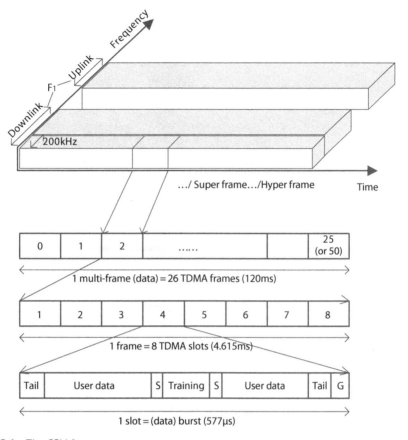

Figure 5.6 The GSM frame structure.

When considering multiple cells, GSM utilises a concept known as frequency re-use distance. Due to higher interference levels, adjacent cells do not employ the same segment of frequency (an example noted as F1 in Figure 5.7). Frequency re-use factors of 3, 4 or 7 are common in GSM, with the numbers referring to distinct frequency bands available to be used in adjacent cells. With a higher re-use factor, the same frequency bands are re-used further away, reducing interference levels. An example of a GSM deployment with re-use factor 3 is shown in Figure 5.7.

The frame structure and frequency planning of GSM is quite simple, but it generates a very robust transmission system. The narrowband nature and higher frequency re-use factor enables GSM cells to cover very large foot-prints, if that is needed. It is not uncommon to find GSM cells spanning more than 30 km in rural regions. This provision of wide coverage at a relatively lower cost is one of the reasons for the longevity of the GSM systems, especially for serving rural areas.

Within the GSM slot allocations, it can be seen that multiple users are allocated to distinct time slots. So GSM and its evolutions in 2G are termed as TDMA (Time Division Multiple Access) technologies. There is also an FDMA (Frequency Division Multiple

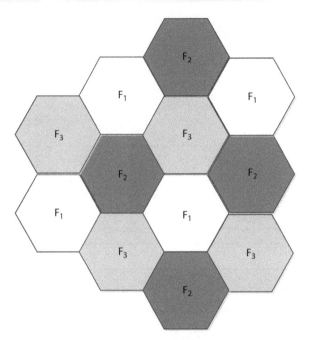

Figure 5.7 Frequency re-use factor = 3 in a GSM deployment.

Access) component in GSM numerology, as multiple frequency channels (spaced 200 kHz apart) can be allocated within a GSM carrier, which is then individually divided into TDMA time slots as detailed above.

5.5.3 GSM (And GPRS) RAN Features

In this section, we will look at unique RAN features and key components of GSM. The more generic RAN features are covered in the sections for later generations, particularly in 4G. A more complete account of GSM features can be found in [7].

5.5.3.1 GMSK Modulation
The digital bits (0 and 1) of the communication source are mapped on to the analogue RF signal through modulation and the basic schemes are covered in Chapter 8. The modulation scheme used in GSM is Gaussian Minimum Shift Keying (GMSK). GMSK is part of the Minimum Shift Keying (MSK) modulation family, where the phase of the analogue signal is varied in-line with the digital information that needs to be transmitted. This signal is filtered with a Gaussian-shaped filter prior to modulation, to generate the GMSK signal. GMSK has some unique features which helped it to be selected as the modulation scheme for GSM. GMSK shows very low out-of-band emissions, i.e. it is a smooth signal in the time domain and the sidebands in the frequency domain are very low. This helps with controlling interference in the GSM numerology where the narrowband frequency channels are positioned 200 kHz apart. Also, GMSK is a constant power level signal, so the peak to

average power level is (or very close to) unity. This feature enables highly efficient power amplifiers to be used in GSM, without the need for a 'back-off' as in 4G LTE. GMSK usage is one reason why the battery life of GSM phones is exceptional, typically lasting for days. GMSK is also more resilient to noise than many other modulation schemes, hence it contributes to increase the coverage range of GSM cells.

5.5.3.2 Synchronisation

In GSM, the mobiles achieve timing synchronisation to the base station by aligning with the synchronisation channel (SCH), carried in the synchronisation burst. This burst contains a long synchronisation (training) sequence (of 64 bits) and also specific information about the frame number and the base station ID. The long training sequence is known by both the base station and the mobile and this can easily be detected with a matched filter at the receiver (the principle of matched filtering is discussed in Chapter 8). Once the mobile aligns with this downlink burst, it can decode the frame and base station numbers as the first step towards completing initial access to this base station. The base station also transmits frequency correction bursts, where the information part of this burst is all set to zero. This ensures a clear frequency signal from the base station with no phase variations. Mobiles can easily align with this signal from time to time, to achieve frequency correction.

5.5.3.3 Handover

Handover is one of the areas where GSM standard developers have worked hard to optimise, so that the mobile user minimally experiences dropped calls. The handover itself can be of several categories. It can be an intra-BTS (within the same BTS), inter-BTS, Inter BSC or inter network (inter MSC) handover. In all these handover scenarios, the mobile station reads the beacon signals of other frequencies of its own BTS (for intra-BTS handover) or the other BTS and reports back to the network. In the intra-BTS handover, the mobile station will be switched to a different frequency channel and a time slot giving a better signal quality in the same BTS. When the mobile station moves out of the current BTS and it reports higher quality beacon signals of a neighbour BTS, the inter-BTS handover will occur. This handover is controlled by the BSC Base station controller and it will assign a new frequency channel and time slot resource to the mobile, before releasing it from the old BTS. If the new BTS belongs to an area controlled by another BSC, the handover is categorised as inter-BSC and resource switching is handled by the MSC. The inter-MSC handover occurs when the mobile moves from a GSM network to another RAT (3G or 4G) network or vice-versa.

5.5.3.4 Frequency Hopping

Frequency hopping is a unique, optional feature of GSM, which in-effect can only be implemented in narrowband systems. As seen in Figure 5.7, the frequency re-use factor can be increased, so interference from neighbour cells can be reduced in GSM. However, in urban areas where the service demand is high, the capacity (or the number of active users to be concurrently served) of GSM has to be increased by reducing the frequency re-use factor. If a given narrowband carrier to a particular user is changed continuously, it is likely that the interference it receives (interference will only come if the same carrier/time slot has active users in the neighbour cells) will average out. This change of frequency is called frequency hopping. It is observed that maximum gains can be achieved when the frequency is hopped in a random manner amongst neighbour cells.

5.5.4 2G Evolutions

We have already looked at how the 2G CN evolved to incorporate packet switching with the inclusion of the GPRS sub-system (GSS). Here, we will briefly look at the GPRS and EDGE evolutions within 2G from a RAN perspective. More details about these 2G evolutions are contained in [8].

5.5.4.1 2.5G GPRS
GPRS stands for Generalised Packet Radio Service. As the name implies, the key feature of GPRS was the inclusion of packet switching to the 2G core. However, it also enabled a significant increase in the 2G data rates, to usher in mobile data services. GPRS facilitated the concatenation of up to eight logical Traffic Channels (TCH) in the access network, allowing maximum data speeds of 171.2 kbps. To put it into context with GSM, GSM typically supported date rates of around 13.5 kbps, which was what the voice codecs required.

5.5.4.2 2.75G EDGE
The next evolution within 2G was the introduction of the EDGE network, known as 2.75G. However, this did not alter the CN topology of GPRS. The changes in RAN were mainly to support higher order modulation (in the form of 8PSK), which can theoretically increase spectrum efficiency by a factor of three, over the GMSK used by GSM. These higher data rates were an attempt to satisfy the huge data demand that was building up with the popularisation of mobile internet.

5.6 The 3G Network

The third Generation of mobile standards were developed to enable networks to handle the anticipated exponential growth in data communications that would follow the success of GSM. After nearly a decade of research and later standardisation activities, the first 3G standard was produced as Release 99, and the resultant first networks came out in Japan in 2001. The W-CDMA air interface based 3G standards were driven by 3GPP, which was also behind the standardisation of later evolutions of GSM.

The 3G Networks included some fundamental changes from the 2G GSM networks from the start and also some gradual evolutions across 3G standard releases. The shift from narrowband TDMA air interface of GSM to the wideband CDMA was a distinct difference. We will look at the W-CDMA adaptation in the 3G RAN section below. The 3G RAN section will also look at other significant improvements with respect to 2G and also at the limitations, which ultimately paved the way for 4G. Initially, the 3G CN remained similar to the GPRS core, with the circuit switching and packet switching core components handling voice and packet data respectively. However, with later 3G releases culminating in HSPA (High Speed Packet Access) standards, the architecture was made flatter (removing the functionalities from some of the nodes), to give a much faster mobile data experience. We will look at these architectural evolutions in the section devoted to HSPA.

The 3G era began with much hype and anticipation of the mobile data revolution that was waiting to happen, on the cusp of the success of 2G GSM. As we discussed in Chapter 1, the mobile operators paid exorbitant license fees for 3G spectrum, anticipating huge demands for 3G services. However, the story did not evolve as planned. There was very slow take up of 3G services initially. This created a vicious cycle as the terminal prices remained high due to low volumes. It was only with the introduction of HSPA that the mobile internet services really picked up and were then given a massive surge by the introduction of Smartphones. Some experts tend to thus discard early 3G as a failure, but, in our view, it was a huge learning experience for the industry and within successive releases of the standards (leading to HSPA) they got many things right. So, it is worth having a look at the technologies that underpinned this generation along with their outcomes. After all, most of the later HSPA generations are still alive today, supporting the mobile broadband services we cannot seem to do without.

The 3GPP 3G system was underpinned by UMTS (Universal Mobile Telecommunications Systems) concept. UMTS was envisaged to encompass multiple technologies (both terrestrial and satellite) to globally facilitate 3G communications in commonly designated frequency bands. However, it was the UMTS Terrestrial Radio Access (UTRA) that became the mainstay and was widely deployed by the operators. The common frequency bands for UTRA were designated in the 1800–2100 MHz range, as both FDD (Frequency Division Duplex) and TDD (Time Division Duplex) modes.

Definitively, the ITU developed the IMT-2000 specifications [9] and the standards which would meet these specifications to be classed as 3G. IMT-2000 produced a vision of an all-encompassing communication model providing indoor, local area, wide area (regional) and global coverage through terrestrial and satellite technologies in a single frequency band at around 2 GHz. It should support a wide array of services from voice, data, multimedia to internet access, using both circuit switched and packet switched transmissions. The specified data rates were up to 2 Mbps in indoor environments, 384 kbps in low mobility and 144 Mbps in high mobility environments [9].

The main changes enacted initially by the UMTS standards were in the radio access network, whilst the CN continued to be much the same from GPRS. A fairly detailed account of the RAN and core functionality of the early 3G releases can be found in [10]. A functional block diagram of the UMTS network architecture is shown below in Figure 5.8, with each of the components described in the sub-sections below. The early 3G architecture (as defined from Release 99) contained the UTRAN, circuit switched core and packet switched core. From Release 5 onwards, the IP Multimedia sub-system (IMS) was added to specifically handle IP traffic. This was a highly successful addition to the 3GPP network architecture, which evolved beyond the third generation releases. In a later Section 5.6.3, we will detail the IP Multimedia subsystem.

5.6.1 The UMTS Terrestrial Radio Access Network (UTRAN)

The basic architecture of the components within the 3G RAN is similar to the 2G RAN, although the terminology has changed. The mobile station of 2G is termed as the User Equipment (UE), to highlight the fact that not only the conventional mobiles, but newer

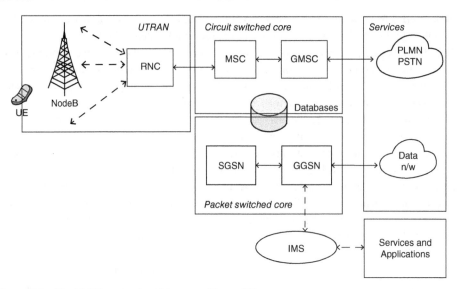

Figure 5.8 The UMTS network architecture. (*Source* [6]).

portable devices (at that time) like PDAs and Laptops can also be connected with 3G. The base station was termed as the Node B, which connects the UEs within its cell to the network through the UTRA air-interface. The RNC stands for the Radio Network Controller, which oversees the radio resource management (RRM) of the multiple Node Bs connected to the RNC.

One of the key changes in 3G is the transition from the narrowband air interface of 2G systems to wideband CDMA. This change was enacted to cater for the then anticipated exponential growth in mobile data. You will see in Chapter 8 that theoretically for very large signal bandwidths, the achievable rate increases linearly with the transmit power, while for smaller bandwidths, this increase is logarithmic. This would mean that the peak user data rates supported by UMTS would be substantially higher. The CDMA principles will be discussed in Chapter 8 and here we will give a brief overview of the structure of how the W-CDMA air-interface is used in 3G UMTS.

5.6.1.1 The UTRAN Air Interface
To realise the promise of much higher data rates, 3G needed a radical change in the air interface from 2G GSM. The narrow band air interface of GSM was seen as a major limitation, even with the relative increase in data rates achieved with subsequent 2G releases (GRRS and EDGE). If you consider similar SNR, the increase in allocated bandwidth per user can linearly increase user capacity or data rate, as will be seen in Section 8.5.1. Wideband Code Division Multiple Access (or W-CDMA) was chosen as the candidate air interface to achieve these very high data rates in 3G UMTS. The technical details of CDMA will be covered in Chapter 8, but here we will look at how this technology was adapted into the air interface of 3G UMTS. The basic concept of CDMA is that users are assigned a unique code (which is of a higher rate than the data rate), which spreads the data into a

wider bandwidth. This is why CDMA is also referred to as a spread-spectrum technique. CDMA allows users to occupy the same frequency and time domain resources, and they are separated by the unique spreading codes.

It is worth reminding (as noted in Chapter 1) that CDMA was already a tried and tested technology for cellular applications in the form of IS-95 and subsequently in CDMA-2000. These were narrowband technologies, with channels occupying a bandwidth of 1.25 MHz. The channel bandwidth was designed to occupy 5 MHz in UMTS. The data 'bits' from the source would be spread to this wider bandwidth by the time domain multiplication with CDMA sequences of 'chips'. This form of CDMA is known as Direct Sequence CDMA (DS-CDMA). Also, these DS-CDMA sequences would be orthogonal to each other, having been derived from the Walsh Hadamard codes [10]. The DS-CDMA sequence lengths can be varied in their lengths, so different basic data rates, from the very low voice codec data to very high mobile video data rates can be adapted. The orthogonal spreading codes can be represented in a code tree, as illustrated in Figure 5.9.

The orthogonality of the spreading codes mean that ideally, the data (spread with these codes) can occupy the same time frequency resources without interference. At the receiver,

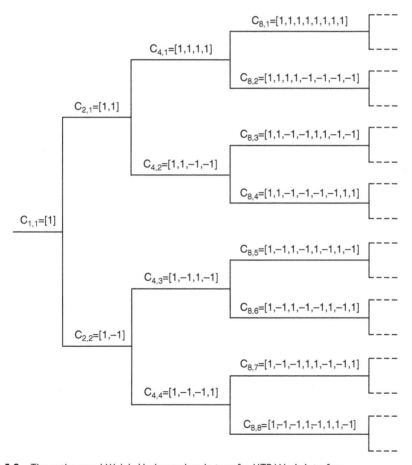

Figure 5.9 The orthogonal Walsh-Hadamard code tree for UTRAN air interface.

the incoming signal is again multiplied with the same sequence (called de-spreading), to recover the data sequence. The interference (i.e. other users' data spread by other DS-CDMA sequences) nullify, as the cross-correlation between these different sequences is zero in ideal conditions.

As seen in the code tree of Figure 5.9, different length spreading codes are available in UMTS. The higher data rates (for applications like mobile video) are spread with shorter codes. There are only a few shorter codes available, so consequently only a few higher rate users can be accommodated in a cell. At the other end, lower data rates (for applications like voice calls) are spread with longer length codes and there are many of these codes available in the code tree. In practise, a mixture of these high and low data rate users will need to be served in the cell and the codes from the tree will likewise be used.

On top of these spreading codes, a scrambling code is applied in the time domain. The scrambling code is derived from PN (Pseudo-Noise) sequences, which have properties very similar to random sequences. Each NodeB would use a unique scrambling code within its cell so, effectively, the UEs within this cell can identify their relevant transmissions from interference coming from the neighbour cells. You will see that the Orthogonal Variable Spreading Factor (OVSF) spreading code tree is finite (for practical spreading factors) and the use of the scrambling code allows each NodeB to re-use the same OVSF codes and yet make its transmissions uniquely identifiable. The orthogonality given by the OVSF codes still holds within a cell, as all transmissions within that cell are scrambled by one single scrambling code [10].

5.6.1.2 UTRAN Frame Structure

The data and control channels in the 3G-UTRA air-interface is arranged in a manner to reflect the W-CDMA operations. At the finest resolution are the 'chips' which denote each bit in the spreading sequence. Each chip would occupy 0.26 µs. 2560 chips make up a slot, which spans a duration of 0.667 ms. The slot would carry a number of data bits, spread by the relevant Walsh-Hadamard sequence. For example, voice codec data spread by a $SF = 128$ sequence will have far fewer data bits in a slot, while high rate data spread by a $SF = 4$ code will carry many more bits in a slot, to support a higher data rate. Fifteen data slots make up a UMTS frame, spanning 10 ms. Then 72 consecutive frames make up a super-frame, spanning 720 ms. The UTRA frame structure for a downlink data channel is shown below in Figure 5.10.

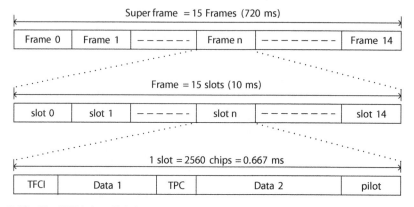

Figure 5.10 The UTRA downlink frame structure.

Within a slot, most of the resources are allocated to data transmission, but there is also essential control information included. The Transport Format Combination Indicator (TFCI) allows the receiver's protocol layers to identify the format combination used and then decode and deliver the data to the correct transport channels. The TPC (Transmit power control) bits allow the NodeB to carry out fast power control at the rate of 1500 Hz, which is an essential requirement for W-CDMA. We will discuss the UTRAN power control features below in Section 5.6.2.1. While this time domain structure looks broadly similar to the GSM frame format, an essential difference is that, in the frequency domain, the physical channels in UTRA are spread to 5 MHz bandwidth by the W-CDMA process. This makes UTRA a wideband technology, with the users multiplexed in the code domain. You will be introduced to the theoretical basics of CDMA in Chapter 8.

5.6.2 UTRAN Features

In this section, we will discuss a few UTRAN features which will show distinct differences from 2G.

5.6.1.1 Synchronisation

The UE synchronisation in 3G UTRA is carried out in several steps, assisted by the Synchronisation Channel (SCH). The SCH contains two parts, the P-SCH (Primary Synchronisation Channel) and the S-SCH (Secondary Synchronisation Channel). The P-SCH allows the UE to obtain slot level synchronisation. P-SCH is a 256-bit long code transmitted at the beginning of each slot, using the first 256 chips. All the NodeBs would transmit the same P-SCH code, but there would be timing differences as the NodeB transmissions are asynchronous. The UE would detect and align with the strongest P-SCH signal, to attain slot-synchronisation. The P-SCH is transmitted without the scrambling operation (unscrambled).

The S-SCH also consists of 256-bit long sequences, transmitted at the start of each slot. S-SCH allows the UE to obtain frame synchronisation and also obtain the scrambling code group number (out of 64 groups). The exact scrambling code used by the synchronised cell can be found by the UE by reading the Primary Common Pilot Channel (PCPICH) as the next step. Once the scrambling code is identified, the UE can decode the BCH (Broadcast Channel) and obtain cell-specific information.

5.6.2.2 Fast Power Control

UTRAN would implement a very tight power control mechanism, to mitigate the 'near-far problem' in CDMA, which we will discuss in Section 5.6.4. The fast power control would enable the NodeB to adjust the UE power precisely, through signalling in the control channel PDCCH. Whilst this would reduce the 'near-far effect' and optimise the power consumption of the active UEs, it would also add considerable signalling overhead to the UTRAN. 3G UTRAN has introduced a two-level power control mechanism, containing a closed loop and an open loop. The closed-loop mechanism delivers the fast power control, where the eNodeB will measure the received signal power levels from the UE and instruct the UE to adjust power levels up or down, in 1, 2 or 3 dB steps. This power control will happen at a rate of 1500 Hz in-line with the TPC commands, as seen in the UTRA frame structure above. The open-loop power control is for the initial power level setting for the UE, when it starts transmitting. The UE will measure received power levels from the eNodeB, which can be coarsely related to the path loss between the UE and the eNodeB. With this knowledge, the UE can set its initial transmit power levels in the uplink.

5.6.2.3 Soft Handover

Soft Handover is a feature unique to 3G UTRAN, which enables both the source Node B and the destination Node B in a handover situation to support the UE through providing active radio resources. Soft handover is possible only when both these NodeBs operate on the same carrier frequencies, so that the UE can receive both signals and effectively combine them. In CDMA, the signals with different path delays can be combined using a technique known as the 'Rake receiver'. In the uplink of the soft handover process however, the received signals of the two NodeBs are not combined, as there are two physically separate receivers in action. The received signals are processed and passed on to the RNC, where the RNC selects the signal with the higher received power, on a frame-by-frame basis. The RNC uses the outer-loop power control information to determine the best received signal power.

There is another version of handover in 3G known as the 'softer handover', where the received signals in both the uplink and downlink can be combined by the use of CDMA Rake receivers. Softer handover is applicable when the UE is transferred between two sectors served by the same NodeB. Then the NodeB has the ability to de-scramble each of the delayed versions of the received signals and apply the Rake combination.

Soft and softer handovers can enhance reliability of the handover process and reduce call dropout rates. However, only a small percentage of handovers fulfil the criteria to be supported by soft or softer handovers. Hence, the customary hard handover option is also available in UTRAN, where the connection between the UE and the current serving NodeB is broken before the new connection between the UE and the destination NodeB is established. Also inter-RAT handovers between UMTS and the 2G counterpart RATs are an important component of hard handovers.

5.6.3 The IP Multimedia Subsystem (IMS)

Although the packet switched domain in the 3G network architecture (Figure 5.5) provided access to the internet services, there was no harmonisation or guarantee of quality in this approach. The IMS was developed by 3GPP to bridge this gap. The IMS would provide a standardised interface between the cellular user and various IP-based internet Multimedia services (VoIP, MMS, video conferencing, web access etc.). In developing IMS, 3GPP endeavoured to use accepted internet protocols, particularly from the IETF (Internet Engineering Task Force) and the use of SIP (Session Initiation Protocol), which governs the IMS-based signalling. This approach helped IMS to be accepted beyond the 3GPP and cellular domain, to the fixed network as well.

The benefits of having this IMS layer between the Internet Multimedia services and the 3G network connecting the user were numerous. Firstly, the IMS could establish a session between the Internet application and the user with a defined level of QoS. This would prevent quality degradation in the middle of a VoIP call, for example, and operators could guarantee the QoS for the duration of the call. They could also establish different charging policies for different QoS levels provided to the user. Different IP-based applications would require different QoS levels and the IMS allows definition and execution of these with ease. The ability to integrate and combine different IP multimedia services provided by different vendors is another main benefit of the IMS. In this way, operators could offer unique customised services to suit the needs of different customers. Also, they would not be tied to a single IP services vendor and their proprietary service platform.

First introduced in 3GPP Release 5, IMS continued to develop across 3G and later 4G releases. IMS was also supported in 3GPP2 technologies, fixed wireless and later WiMAX systems. IMS is an integral part of Rich Communication Services promoted by GSMA [11], as an alternative to the OTT voice and video calling services that proliferated in the period of early 4G (discussed in Chapter 1.3). An in-depth technical coverage of this IMS topic can be found in [12].

5.6.4   Issues with the UMTS Air Interface

The W-CDMA was seen as a powerful solution to enable mobile broadband communications, but it was not without its shortcomings. One of the early issues identified in the research phase was the 'near-far problem'. A user near to Node B would be received with much higher power than a user further away from Node-B. Thus, the received signal from the further out user can be masked by the higher power of the nearby user. The solution to this near-far problem is to control the transmit power of the users in a fast and accurate manner. UMTS includes dynamic power control with very fine resolution. Node B would instruct users to increase or decrease the power levels in frequent power control messages, as we discussed earlier in the 3G RAN procedures.

Another issue that arose in UMTS was the intra-cell (within a single cell) interference due to the spreading codes losing their orthogonality. If you recall how the OVSF codes are constructed (Figure 5.9), they would show perfect orthogonality when perfectly aligned (i.e. zero cross-correlation at zero lag). However, the radio channel introduces multi-path delays and some delayed versions of the OVSF codes overlap with each other at each receiver. These delays come from the multiple reflective and diffractive radio signals in a typical environment (called the scattering environment). These OVSF code distortions introduce some interference at the receiver, which limits the overall number of users and their data rates that can be effectively supported in a UMTS cell (i.e. the cell capacity). It has been shown that this effect can be quantified in a code orthogonality factor, which is related to the level of multipath delay (the 'delay spread') of a given radio channel [13]. This effectively meant that the full cell capacities theoretically shown to be possible with W-CDMA could not be achieved in 3G.

The UMTS air interface and the W-CDMA spreading code structure was fixed to a 5 MHz bandwidth. It later became evident that, for very high data rates, this bandwidth would not be sufficient and the flexibility to operate in multiple bandwidths would be a key necessity. This was also driven by the lack of availability of contiguous spectrum bands, due to the high demand on spectrum for cellular and various other uses (we will discuss spectrum issues in Chapter 7). With 4G standardisation, OFDMA emerged as the preferred air-interface, which uses the frequency domain multiplexing of users, and where the bandwidth can easily be adapted with high flexibility. We will look at the OFDMA technology in detail under 4G RAN.

5.6.5   3G Evolution to HSPA

As mobile broadband services began to take traction, it soon became evident that the capabilities offered by early releases of 3G were not sufficient to meet performance demands. The early 3G versions would support maximum downlink data rates of 2 Mbps and with many video downloading applications becoming popular, this limit would soon become a performance barrier. Also, the higher latencies and power consumption issues in the UE needed to be addressed to provide a true mobile broadband experience.

The packet-routing mechanisms over 3G RAN and the core were modified in a number subsequent 3GPP releases to overcome the above issues. In Releases 5 and 6, the High Speed Downlink Packet Access (HSDPA) and High Speed Uplink Packet Access (HSUPA) variants were introduced, which became commonly known as High Speed Packet Access (HSPA). HSPA was further evolved in Release 7 where a flat architecture, similar to LTE, was introduced. This became known as HSPA+ and further refinements of HSPA continued up until 3GPP Release 11.

We will first look at the architectural changes introduced over the evolution of HSPA. The Architecture of Release 6 HSPA would be similar to the earlier 3G versions, with both the control and data planes going through the Node B and RNC in the RAN and SGSN and GGSN in the core. In the later versions of Release 7, the HSPA+ architecture has been made much flatter, with the RNC functionality incorporated into the Node B. The data plane would connect directly to the GGSN (called the direct tunnel solution bypassing the SGSN) and the control plane would go through the SGSN and then to the GGSN from Node B. The architecture comparison is shown below in Figure 5.11 (from [14])

By making the HSPA+ architecture flatter with fewer nodes, latencies with packet access were significantly reduced. The latencies of over 100 ms for one-way packet transmission

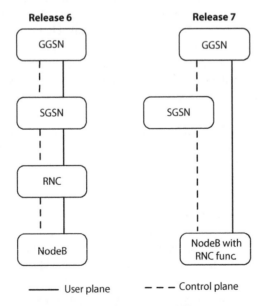

Figure 5.11   The architecture evolution in HSPA. (*Source* [14]).

in early 3G releases dropped to below 25 ms for HSPA+. Also, the cost of data packet transmissions could be reduced, enabling lower cost per bit for the operators, who were beginning to experience exponential growth in mobile data volumes around this time period. The architecture evolution in HSPA is designed to retain backwards compatibility, i.e. existing UEs could operate in this newly evolved architecture as the functional split points between the RAN and the CNs were retained as the same. Release 7 HSPA+ architecture is basically similar to the flat LTE architecture which was introduced in Release 8 onwards. The handling of the user plane in the GGSN would be similar to the functionality of the SAE Gateway of LTE (detailed in Section 5.7 below). It was envisioned that such similarity would enable the HSPA+ node B to later connect to the LTE core in the user plane, enabling cost-efficient evolution paths for HSPA+ networks [14].

HSPA evolutions brought about three significant improvements in 3G networks: lower latencies, higher data rates and lower power consumptions in the UE. We have looked at how architectural changes helped reduce latencies. In terms of improving data rates, there were some significant changes in the HSPA RAN. The air interface retained as W-CDMA in HSPA, with a channel bandwidth of 5 MHz. The highest modulation order in the downlink was improved in HSPA to 64 QAM, allowing 6 bits per data symbol. Also, the 2 × 2 MIMO schemes were introduced. These changes would triple the peak data rate of HSPA (from the previous 3G releases) to 42 Mbps in the downlink. The uplink modulation order was also improved to 16 QAM in HSPA (Release 7), enabling peak data rates of 11.5 Mbps. HSPA in Release 8 also introduced the dual-cell operation in the downlink, enabling two in-band 5 MHz carriers to be used in joint resource allocation through scheduling. This was further extended to four carriers in multi-bands, enabling a peak data rate of 168 Mbps in the downlink. For voice services, HSPA introduced circuit switched (CS) voice mapping over HSPA channels. This is a combination of VoIP (Voice over IP) in the radio channels and CS in the CN [14].

Improving the UE power consumption was another key benefit brought on by HSPA. This was enabled by introduction of discontinuous transmission and reception by the UE, at idle periods of data communication. For example, in previous 3G releases (Release 6 and before) the control channels need to be received continuously even in periods when there was no data download in the UE. The new HSPA updates from Release 7 would make these receptions (and transmissions in the UL) dis-continuous and limited to only short time periods. This benefit was also achieved with the CS voice mapping over HSPA solution for voice calls detailed above.

HSPA continued to evolve within the LTE releases, up until Release 11, providing a much needed baseline for mobile data networks in the early days of 4G evolution.

5.7 The 4G Network

The proliferation of internet broadband access through improved wired connections opened up numerous possibilities of a connected world. Soon, the potential of mobile internet also became apparent with the realisation that data rates needed to catch up with those offered by the ever-improving wired connections, to achieve the same user experience. With the first deployments of 3G UMTS networks offering only 384 kbps maximum downlink data rates, there was the need to push the data rates much higher to the 100's of Mbps realm. Whilst the work to improve the 3G UMTS standards continued, the need to develop an entirely new, mobile data-oriented 4G standard became imperative. Thus, the work on the 3GPP LTE standard began in 2004, with similar activity on the IEEE 802.16 WiMAX standard having begun a couple of years earlier.

In looking at the evolution of the 4G networks, we will focus here on the 3GPP LTE standards. As noted in Chapter 1, initially there was competition from the IEEE 802.16 WiMAX to become the flag bearer for 4G, but the markets chose the LTE evolution path. We will also discuss the key evolutions from the first LTE standard 3GPP Release 8 to current Release 15, to track the rapid advancements enabling all IP mobile data networks. Generally, all LTE releases are considered as 4G networks (from 3GPP Release 8 onwards), although more precisely it was 3GPP Release 10 which first met the IMT-Advanced requirements [15] to be categorised as a 4G network. The IMT Advanced specifications stipulated that 4G networks should achieve 100 Mbps in high mobility conditions and up to 1 Gbps in low mobility conditions.

LTE was the first fully packet-switched cellular network in the 3GPP series, replacing the circuit-switched voice component of the previous generations with VoLTE (Voice over LTE). LTE also introduced a flat architecture, removing many of the control nodes of previous generations. From the RAN aspects, LTE introduced a new air interface, OFDM (Orthogonal Frequency Division Multiplexing), which we will study in some detail in Chapter 8. Unlike previous 3G UMTS, where the channel bandwidth was fixed at 5 MHz, LTE offered flexibility to network operators to select from a number of bandwidths ranging from 1.4 MHz to 20 MHz. This adaptation is recognition of the fact that, in many countries and regions, the spectrum has become fragmented and utilising contiguous, fixed bandwidths was not feasible. In later LTE releases, carrier aggregation was introduced, where disparate frequency bands could be combined together to provide wider bandwidths, to facilitate even higher data rates.

5.7.1 LTE System Architecture

The system architecture in LTE is one of the key areas that shows significant differences from previous cellular generations. LTE CN is the first in the 3GPP generations to run entirely as a packet-switched network. Some of the features, like moving towards a flat architecture, are evolutions from the features introduced in Releases 6 and 7 of 3G UMTS. The architectural changes in LTE are generally categorised as System Architecture Evolution (SAE). The flat architecture is achieved mainly by increasing the functionality of eNodeB, as we discuss below. Also, there is significant separation in the control plane and user plane paths. This flat architecture enables LTE to achieve significantly lower latencies (up to 10 ms) and helps it to deliver real-time data applications.

The general LTE architecture is schematically shown in the block diagram of Figure 5.11. Some of the main interfaces between the functional blocks are also shown and named here, with separate paths of the control and user planes illustrated. We will detail some of the main functional blocks and interfaces in the paragraphs below. A detailed description of the 4G CN (the evolved packet core –EPC) and SAE can be found in [16].

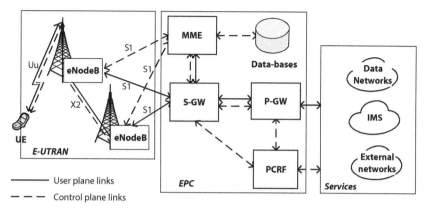

Figure 5.12 The LTE network architecture, as captured in SAE. (*Source* [14]).

The eNodeB plays a central part in the LTE system architecture. It not only handles the many radio interface-related functions in communicating with the UE, but also provides the functionalities previously supported by the BSC (in 2G) and the RNC (in 3G). For example, the RRM (radio resource management) control functions of user scheduling and QoS-based radio resource allocations are now conducted by the eNodeB. Exemplifying the flat architecture, there is now a direct interface (termed as X2) between the eNodeBs, which can support the user and control plane data transfer in handovers. The eNodeBs control many of the aspects of mobility management, including the conduct and analysis of RSRP and RSRQ measurements, making handover decisions based on these measurements and handover-related signalling with other eNodeBs (through X2 interface) and with the MME (through S1 interface). In LTE, there is no concept of soft handover as in 3G, so a given UE is only connected to one eNodeB at a time. The eNodeB passes the control plane information to the MME in a handover. A given UE is associated with only one MME at a given time, and if this needs to change through UE mobility, the eNodeB has to initiate that change. Also, the eNodeB is responsible for connecting the user plane for a particular active UE to the external networks (for DL and UL data delivery) through user plane tunnels to the S-GW (Serving Gateway).

The MME (Mobility Management Entity) is the main control element in the LTE CN, the Evolved Packet Core (EPC). It has control plane interfaces to the gateways, databases (HSS) and to the eNodeBs. As the name suggests, the MME is responsible for the mobility management of the UEs. The MME will track the location of the UE in both active and idle modes. In the active mode, the MME will track the eNodeB the UE is attached to, and if this changes (through handover), the control signalling during the handover is handled by the MME. In some handover situations, the S-GW and the MME itself will need to change, in addition to the eNodeB, and the control signalling for these changes are also carried out by the MME. In the idle mode, the MME will track the UE, based on the tracking area (TA), where the TA consists of a neighbouring group of eNodeBs. The MME also handles the authentication when a UE first joins or re-joins an LTE network. The authentication is done through checking the UE credentials in the HSS (Home Subscriber System) registries. For roaming users, the MME will retrieve user information from the user's home network HSS, to execute authentication. In the protocol stack for LTE, the Non-Access Stratum (NAS) protocol layer (discussed later) represents the link-up between the UE and the MME.

The Serving Gateway (S–GW) is responsible for the relaying of data (IP packet) flows between the eNodeB and the PDN Gateway (P-GW), for all UEs in the connected mode. If data (IP packet) flow intended for a UE currently in the idle mode comes through from the P-GW, the data is buffered at the S-GW and the S-GW will request the MME to initiate paging for the UE. Only when the UE transfers to the RRC connected mode will that data flow be tunnelled by the S-GW to the eNodeB. Thus, the S-GW functions as the IP router/buffer for the UE, as per its RRC connected or idle states. During the handover process for a UE (when it changes the serving eNodeB), the S-GW will function as the anchor point. The S-GW will switch the user plane tunnelling from the current serving eNodeB to the target eNodeB as the handover completes. During handover to/from LTE to another RAT (2G or 3G), it transfers data packets to the SGSN. A given UE will be served by a single S-GW at any given time. During some cases of UE mobility, the S-GW itself will have to change and this is done under the control of the MME.

The Packet Data Network (PDN) Gateway (P-GW) provides the routing of data packets between the external networks and the S-GW (ultimately linking the UE). The data flow between the external data networks and the P-GW is in the form of IP packets. The data flow between the P-GW and the S-GW is through tunnels and, in many cases, these tunnels are based on the protocol GTP (which we discussed earlier). The P-GW does the mapping between the IP data flows and these GTP tunnels. The P-GW is responsible for allocating IP addresses to the UE, which the UE then uses to connect and communicate with the internet. The P-GW can act as a final mobility anchor, when a UE moves from an area controlled by one S-GW to another. Different service policies can be set for a UE and the P-GW will set up the gating/filtering functions to reflect these. It will also collect the related charging information. The P-GW can also monitor the data flows for lawful interception.

The policy and charging resource function (PCRF) is responsible for deciding upon the level of QoS (Quality of Service) to be provided for each data link. The PCRF relays this information, known as the Policy and Charging Control (PCC) rules to the P-GW. The PCC rules are relayed by the PCRF whenever a new 'bearer' is set-up, for example when a UE is initially attached to the network.

5.7.2 LTE Protocol Layers

The protocol layer descriptions for the user plane and control plane of LTE is provided below. Let's recall that the data plane is concerned with the transfer of the user data from the application down to the physical transmission layer. The control plane is concerned with the passing down the control information related to each of the processes in the protocol stack. For example, in the physical layer, it is important that each user is made aware exactly where in the radio frame the user data related to that user is positioned. A control channel in the downlink, named the PDCCH (Physical Downlink Control Channel) is used for this purpose.

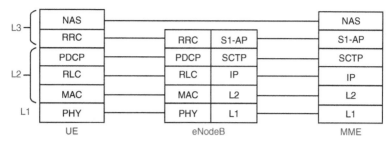

Figure 5.13 Control plane protocol layers between UE and the MME.

We will discuss (particularly) the lower layer protocol stacks of LTE, which helps to highlight how the control plane signalling and user plane data traffic are routed quite differently in LTE. We will start with the control plane where, as stated above, the MME plays the central role. The protocol layers for the LTE control plane is shown below in Figure 5.13, for the signalling between the MME, eNodeB and the UE. It should be noted that the horizontal lines represent logical links, except for the physical links of PHY to PHY between the UE and the eNodeB and of L1 to L1 between the eNodeB and the MME. The logical links between similar protocol entities indicate that the related information is inserted at one end of the link and then extracted and utilised at the other end of the link.

As noted above, the MME handles the mobility management of the UE, both in the connected and idle modes. Although physically the related control signalling to this is transmitted through the eNodeB, the eNodeB does not decipher or take actions on these control messages. So the MME and UE are logically connected 'directly' by this NAS protocol layer as shown in the diagram. The main functions of the NAS layer are to manage the mobility of the UE and also support the establishment of communication sessions between the UE and core. All protocol layers below this NAS link the UE and eNodeB for access related control signalling, and this is called the AS (Access Stratum). The RRC (Radio Resource Control) layer, as the name implies, controls usage of radio resource. RRC is responsible for many kinds of signalling to the UE, like the broadcast messages and paging. All UEs should establish an RRC connection before any data transmission can occur. In LTE, a UE has to be in either RRC-connected or RRC-idle states. The more complex layer 2 (L2) is divided into three sub-layers. The PDCP sub-layer is in charge of ciphering/deciphering and integrity protection in the control plane. The RLC sub-layer conducts error correction with ARQ. It also segments and concatenates the data units coming from the PDCP layer, making it suitable for radio transmission. The MAC layer is responsible for handling multiple requests for the physical resource through scheduling, priority allocation of UEs and error corrections through the HARQ procedure. Recall that scheduling is a key RAN-related procedure as we discussed in Section 5.2.5. For LTE, the MAC scheduler can select from dynamic, persistent and semi-persistent options. In dynamic scheduling, the data availability for each active user and their channel conditions are checked at every transmit time interval (TTI) and the schedule is created. This schedule has to be informed to the UEs through the control channels at every TTI, creating a greater overhead. In persistent scheduling, allocations are fixed for the duration of the whole transmission (eg: duration of a call) and the schedule has to be informed to a user only once through the control channels. However, the performance can be sub-optimal as the user channel conditions can vary during this period and the allocated MCS (for example) may not match. The semi-persistent scheduling is a hybrid of both above schemes, as the name suggests. The lowest layer (L1) is the physical (PHY) layer and this is basically the radio interface-related functions between the eNodeB and UE. The physical layer receives the transport channels from the MAC layer and converts them to physical channels, which are carried over the radio interface.

There are control signals passing to/from other CN nodes to/from the MME, but we will not discuss these for brevity. One point to note is that the GTP-C protocol is used in the higher layers for this control signalling.

Due to the flat architecture defined in LTE SAE (the removal of radio control nodes such as the RNC in 3G), all protocol functionalities from PDCP layer and beneath in the data plane (and also from RRC layer and beneath in the control plane) are carried out in the eNodeB and the UE respectively. The IP layer sits above the PDCP layer in the UE and logically connects directly with the IP layer in the P-GW. The user mobility (to a different eNodeB or even a different S-GW) is supported by the GTP tunnelling protocol used in LTE, as discussed earlier. There are basically three levels of IP layers used within this configuration. A user IP layer, which feeds the user application at the UE and the corresponding application layer at the external server. Public IP addresses are used for this layer. Then an LTE application IP layer is created between the P-GW and other network elements to carry out the GTP-U based tunnelling of the IP packets, with private IP addresses. Thirdly, an IPsec layer is available optionally, to increase data security.

The PDCP layer in the user plane is responsible for header compression/de-compression of the IP packets. This reduces overheads in terms of radio-resource usage and improves efficiency, especially in the case of smaller IP packets in VoIP. As in the control plane, the PDCP layer also carries out ciphering/deciphering. There are also specific PDCP functions in the case of LTE handovers utilising the X2 link. These functions ensure that all non-delivered packets are sent to the new source eNodeB prior to completion of the handover with the late path switch. The RLC layer in this data plane carries out similar functions to that in the control plane. The MAC layer, in addition to the scheduling and HARQ functions mentioned in the control plane, also conducts transport channel format selection which help with the link adaptation functionality. In the PHY layer, transport channels are mapped to physical channels. The user data in the PHY layer (along with some control data) is carried in shared channels between multiple active users – the PDSCH (Physical Downlink Shared Channel) and PUSCH (Physical Uplink Shared Channel). We will look at how this multiple-shared access is provided in the air interface in the next section.

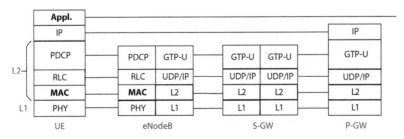

Figure 5.14 User plane protocol layers between UE and the eNodeB and core.

5.7.3 LTE Multiple Access Schemes

LTE uses two distinct multiple access schemes for the downlink and uplink and is the first cellular generation to have such a division. The downlink uses OFDMA and the uplink uses SC-FDMA (Single Carrier Frequency Division Multiple Access). We will look at these multiple access schemes in detail and also what prompted LTE to adapt these different schemes.

More in-depth information about the processes in digital communications are provided in Chapter 8 including generic details about OFDMA and SC-FDMA. Here, we will focus on their application to the LTE and the reasons why they were selected.

OFDMA is essentially separating data streams in the frequency domain. The frequency fingers, on which the data is modulated, are called sub-carriers. In LTE, these sub-carriers are spaced 15 kHz apart. In a traditional FDMA system, the sub-carriers (or frequency fingers) cannot be this close together, as there needs to be guard bands to control interference on the data modulated on one sub-carrier from the adjacent ones. What makes OFDMA special is that the sub-carriers are orthogonal to each other. Looking at Figure 8.24 in Chapter 8, you can visualise what this orthogonality means in practice. When each of the sub-carriers is sampled at the peak of its signal, all other sub-carriers have zero value. Hence, no inter-carrier interference is sampled into the data symbol. 'Symbol' is the term used to represent the basic time unit of a data bit in LTE.

Due to multiple sub-carriers individually carrying modulated data symbols, OFDMA is known as a multi-carrier system. OFDMA has become the de-facto multi-access scheme for modern communication systems from DVB in TV broadcasting to LTE (and WiMAX) to IEEE 802.11 for Wi-Fi. Before looking at the major benefits of OFDMA which prompted its selection for LTE, we will briefly look at how the transmit and receive chains operate.

Regardless of bandwidth of the LTE system, the sub-carriers are spaced 15 kHz apart and are modulated with the data bits separately. The time domain composite transmit signal is generated through a process called Inverse Fast Fourier Transform (IFFT). You will study about the Discrete Fourier Transform in Chapter 8, which is essentially a conversion of a signal from the time domain to the frequency domain. The inverse transform carries out the reverse operation. The 'Fast' term represents a more efficient implementation of the transformation, with 2 N points. The modulated signals (on sub-carriers) are parallelised and are fed together to the IFFT block, which outputs the time domain signal, which is ready to be transmitted. One instance of the IFFT output signal (in time domain) generates a symbol. However, a cyclic prefix is added to the symbol (where the tail part of the symbol is copied and added to the front – as a prefix) before transmission. The purpose of the cyclic prefix is to counter the inter-symbol interference (ISI) as explained in more depth in Chapter 8.

By adding the cyclic prefix (CP), no new information is carried in the first part of the symbols, and this can be now discarded, nullifying ISI. In LTE, two basic CP lengths are specified, to be selected based on the level of radio-channel dispersion. For short delay spread in smaller cells (typically in urban areas) a normal CP length of 4.7 μs is used, while for larger cells an extended CP length of 16.7 μs is recommended. The CP addition incurs a 7.5% overhead in the normal CP case, but enables an elegant reception of a series of narrowband modulated signals as explained more in Chapter 8. The removal of CP makes the

OFDMA received symbol detection just a multiplication with the (conjugate of the) main signal path, and not a complex de-convolution operation. In Figure 5.15, the multi-path radio channel and the CP insertion as per the LTE specifications is shown.

The LTE transmit chain thus involves modulation of the data bits onto each of the sub-carriers, a serial-to-parallel conversion to feed these into the IFFT operator, the CP addition to each symbol, the pulse shaping (windowing) to limit the out-of-band emissions of a sharp-edged signal, digital to analogue conversions and RF transmission stages. The LTE receiver conducts the reverse operations to extract the data bits from the multi-carrier transmit signal. After the CP removal, there is a serial-to-parallel conversion to feed to the N-point FFT operation to convert the signal into the frequency domain, where the modulated sub-carriers are recovered. The channel equaliser takes away the effect of the radio channel through a simple multiplication, as noted above. The LTE OFDMA transmitter and receiver chains are shown below in a block diagram form in Figure 5.16.

The OFDMA scheme brought multiple benefits to the LTE radio layer. From the inception, LTE aimed to support multiple bandwidths, to reflect different band availabilities across the world. The number of sub-carriers in the OFDMA scheme can be varied by only changing the IFFT and corresponding FFT block sizes in the above diagram corresponding to the selected bandwidth. OFDMA is an efficient collection of multiple narrow-band channels and this enabled each of the sub-carriers to be less affected by the multipath nature of the channel. The multipath delay spread in the radio channel causes fades in the frequency response (again a Fourier transform relationship) and having narrow band sub-carriers capture only a low amount of this frequency selective fading. As noted in the CP description, the channel equalisation in the receiver amounts only to a simple conjugate

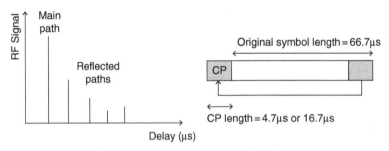

Figure 5.15 The inclusion of Cyclic prefix in OFDMA.

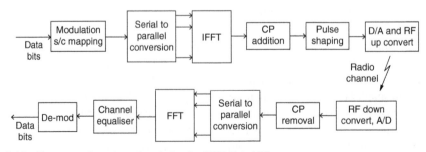

Figure 5.16 The transmit and receive chains for OFDMA in LTE.

multiplication in OFDMA. Also, OFDMA offers the opportunity for users to report the best sub-band (or group of contiguous sub-carriers) through measurements to the eNodeB, depending on their radio channel conditions. This frequency diversity (sometimes called multi-user diversity) enables the best sub-bands to be utilised for each user, increasing the system data rates (or throughputs). OFDMA allows users to be scheduled in both frequency and time domains as we will see in the LTE frame structures later.

The OFDMA scheme also brings in some challenges. One of the main challenges is the high levels of PAPR (Peak to Average Power Ratio) for the RF amplifiers. From Figure 8.24, you could see that the OFDMA waveform is a collection of sinusoids (sine wave shaped sub-carriers), modulated with their data bits. In this composite sinusoidal wave, there are large variations of average and peak power levels. Discussed in Chapter 8, the RF power amplifier in the LTE transmitter needs to accommodate average and peak power levels within its linear operation region, with a large back-off level. This would mean more linearity and power consumption requirements on the RF amplifier in the OFDMA transmitter. Whilst these requirements can be met in eNodeB, for the user devices, the power amplifier needs to be simpler and more power and cost efficient. Also, as the bandwidth in the uplink is shared amongst active users, the sub-carrier segments will be generated by different frequency oscillators. Invariably, these oscillators will have slight variations and it would be impossible to maintain the sub-carrier orthogonality of the entire uplink band, leading to inter-carrier interference (ICI). Due mainly to these reasons, OFDMA was not implemented as the multi-access scheme in the LTE uplink. A single-carrier version, SC-FDMA was selected and its functionality is explained below. Again, a more in-depth discussion on SC-FDMA and other single-carrier modulation schemes can be found in Chapter 8.

In SC-FDMA, the modulated signal (in the time domain) is first converted into the frequency domain by a Discrete Fourier Transform (DFT) operation. The rest of the transmitter chain is equivalent to the OFDMA transmitter. Different users, who are allocated separate parts of the uplink bandwidth, generate their transmit signals in this same way. The frequency domain generation of the signal (through DFT) retains the OFDMA property of good frequency separation between the users. Hence user bands can be packed closer together, without the need for guard bands. Within each user's band however, the DFT operation 'spreads' the signal throughout the band. Hence SC-FDMA is also termed as DFT-spread-OFDM (DFT-s-OFDM).

The CP addition however, is for a block of symbols in SC-FDMA. Within this block of symbols, there can be Inter-Symbol Interference (ISI). Thus, the SC-FDMA receiver (at the eNodeB) needs to employ more complex equalisation schemes. Some of these are described in Chapter 8. On the whole, the receiver operation is similar to OFDMA, apart from the addition of the inverse DFT (IDFT) operation before symbol de-modulation. The SC-FDMA transmit and receive chains are depicted in Figure 5.17.

Finally, on this topic of LTE multiple access schemes, let's look at how multiple users share the LTE bandwidth. In the downlink OFDMA, there are orthogonal sub-carriers across the bandwidth and each user is assigned a contiguous portion of these sub-carriers (called localised allocation – see Chapter 8). In the uplink SC-FDMA, there are separate, adjacent bands allocated for each user, where the OFDMA like sub-carrier mapping is spread by the DFT operation. Thus, within a user-allocated band in the uplink, a single-carrier waveform is

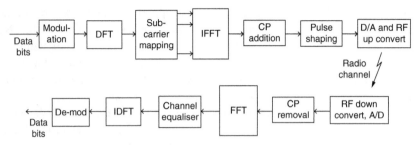

Figure 5.17 The transmit and receive chains for SC-FDMA in LTE.

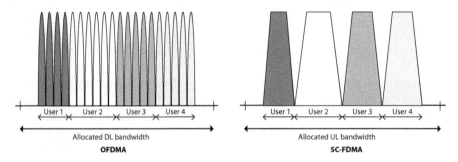

Figure 5.18 The band allocation for users in LTE downlink and uplink.

generated. In both the downlink and uplink, the entire bandwidth is not usable for user allocations. To avoid out-of-band interference to other operators in adjacent bandwidths, there needs to be guard bands on either side for which some sub-carriers need to be reserved. The basic allocation unit is a Physical Resource Block (PRB), which contains 12 contiguous sub-carriers and a 1 ms time unit (of either 12 or 14 symbols). We will look at the frame structure and PRB in more detail in the next section. The LTE downlink and uplink bandwidth allocations are shown in Figure 5.18.

5.7.4 LTE Frame Structures

Although the radio waveforms in the LTE uplink and downlink are designed differently, the same basic frame structure is employed in both links. The radio resources are dimensioned in both time and frequency domains in LTE. The time domain is divided into a number of frames, sub-frames, slots and then the basic unit of symbols. The frequency domain is represented by the sub-carriers. The basic frame structure is depicted in Figure 5.19.

The LTE radio frame is made up of 10 sub-frames, each with 1 ms duration. Each sub-frame contains two slots, which are of 0.5 ms duration. Each slot contains a number of symbols (with their CPs), as shown in Figure 5.19. Depending on the length of the CP, there can be six or seven symbols per time slot. In the uplink, however, the CP is added at the end of a symbol block and not in the way depicted in this figure. However, the numerology as per the number of symbols per slot is the same as the downlink. The frequency dimension, which is usually represented by the vertical axis, contains a given number of sub-carriers

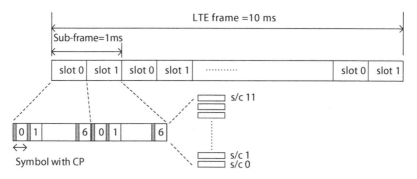

Figure 5.19 The basic units of the LTE Frame structure.

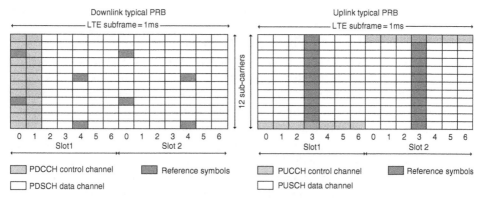

Figure 5.20 Typical PRB structures for the downlink and uplink of LTE.

as per the allocated bandwidth. For basic resource allocation, the PRB unit contains two time slots and 12 contiguous sub-carriers.

The PRB will have the same structure in the uplink and downlink, but the radio channels and reference symbols carried by the PRB in two directions will obviously be different. A time-frequency instance in the PRB is called a resource element (RE) and there are $14 \times 12 = 168$ REs in a PRB with a normal CP. These REs can be used to transmit data or control channel information and also reference symbols. The PRB is the basic unit of resource allocation to the users in LTE. Typical examples of the LTE downlink and uplink PRB configurations are shown in Figure 5.20. They show common usage patterns for control, data and reference signals, but there can be other channels inserted at given times (e.g.: the uplink random access channel PRACH, or downlink synchronisation signals PSS and SSS). The PSS and SSS for example, can span multiple PRBs in the frequency domain at given symbol positions periodically, in given sub-frames.

Comparing the frame structure of 2G GSM (in Figure 5.6) and 4G LTE, the flexibilities provided by the 2D time/frequency allocations of the LTE frame are quite obvious. For example, the LTE frame can easily support multi-antenna (MIMO) adaptations, by simply adding more reference signal slots in the PRB structure. An example of a Release 8-9 LTE 2×2 MIMO configuration is discussed in the network planning section of Chapter 6.

5.7.5 LTE Reference Signals

The reference signals in LTE have seen significant evolutions over the releases. There are two broad types of reference signals utilised in LTE. The first type are cell-specific reference signals, which are transmitted at designated positions in the resource blocks of the downlink. Any active UE can read these reference signals in its allocated PRB and use it to decode the channel and demodulate its data and control information. The second type is user-specific reference signals, which are intended for a specific user and transmitted only in the related PRBs allocated to that user.

In LTE Releases 8 and 9, the dominant reference signal is the cell-specific, Common Reference Signal (CRS). The CRS would enable the channel state information (CSI) measurements and data demodulation. As these initial LTE releases supported up to four transmit antennas in MIMO configurations, the resource elements used for CRS would multiply accordingly in a mutually orthogonal manner. This means, that the CRS locations (REs) for one-transmit antenna in the PRB would be kept vacant in other transmit antennas. An example for a 2 × 2 MIMO system is shown in Figure 6.8 of Chapter 6.

The major drawback of this type of cell-specific reference signal is that, as the overhead for the CRS grows with the highest supported MIMO configuration, not all UEs would benefit from this. For example, the cell-edge UEs would not see a benefit in MIMO as their SINR levels are typically low, but they too will face the same CRS overhead as configured to the highest number of transmit antennas in the system.

In LTE Release 10, the supported MIMO configuration was extended to eight transmit antennas. A simple extension of the CRS would create an unwanted overhead for most of the UEs which cannot benefit from this MIMO configuration. Also, an issue of backwards compatibility for the Releases 8 and 9 UEs would occur, as these would not be aware of the new CRS positions to support eight transmit ports. The solution developed in Release 10 was to split the tasks of CSI measurements and data de-modulation into two separate reference signals. The CSI measurement and reporting by the UE is aided by new CSI Reference Symbols (CSI-RS). These need to be positioned at much lower density then CRS, as only the channel variations can be tracked with a lower rate pilot pattern. A user-specific reference signal, called the De-Modulation Reference Symbols (DMRS) was introduced to support the data demodulation from up to eight transmit ports. By having the user-specific nature, this DMRS overhead would be used to target only the higher channel rank UEs who could benefit from higher order MIMO. Also, these DMRS would be pre-coded as data bits, so the transmitter does not have to stick to a pre-selected pre-coder matrix. This feature is seen as a key enabler for MU-MIMO, where the composite channel of multiple users need the ability to flexibly select and change the pre-coder. Having the DMRS pre-coded makes it aligned with the radio channel and gives better chance of accurate decoding and hence improves the UE link performance.

5.7.6 LTE Main RAN Procedures

We will detail the main RAN procedures in this section, giving focus to the Physical layer procedures. An in-depth reference to the LTE downlink and uplink Physical layer can be found in [17].

5.7.6.1 Synchronisation

The UE synchronisation in LTE is carried out in two steps, involving two specific signals. These are the PSS (Primary Synchronisation Signal) and the SSS (Secondary Synchronisation Signal). The aim of providing two signals is to reduce the complexities of the synchronisation and the cell search procedures for the UE. In the LTE (FDD) radio frame, the PSS is mapped to the middle 72 sub-carriers of the last OFDM symbol in the first slot and this is repeated in every fifth sub-frame (at 5 ms periodicity). The PSS sequence itself is a ZC (Zudoff Chu) sequence showing peak auto-correlation when aligned and zero auto-correlation with any offset. The UE will correlate the received sequence with one generated internally and thus achieve frequency alignment. The PSS also carries one part of the cell ID (the group cell ID).

The SSS generation is based on maximal length sequences (also known as m-sequences). The m-sequences are scrambled with two different codes to generate two distinct SSS sequences. These are transmitted in the symbol immediately before the PSS, in slot 0 and slot 5, again occupying the middle 72 sub-carriers (or 6 PRBs). The two SSS sequences are interleaved in the two slots of the sub-frame, allowing the UE to attain timing synchronisation. The SSS also carry information on the second part of the cell ID (physical layer cell identity group). Combining these two (PSS and SSS) cell ID information, the UE is able to work out the unique cell ID of the transmissions. This will have a value ranging from 0 to 503.

Once initial cell synchronisation is complete and the UE obtains the cell ID, it can read the critical information about the cell contained in the MIB (Master Information Block) and the SIB (System Information Block). The MIB carries critical information like the system bandwidth and the system frame number. The MIB is carried on the Physical Broadcast Channel (PBCH) and this information is repeated on every radio frame (10 ms periodicity). The SIB carries more detailed system information and is carried in the control channels of the Physical layer.

5.7.6.2 Random Access

The LTE random access procedure is versatile enough to cater for a number of different radio propagation scenarios that can occur in the uplink. As with any uplink LTE transmission, the SC-FDMA designed RACH signal contains a CP for a block of symbols. This CP length can change within a number of possible RACH formats in LTE, to reflect the different delay spreads of the radio channel. Also, due to the uplink timing uncertainty (discussed earlier in the chapter) a guard period is maintained after the RACH sequence, up to the RACH slot length border. The RACH sequence can be repeated several times in the transmission, to support weaker signal reception scenarios. The sequence lengths can be set to longer and shorter sequence formats, again to cater for longer or shorter delay spreads and propagation distances. A generic format of the RACH signal (known as PRACH channel) is shown in Figure 5.21.

The LTE RACH sequence (or preamble) is designed with the Zadoff- Chu (ZC) sequences, which show very good auto-correlation properties. As discussed earlier, this good auto-correlation allows quick sequence detection and timing estimation at the eNodeB receiver. A cyclic shift of the ZC sequence will produce another ZC sequence. Thus, LTE defines a family of preamble ZC sequences, which a user in a cell can randomly select and use. The

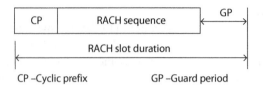

Figure 5.21 The RACH signal format for LTE.

RACH transmission from the UE occurs under the control of the MAC layer, at the designated time slots of the uplink frame. The normal random access process is contention based and a contention occurs if two users select the same ZC sequence and transmit to the eNodeB in the same time slot. Then, the eNodeB has to carry out a contention resolution process, which involves the users not acknowledged with a resource grant for their intended uplink trying again with a RACH sequence, after a set wait period.

There is another form of random access specified in LTE – a contention-free process operable in a handover. A certain portion of the RACH sequences allocated for a cell is reserved for these contention-free allocations. Once a UE enters a handover process, it is granted these RACH sequences, to acquire the timing synchronisation and the uplink resources for the destination cell. As the RACH sequence is a dedicated allocation for the particular UE in the handover, it is always guaranteed to connect to the destination cell (in RRC-connected mode) without contention.

5.7.6.3 Power Control

In the LTE uplink, an active user, in a single transmission session, can be allocated a different number of PRBs (implying different transmission bandwidths). The uplink transmit power needs to change with changing bandwidth allocations, to maintain a constant power spectral density (or power per 1 Hz of unit bandwidth). The uplink power control in LTE is thus aimed at regulating the power spectral density (PSD), rather than the actual power itself.

As in 3G UMTS, the LTE power control is made of two components. The open-loop power control part is enacted by the UE itself, where the UE estimates the downlink path loss, assumes a similar loss in the uplink and regulates the uplink PSD to reflect this. A dynamic power offset is added to this, which is based on the closed-loop power control commands provided by the eNodeB, in the PDCCH channel. The difference from UMTS is the much slower rate that this dynamic component is updated, which is around 100 Hz, or once in every 10 ms (once per LTE frame duration). This slow power control is sufficient due to the much better orthogonality properties shown by OFDMA in LTE.

5.7.6.4 Handover

The LTE handovers are distinctly different from the 3G UMTS in that there is only the option similar to the hard handover available, with clear transition of the user from the source eNodeB to the target eNodeB. The LTE handover is entirely controlled by the RAN, with the CN updated only after the completion of the handover. Let's discuss the full handover procedure in LTE.

All UE's in LTE would be configured to take measurements of the signal strengths of neighbour eNodeBs. The UE could send these reports periodically to the serving eNodeB or the serving eNodeB can set a threshold value for signal strengths, above which the UE would report them. Once this threshold is crossed, the serving eNodeB would select the neighbour with the highest reported signal strength as the target eNodeB and would send a handover request to it. Once the target eNodeB acknowledges this request, the source eNodeB will push a handover command to the specific UE. At this point, all downlink data packets that the user has not acknowledged would be transferred to the target eNodeB through the X2 interface. This process may cause some overheads as some packets (for which the ACK from the user is on the way) may become duplicated, but it effectively ensures a lossless handover. This is an important feature which was enabled by the newly introduced X2 interface in the 4G network architecture. The UE would make timing synchronisation and access the uplink resources through the contention free RACH procedure explained in random access section above. Once this RACH process is complete, the UE would send the handover complete message to the target eNodeB.

The target eNodeB would inform the MME of the CN about the handover only at this stage. This is called a late path switch and effectively it does not affect the QoS the user is receiving, as the incoming packets up to this stage have been transferred to the target eNodeB through the X2 interface. The path switch message will cascade to the serving gateway (S-GW) and it will now switch the UEs data path to the target eNodeB. Finally, the target eNodeB will send a release resource message to the source eNodeB for it to release the data and control plane-related resources associated with this specific UE.

5.7.6.5 HARQ

The HARQ procedure in LTE is central to increasing the robustness of the data packets against transmission and propagation (channel) impairments. The LTE HARQ process is based on the 'stop and wait (SAW)' concept, where once a packet is transmitted, it is buffered until a feedback (ACK/NACK) is received. The LTE HARQ procedure is applied in both the uplink and downlink. In order to keep the physical channel fully occupied, a number of consecutive data packets are transmitted and buffered, while waiting for the ACK/NACK feedback messages. In LTE, this number is set to eight packets, allowing an 8 ms timeframe for the first re-transmission (for NACK feedback) or removal of the packet from buffer (for ACK feedback). This is depicted in Figure 5.22.

The one occasion in LTE where the SAW HARQ process is altered is when transmitting VoIP packets in the uplink, using TTI bundling. The VoIP packet interval (for AMR coding) is 20 ms and this is far too long for the 1 ms LTE packet sizes. In TTI bundling, one VoIP packet is repeated four times (without waiting for HARQ of a single packet), to create a 4 ms TTI. The HARQ procedure is conducted for the TTI bundle, and there is a far greater chance of responding with an ACK for a TTI bundle than a single 1 ms packet, in cell-edge conditions. This can effectively reduce the 8 ms delay associated with HARQ (as shown in Figure 5.22) as this average delay will be unacceptable for VoIP communications. Also, by packing the uplink transmissions with 4 ms of TTI bundle and possible re-transmissions, the VoIP packet interval of 20 ms is filled more with actual transmission time. This helps to increase coverage of the uplink, and indirectly benefits the cell-edge VoIP users. The uplink power budget can be improved by around 4 dB by the use of TTI bundling and associated HARQ [14].

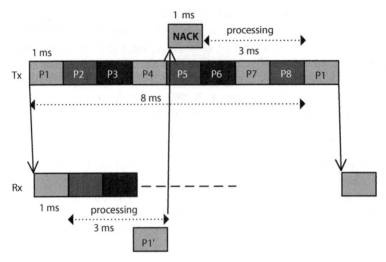

Figure 5.22 The stop and wait HARQ procedure in LTE.

5.7.6.6 MIMO Evolution within LTE

The MIMO techniques, introduced as 2×2 MIMO in 3G HSPA, saw significant evolutions within LTE. LTE Releases 8 and 9 saw the evolution of MIMO to four transmit ports and the corresponding changes in the CRS reference signalling (as covered in previous sections). We also discussed how the reference signalling included the user-specific DMRS to accommodate the eight port MIMO configurations in Release 10 LTE. In this section, we will look at the code book design in these LTE releases.

Release 8 and 9 LTE versions included single code books for different configurations (up to 4×4) and different channel ranks. LTE Release 10 saw the introduction of MU-MIMO, in addition to the support for Single User MIMO (SU-MIMO), and the extension of transmit ports up to eight. With the larger number of antenna ports, different antenna spacings and polarisations would give different radio-channel conditions and correspondingly different channel ranks. Rather than developing a very complex single code book to capture all possible variations, the Release 10 code book design consists of two distinct code books. The code book C1 accounts for the long-term wide band channel properties while the code book C2 account for the short-term narrow band channel properties. The resultant precoder matrix C is calculated as C = C1xC2. As C1 varies at a much slower rate than C2, the update rates for C1 can be longer than for C2. This double code book design in Release 10 offers the flexibility to support both SU-MIMO and MU-MIMO operations under a variety of channel conditions.

5.7.7 Main Features of Subsequent LTE Releases

We have so far looked at the main features of LTE and LTE-Advanced systems as defined by the standards up to Release 10. As noted before, 3GPP standardisation is a continuous process and new features are studied and added to subsequent releases all the time. We will briefly analyse the main features that were defined in Releases 12, 13 and 14 of 4G LTE.

In Release 12 (finalised in late 2014), one of the key themes was the support for small cells (SCs). There is a dedicated Chapter 10 in this book on the topic of SCs, where finer details can be found. The essence for SC support in LTE was the recognition that through the denser re-use of the radio spectrum, much higher data capacities per area can be achieved. This was seen as a key solution for the x1000 more capacity growth that is fuelled by Smartphones and other data hungry devices. One of the key issues that needed containment was the interference effects between the SC(s) and the macro cell, which will utilise the same radio spectrum. In Release 12, a number of solutions themed under eICIC (enhanced Inter-cell Interference Co-ordination) has been provided. The ability to configure Almost Blank Sub-frames (ABS), where the overlapping sub-frames of the macro cell can be made blank of data channels (to only carry control channels) is an example. The concept of Phantom cell [18], where the SC(s) only carry data channels and the wider macro cell carries the control channels has been widely discussed during Release 12. A similar feature was standardised as dual connectivity (DC) in Release 12 and this continues to have wider implications even in 5G Release 15 (as we will study later in this chapter).

Device to Device (D2D) communications is another main feature introduced in Release 12. This feature enables mobile devices to communicate directly with other nearby mobile devices, without intervention of the eNodeBs. D2D communications can be very useful in disaster and emergency situations where the serving eNodeB can be made dysfunctional. In fact, development of D2D communications (also known as Proximity services, ProSe) started as a step to enhance the LTE specifications making it suitable for use in mission critical communications (including for emergency services). D2D has other commercial uses such as finding friends and restaurants and advertising targeted content directly to nearby users. The D2D-related standards cover two main scenarios. First, the autonomous D2D communications without any support from the LTE network and second, communications supported by the network for device discovery. D2D communications can extend the coverage range of a network as well as increase the capacity for a given bandwidth and reduce the network load. As with many aspects of the 3GPP standards, the D2D features continue to be developed in subsequent releases.

In LTE Release 12, the capabilities of Carrier Aggregation (CA) were further improved, to accommodate joint operation of FDD and TDD in component carriers. Also, Release 12 saw major improvements to security features, partly to enable LTE to function in the mission critical communications domain.

LTE, as defined by Releases 13 and 14 is termed as LTE-Advanced Pro, to indicate the enhanced and additional features provided by these releases. Release 13 (completed in June 2016) saw the first introduction of 3GPP standards to support massive machine-type communications (mMTC). The narrow band Internet of Things (NB-IoT) and the LTE category M (also known as LTE-M) are both components designed to support mMTC. The NB-IoT strand occupies 180 kHz bandwidth (width of a LTE RB) and can operate in standalone, in-band or guard-band modes. Operating within the re-farmed GSM systems is also possible, to occupy the 200 kHz narrow band carriers of GSM. The maximum path loss (we will study path loss in Chapter 6) has been extended to 164 dBm in NB-IoT, to support wide area coverage. It is suited for low throughput, delay tolerant and low mobility use cases such as operation of Smart meters. LTE-M, on the other hand, operates in a wider band of 1.08 MHz and can support higher data rates up to 1 Mbps. LTE–M has been designed to support 155.7 dBm path loss to provide sufficiently wide coverage. LTE-M can co-exist with other types of LTE traffic in a single system.

As noted above, the LTE standard has been viewed by many emergency and mission critical services as a potential technology to complement or even replace the dedicated/proprietary communication technologies used in this field. Up until Release 13, LTE had been developing enabling technologies for this domain, like the D2D. The Mission Critical Push To Talk (MC-PTT), introduced in Release 13, was the first dedicated service for the MC communications, offered by 3GPP. The MC-PTT is a voice communication solution which provides features like clear audio quality, high availability/reliability, lower latency, support for group and one-to-one calls, talker identification, D2D communications, emergency calling etc. We will look at the MC thread of standards in the 5G support for the emergency services vertical in Chapter 12.

Release 14 (completed in mid-2017) saw the extension of LTE features towards the vehicle to everything (V2X) domain, a key area for future growth. V2X includes Vehicle-to-Vehicle (V2V), Vehicle to Infra-structure (V2I), Vehicle to Pedestrian (V2P) and other communication possibilities. These will enable the automation of a number features in the automotive sector, leading to assisted driving and finally self-driving vehicles. The capabilities of Release 14 can be utilised to develop the first C-V2X (Cellular V2X) systems. We will discuss this automotive vertical in detail in Chapter 12.

Another area that saw significant development in Release 14 is the expansion of LTE systems into the unlicensed spectrum. Traditionally, the unlicensed spectrum has been utilised by Wi-Fi, Bluetooth and other similar local area and short-range communication systems. Due to the pressures of spectrum scarcity, LTE has started to look at expanding into the unlicensed spectrum since release 13. In Release 14, mechanisms to aggregate unlicensed spectrum carriers with LTE licensed spectrum carriers using LAA (Licensed Assisted Access) and LTE-U (LTE- Unlicensed), have been enhanced. Another related work area in LWA (LTE- Wireless LAN Aggregation), where the operator deployed Wi-Fi (known as carrier Wi-Fi) resources can be better managed by an LTE anchor eNodeB, has also been enhanced. Section 11.1.4 takes a more detailed look at LTE-U and the ways of amalgamating LTE and WiFi technologies.

Release 14 also saw the completion of MC-data and MC-video components of the mission critical communications domain, making a full suite of MC services ready. There were also significant developments in eMBMS (evolved Multimedia Broadcast Multicast Services), where point to multi-point communication services are optimised to deliver IPTV content on mobiles for example. The mm-wave channel model, which spans from 6 GHz up to 100 GHz and will be useful in simulating and understanding the 5G higher frequency systems, was also completed in Release 14. This is the last release to be classified within 4G LTE domain, and we will look at 5G networks from the next section.

5.8 The 5G Network

The 5G network is envisioned to support many more applications than simply increased data rates. The network design will try to capture the requirements of many other verticals, like automotive, industry automation, Smart city, eHealth etc., where different strands of 5G services are likely to emerge. Thus, the 5G network will be multi-dimensional, making

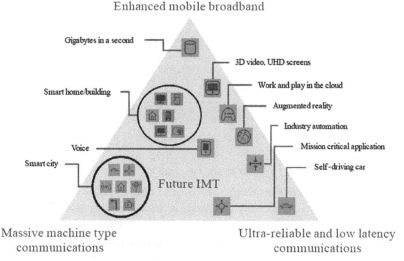

Figure 5.23 The 5G key services triangle. (*Source* [19]).

the network design complexities even greater. We will talk about these vertical applications and the likely impact they will have in Chapters 11 and 12, but here we will simply look at the technical features of the emerging 5G network.

It is now widely accepted that 5G networks will have to support three core service types. These are the provision of very high data rates, as termed by extreme mobile broadband (eMBB), the provision of ultra-reliable, low latency communications (URLLC) and supporting millions of connected devices as massive machine type communications (mMTC). These three pillars of 5G services are usually depicted in triangular form, with various 5G applications falling within this triangle, with their position representing the degree of reliance upon these core requirements. Such a descriptive diagram is shown below in Figure 5.23, from [19].

Developing a 5G standard to encapsulate all these requirements in one go is a near impossible task. Thus, 3GPP has taken a stepwise approach to 5G standardisation. The 3GPP Release 15 (known as 5G phase 1) will contain first components of 5G, known as New Radio (NR). This 5G-NR Release 15 standard details a number of deployment options for 5G, in corporation with LTE or as a stand-alone network. We will look at these options and the rationale behind them in the section below. Release 15 will mostly support eMBB applications, with some features for URLLC support. Currently, the NB-IoT and LTE-M strands of the Release 13 standard are expected to support the emerging mMTC deployments. The non-stand alone component (where it operates on the support of a LTE network) of Release 15 was completed in December 2017, and the stand-alone component was completed by June 2018. The development of 5G phase 2 standardisation, in the form of 3GPP Release 16 is on-going at the time of writing, in January 2020. This is expected to support even higher mm-wave frequency bands and provide more features to enable URLLC and mMTC applications.

As with the previous generations, the International Telecommunications Union (ITU) will develop a set of key requirements and related evaluation criteria that a standardised technology has to satisfy in order to be classed as 5G. These are currently being developed, and are termed as IMT 2020 evaluation criteria [19]. While Release 15 contains the first 5G

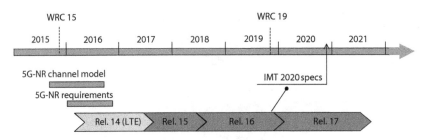

Figure 5.24 The timeline for 3GPP 5G standards development.

capabilities, the aim for 3GPP is to make Release 16 fully compliant with IMT 2020 specifications for Release 16 to be formally labelled as a 5G technology. You can see how these 3GPP 5G and IMT 2020 timelines align in Figure 5.24. It also contains reference to WRC (World Radiocommunication Conference) meetings, where the decisions on which spectrum to be used for 5G and other terrestrial/satellite applications) are taken by the ITU. More details of this WRC process can be found in Chapter 7.

In terms of terminology, 5G refers to the overall 5G system currently developed by 3GPP, as there is no other 5G standardisation process. We use 5G-NR to refer to the 5G radio layers and procedures, as this is the common practice.

5.8.1 5G Deployment Options

The 5G standards have acknowledged that different operators would prefer different timeframes and pathways to migrate from 4G to 5G. It is likely that most operators would want to utilise the 4G infra-structure for some more time, considering the huge investments they made in 4G. So, some of the 5G architectural options are built on the possibility of dual connectivity (DC) for UEs in the RAN and the ability to connect to either the 4G core (EPC) or 5G core. Dual connectivity is a feature which allows a UE to maintain two radio links concurrently and was introduced in Release 12. The architectural options further include options for UEs to utilise 4G RAN but connect to the 5G core (EPC) as well as utilise 5G RAN and 5G core. Thus, there is a clear migration path preferable for most operators, starting with non-stand-alone systems (5G UEs with the support of 4G RAN and core) to fully stand-alone systems (5G UEs using entirely 5G RAN and 5G core).

The most cost-efficient approach to start 5G services will be to deploy the 5G RAN at specific high-demand locations, with the support of 4G RAN (as an underlay network) and connect to the 4G core (EPC). The 5G cells or gNBs- as they are known in 3GPP standards, will only provide the data connectivity or the data plane. The 4G cells or eNBs, will provide the sole control plane and (optionally) a data plane. This is a non-stand-alone deployment, supported by the first phase of Release 15. For operators, this approach helps to deploy 5G cells (or gNBs) to the most likely hotspots where the demand for 5G eMBB services would be the highest, and support these cells with the existing 4G network, where the investment has already been made. This approach in known as option 3, in the 3GPP literature [20].

Option 3 has further three variations, as shown in Figure 5.25. The first variant has both data plane and control plane connectivity to the EPC entirely through 4G eNB. 5G gNB only provides data plane connectivity through 4G eNB. In the second variant, the data

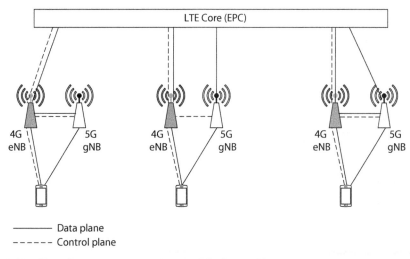

Figure 5.25 The NSA architectural options for 5G. (*Source* [21]).

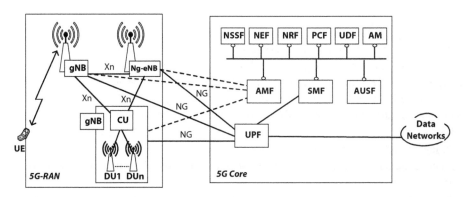

Figure 5.26 The 5G system architecture.

plane of the 5G gNB can connect directly to the EPC. In the third variant, the 5G gNB data plane has a direct link and a link through the 4G eNB to the EPC. In all three options, a data plane is also provided by the 4G eNB [21].

5.8.2 5G System Architecture

5G is in the process of developing the 5G system architecture with 5G RAN and 5G core, enabling the stand-alone mode of deployment. The 5G system concept will enable not only the 5G-RAN, but evolved LTE RAN and other wireless access technologies to connect to the 5G core. The 5G core itself is significantly different from the LTE core and is designed to enable many of the new requirements and emerging networking concepts, which we will look at in Section 5.9. The 5G system architecture is depicted in Figure 5.26. A useful reference to the new 5G architecture and layer 1/2/3 aspects is [22].

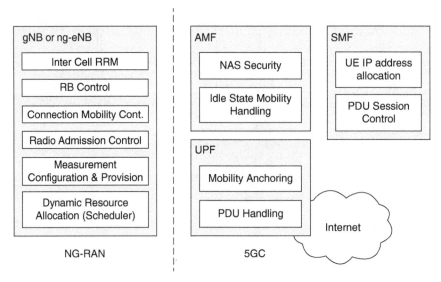

Figure 5.27 5G RAN and Core functionalities. (*Source* [23]).

Under the overall 5G architecture as specified in 3GPP TS 38.300 [23], both 5G gNB and ng-eNB access nodes can connect to the 5G core. The ng-eNBs can provide radio access through the E-UTRAN, i.e the LTE RAN. The radio nodes inter-connect through the Xn interfaces while the nodes connect to the 5G-core through the NG interfaces. The RAN nodes can optionally have a centralised architecture, where a central unit (CU) does the higher layer processing for a number of distributed units (DU), which can perform only the lower layer functions (for example, only the radio transmission). The centralisation of RAN offers many benefits and is a current hot topic in the industry. We will look at this aspect in detail in the next section.

The design of 5G core enables a number of new network concepts to be executed, paving the way for a service based, virtualised 5G architecture. The 5G core is also designed to enable fixed wireless and other access technologies to connect to it, paving a future evolution path for mature 4G networks as well. We will study 5G core concepts and features in a bit more detail in Section 5.8.8. The main functionalities of 5G RAN and 5G core are identified by 3GPP in Figure 5.27. In 5G Core, only the three main units of AMF (Access and Mobility Function), UPF (User Plane Function) and SMF (Session Management Function) are shown.

5.8.3 Spectrum Options for 5G

From the very onset of 5G research, it was identified that the diverse, technically demanding set of 5G use cases (examples in Figure 5.23) will require a very large amount of spectrum resources. The lower frequency bands are now particularly congested with a lot of cellular and non-cellular applications and naturally 5G research started to look for higher frequencies, known as millimetre wave (mmWave) spectrum (a common denominator is to treat spectrum from 6 GHz to 100 GHz as mmWave). The approach in 5G has been to cater for operations in high range (mmWave), mid-range (between 1 GHz and 6 GHz) and low range (below 1 GHz) frequencies viable in the networks. The advancements of RF circuitry and chip design would mean that 5G devices would be able to hop between cells operating these frequencies and also use these spectrum concurrently to increase the data rates, in carrier aggregation and RAT interworking solutions.

The availability of wide contiguous bandwidths was an appealing feature in mmWave, but they also provide challenges like the high free space loss and high blockage due to foliage and obstacles. Beamforming with a very large number of antenna elements is a viable solution, especially as the higher frequencies make the antenna array sizes smaller. 5G-NR supports beamforming operations in initial access, as detailed in Section 5.8.6. Within mmWave, the spectrum around 28 GHz has become a primary band, particularly for 5G fixed wireless access (FWA). Generally, mmWave spectrum can be used in 5G for smaller cells with very high capacity, in particular to drive the NSA (Non-Stand-Alone) deployment option 3, with support of larger LTE cells.

5G would also support the mid-frequency bands, especially to ensure sufficient coverage as deployments mature. The 3.5 GHz band has become popular in this range, where in many countries sufficiently large bandwidths (exceeding 100 MHz) are available. In this spectrum, both coverage and capacity can be met at sufficiently high levels. The 3.5 GHz is the primary band used in many of the first 5G networks that went live in 2019.

The low range of frequencies would be important for 5G in the longer term, when standalone networks with wider coverage will start to be rolled out and for future MTC based IoT networks. In some countries, there are spectrum bands available from the switch to digital TV from analogue systems and these can be utilised for 5G. Also, in some countries, 2G and 3G systems are being switched off by some operators, and these spectrum can be effectively re-farmed for 5G.

5.8.4 5G-NR Protocol Layers

In looking at the protocol layers of 5G-NR, the layer denomination is similar to LTE, particularly in the control plane. However, the functionality of some layers has been extended to support complex 5G requirements. We will look first at the control plane 5G protocol layers and then at the data plane protocol layers.

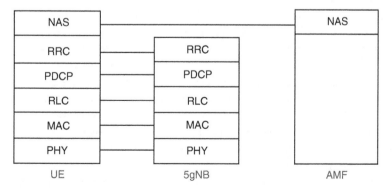

Figure 5.28 Control plane protocol layers for 5G-NR.

Similar to LTE, the NAS layer in 5G-NR logically connects UE directly to the 5G core, bypassing the gNB. This functional unit in the core is termed as AMF. Similar to LTE, the NAS protocol layer is responsible for managing mobility and supporting communication sessions between the UE and 5G core.

The RRC layer is further enhanced from LTE, to support numerous new requirements in 5G-NR. The RRC layer (in essence) serves the same basic function of the establishment, maintenance and release of and RRC connection between the 5G-RAN and UE. This basic function is adapted in 5G-NR to support the DC and CA features. To meet the new requirements in dual connectivity, the secondary node (the 5G gNB in option 3 discussed above) can also transmit the RRC message in addition to the master node (the 4G eNB in option 3) enabling some diversity gain. However, the release of the RRC connection can only be done by the master (LTE) node. The RRC layer of the 5G-NR also has a new 'RRC inactive' state introduced. This will function in between the RRC idle and RRC connected states of 4G LTE (and the previous radio generations). In 4G LTE and before, the UE has to be moved to the RRC connected state for any kind of data transmission. The transition from RRC idle to RRC connected states will involve a number of message transfers and hence a sizable delay and energy consumption. New applications like mMTC need to be highly energy efficient and URLLC need to have ultra low latency. Both these domains involve short-packet transmissions. Thus, the RRC inactive state is introduced, where the UE can move back to the RRC connected state (for the packet transmission/reception) with the minimum number of 'wake-up' signalling messages.

This RRC inactive state is governed by the new 'suspend' and 'resume' procedures for the UE. When the UE moves to RRC inactive state, the 'suspend' procedure is enacted and both UE and RAN store all the information needed to 'resume' the UE back to the RRC connected state. These include the RAN Notification Area (RNA), which the UE is allowed to move before updating the CN (AMF) and the UE security information. Thus, when there is a need to 'resume' the UE to the RRC connected state, this could be done with the transfer of only three messages (on average) entirely between the UE and RAN. If the UE is to 'wake-up' from the RRC-idle state, this would require seven messages, with both RAN and CN involvement. If there are infrequent short-packet transmissions, then moving the RRC state to RRC inactive can significantly reduce latencies, signalling overheads and power consumption.

In layer 2, the PDCP sub-layer in the control plane is responsible for ciphering/deciphering and integrity protection, as in LTE. However, in the data plane, the PDCP sub-layer plays a much wider role as discussed later. The RLC sub-layer plays the role of ARQ retransmission. In 5G-NR, the RLC configuration per logical channel does not depend on numerologies or TTI durations of the physical (PHY layer) channels. The MAC sub-layer, as in LTE, multiplexes and de-multiplexes MAC SDU (belonging to different logical channels) to form transport blocks. MAC also carries out the HARQ process, based on these transport blocks and is responsible for the scheduling procedure as in LTE. One of the differences in 5G-NR scheduling is that the resource unit can be even smaller, with mini-slot scheduling made possible to accommodate short-packet transmissions.

In the PHY layer, the physical control channels are configured from the transport channel information coming from the MAC SDUs. These channels include PDCCH in the downlink, PUCCH and PRACH in the uplink, for example. These physical control

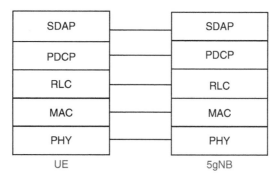

Figure 5.29 User plane protocol layers for 5G-NR.

channels are not dissimilar to LTE. However, the sidelink transmissions have evolved significantly from the beginnings in LTE Release 12 (to support device-to-device, D2D, communications) to support the V2V communications in 5G-NR. The PSCCH (Physical Sidelink Control Channel) is developed particularly in 5G-NR Release 16, to carry the control information for uni-cast or multi-cast V2V situations, with low latency and high reliability.

The user plane configuration of the 5G-NR protocol layer in the RAN is shown in Figure 5.29.

At the top of the protocol layers is the new L2 sub-layer Service Data Adaptation Protocol (SDAP). The function of the SDAP layer is the mapping between QoS flows and data radio bearers. In 5G-NR, the QoS is defined in terms of flows, whereas in 4G, QoS is defined in terms of bearers. QoS flows in 5G are configured from the 5G core down to the UE. A QFI (QoS Flow Identifier) is identified within a PDU session. The SDAP sub-layer also performs the marking of QFI in both DL and UL packets.

Primary functions of the PDCP sub-layer in the data plane are sequence numbering, detection and removal of duplications and header compression. In the 5G-NR PDCP sub-layer, data PDUs can be duplicated to travel through two different RLC paths. First, duplication has to be configured in the RRC layer (in the control plane) and a secondary RLC entity and a logical channel is then added. This feature increases reliability in 5G-NR, to enable URLLC implementations. The PDCP control sub-layer (discussed above) PDUs are not duplicated in this way and are always submitted to the primary RLC entity. The RLC layer performs the ARQ function and also carries out a sequence numbering procedure (separate to that of the PDCP sub-layer). For the dual connectivity option in 5G-NR, two MAC entities can be created in the UE, one to support the master node connection and one to support the secondary node connection. As in the control plane, in the data plane the MAC sub-layer is responsible for creating the transport blocks (TB). The 5G-NR MAC entity can support multiple numerologies and TTI.

The PHY layer in the data plane is responsible for creating and transceiving physical data channels in the DL and UL (PDSCH and PUSCH). This is similar to operations in LTE. However, the data channels in 5G-NR can be configured to support multiple numerologies and TTIs, as we will discuss in the section below. The gNBs can enact these multiple options simultaneously. On the UE side, the PHY layer can be configured to operate in one or more bandwidth parts (BWP), which can be less than the total system bandwidth. In sidelink operations, the PSSCH is dedicated to carry data, primarily for the V2V communications as developed in Release 16.

5.8.5 The 5G-NR Air Interface

Let's discuss some of the key air interface features of 5G-NR. As mentioned above, 5G-NR is developed as an evolution of the LTE-Adv standards, so many RAN features are common with LTE-Adv. The air interface employed is CP enabled OFDM (CP-OFDM), as in line with 4G LTE. However, different numerologies of CP-OFDM are made available in 5G-NR. As we discussed in the 4G RAT, LTE air interface has a fixed 15 kHz sub-carrier spacing, which corresponds to 0.067 ms symbol durations. The bandwidth options in LTE vary from 1.4 MHz to 20 MHz, but the sub-carrier spacing (SCS) remains fixed. The 5G-NR is supporting much wider bandwidths and with a scalable SCS, as a function of 2^nx15 kHz (n = 0, 1, 2, ...). For carrier frequencies below 6 GHz (known as FR1) the SCS is adaptable as 15 kHz, 30 kHz, 60 kHz. For carrier frequencies above 6 GHz (known as FR2 or mm-wave bands) the SCS is adaptable as 60 kHz, 120 kHz or 240 kHz. Consequently, the higher SCSs will correspondently shorten the symbol duration, and the TTI. This is beneficial for short-packet, low latency communications. Wider SCS allows more tolerance for the Doppler spreads (i.e. faster UE speeds than in LTE) and also for phase noise, which is a common problem in higher mm-wave frequencies. These mm-wave frequencies (FR2) are expected to be deployed as SCs, with less propagation delay, so the smaller symbol durations will not be impacted by longer delay spreads. By allowing a scalable SCS in 15 kHz multiples, 5G-NR has retained compatibility with LTE and also meets the needs of wider bandwidth, low latency communications.

Next, we will discuss the frame structure of 5G-NR. As in LTE, the NR frame is made up of 10 sub-frames, with each sub-frame having a duration of 1 ms. You may re-call that the LTE sub-frame is further divided into two slots of 0.5 ms duration, with each slot having six or seven symbols (depending on the length of the CP). In 5G-NR however, the number of slots per sub-frame varies, depending on the SCS utilised. The NR slot is defined as consisting of 14 symbols (for the normal CP). For the 15 kHz SCS, this would mean the slot would span for 1 ms, i.e. only one slot would occupy the sub-frame. As the SCS is increased as 2^nx15 kHz, the symbol durations become shorter as will slot durations. Slot durations will reduce at the rate of $1/2^n$ ms (n = 1, 2, 3 ...) and correspondingly the number of slots per sub-frame would increase. This is depicted in Figure 5.30, for three SCS selections. The CP is not shown here, but this CP is appended to the symbols as in LTE.

In LTE, the granularity for user allocations was based on PRBs, made up of 12 sub-carriers in the frequency domain and 12 or 14 symbols (2 LTE slots) in the time domain. This kind of slot-based scheduling is also possible in NR, where the slot would now contain 14 symbols. One difference in NR is that the symbol durations would reduce as SCS is increased and, as such, time durations of PRB allocations would also reduce. Also, 5G-NR specifies a mini-slot based allocation option, where a single-user allocation can contain two, four or seven symbols only. This kind of mini-slot allocations would reduce the time required for a single transmission to a user, even with 15 kHz SCS. Such types of allocations are useful in URLLC applications, to ensure low latencies approach the 1 ms threshold.

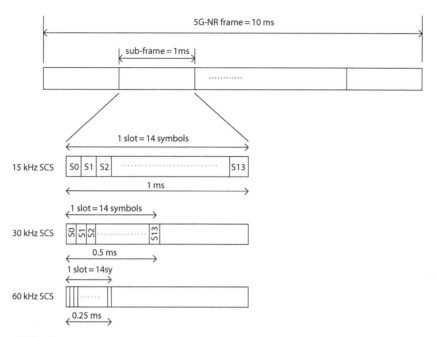

Figure 5.30 The frame structure options for 5G-NR.

Also, for TDD applications (which is likely to dominate the first mm-wave 5G-NR systems), having a low transmit period per user means the DL and UL can be multi-plexed in very quick succession. This enables channel information in one direction to be effectively used in the other direction as well, even for very fast moving users. It also enables accurate, fast-beam sweeping – a likely requirement in mm-wave systems (we will discuss this below, under reference signals).

Another difference in 5G-NR (from LTE) is the introduction of the bandwidth part (BWP) concept. This BWP is defined as a contiguous block of resource blocks (RB), which can be less than the total bandwidth. A UE can operate at a single BWP, so effectively in a selected narrowband segment in a wideband channel allocation. A UE can be configured with up to four carrier BWPs, with only one BWP active at a given time. This enables UEs to adapt the bandwidth according to the data rate (and latency) needs, allowing energy efficiency as well.

To increase spectrum efficiency of radio access, a vast array of filtered OFDM and also non-orthogonal waveforms are being studied. The aim of filtering OFDM is to reduce the out-of-band emissions in the frequency domain. Different filtered OFDM variants are derived based on the granularity of applying the filters (i.e. per sub-carrier level or per group of sub-carriers). Filtered OFDM is still very much in the research domain, and is unlikely to be discussed within 3GPP soon. The non-orthogonal multiple access (NOMA) is also a significant area of 5G radio-access research. The increasing processing power means that the receivers are able to cancel out any interference from users who occupy the same time-frequency slots. In the simpler schemes of NOMA, two users are allowed to occupy the same time-frequency slots, but they will be transmitted at different power levels. This ena-bles the receiver to use Successive Interference Cancellation (SIC) techniques to recover the necessary signals. By allowing more than one user in a time-frequency slot, the overall spectral efficiency can be improved. However, 3GPP has decided not to consider NOMA in the current releases of 5G-NR standardisation.

5.8.6 5G-NR RAN Procedures

Many of the 5G-NR RAN procedures are based on their LTE predecessors, so there are a lot of similarities between them. We will focus on one main area where 5G-NR has developed distinctly new procedures, i.e. to support beamforming. As noted before, 5G-NR systems can operate in mm-wave frequency bands (above 6GHz, denoted as FR2), where the signal attenuation can be very high. To compensate this, the transmitter (and also possibly the receiver) can be configured to have multiple narrow beams, which makes the cell coverage rather different to LTE and previous cellular systems. We will look at how the 5G-NR RAN procedures can support these beamforming systems.

5.8.6.1 Synchronisation

In LTE and its predecessors, the synchronisation signal set is broadcast over the entire cell with wide coverage. With the 5G-NR option to cover a cell with multiple beams, the UE needs to determine which of the beams it can select to receive with the best signal quality. The UE may itself apply beamforming in the uplink and the gNB needs to be aware as to which of its beams can receive the UE signal with the best signal quality. To facilitate these selections, synchronisation signals are transmitted by the gNB in a procedure known as 'beam sweeping'.

'Beam sweeping' involves the synchronisation signal blocks (SSB) transmitted from each of the beams in consecutive symbols (in the time domain). This transmission is called a Synchronisation Signal Burst (SSB). The SSB itself contains the primary synchronisation signal (PSS), the secondary synchronisation signal (SSS) and the physical broadcast channel (PBCH). In the frequency domain, the SSB occupies the middle 20 PRBs (or 240 sub-carriers). This beam-sweeping procedure and the format of the SSB is depicted in Figure 5.31.

Figure 5.31 shows how four consecutive slots (in the time domain) are used to formulate the SSB for each beam and how these blocks are positioned consecutively to produce the SSB. The SSB can be configured to have one or more bursts within the 5 ms half frame window of 5G-NR, to configure an SSB set. The periodicity of the SSB set is configurable and typically set at 20 ms. The PSS, SSS and PBCH components within the SS block carry out similar functions to LTE but, additionally, the PBCH also carries DMRS (De-Modulation Reference Signals). As detailed in Section 5.8.7, the DMRS helps with the demodulation of a downlink signal specifically targeted for a user. In the PBCH, the DMRS helps the UEs estimate signal strengths of each of the beams and then to select the best beam for its communications. The best beam is identified as a time index within the SSB.

How would the gNB know of the beam selected by a specific UE? The UE would read the synchronisation signals to identify the gNB and sync with this. When the UE wants to communicate, it has to request resources through the random access procedure. The PRACH of the 5G-NR in this beamforming scenario would have a sub-set of the RACH pre-amble codes to use depending on the best beam (time index) selection. The UE would only use a

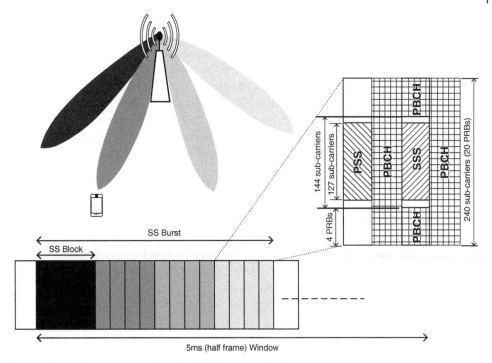

Figure 5.31 The beam sweeping for 5G-NR synchronisation.

pre-amble from this sub-set in the random access procedure. This pre-amble selection will indirectly inform the gNB of the beam selection of the particular UE. If the UE is granted access, the gNB will continue communication with the UE on this selected beam.

5.8.7 5G-NR Reference Signals

In terms of reference signals, there is a move towards user-specific reference signals in 5G-NR, from the LTE norm of cell-specific reference signals. This new feature enables greater energy efficiency, as radio frames where there are no data transmissions can now be kept virtually empty. The cell-specific C-RS of LTE is not present in 5G-NR. The user-specific demodulation reference signal (DM-RS) is present in both DL and UL for 5G-NR. The DM-RS can be configured on-demand and also as a 'front loaded' signal at the start of the user-slot allocation (in the time domain). Having DM-RS front loaded helps to identify this signal early and helps with the demodulation of both data and control channels quickly. This front loading is yet another feature in 5G-NR to facilitate low latency communications. Also, with its user-specific nature, DM-RS can be made available within a specific radio beam and will assist with the demodulation in beam-formed systems (in higher frequency operations).

The CSI-RS (channel state information reference signals) is used for downlink channel estimation and codebook selection for MIMO operations. This functionality is similar to that of LTE. CSI-RS can also help with the finer tracking of carrier frequency and timing offsets. In 5G-NR, the CSI-RS can be configured for more than 32 antenna ports.

A new type of reference signal for 5G-NR is the PT-RS (Phase Tracking Reference Signal), which helps to track the phase variations between the transmitter and receiver in high-frequency operations. Phase noise is generated when radio oscillators in the transmitter and receiver have separate and independent frequency drifts. The phase error component corrected by PT-RS is known as the common phase error (CPE), which is a common offset for a number of sub-carriers in the frequency domain. The CPE will only impact the very high MCS (Modulation and Coding Schemes) and has more impact on shorter PRB allocations. Thus, the PT-RS densities for a specific-user allocation can be configured, depending on the MCS and the PRB length used. PT-RS is configurable both in the DL and in UL and QPSK reference symbols are used in the PT-RS.

The positioning reference signal (PRS) is being specified in 5G-NR to support positioning and localisation in the downlink, to support emergency and commercial use cases. As in LTE, PRS is a configurable reference signal with different periodicities. The PRS is primarily useful for multi-lateration based positioning methods, and its accuracy is basically reliant upon the frequency sampling rate. With the use of wider bandwidths and large sub-carrier spacings, the sampling rate increases and positioning accuracy can be much improved from what is achieved in LTE systems.

The last reference signal we will note here is the SRS (Sounding Reference Signal), which is used on the UL. It helps the gNB detect the CSI (Channel State Information), for each radio channel from active users in the UL. The SRS is also being considered as the primary reference signal to support uplink positioning mechanisms.

5.8.8 5G Core – Concepts and Functionalities

To understand the differences of the 5G core from previous generations, the concept of Service-Based Architecture (SBA) needs to be explained. In previous generations from 2G to 4G, the CN followed a 'point-to-point' architecture, where specific functional elements were connected to each other through specific interfaces. This kind of static connections are suited for networks delivering only a few services (voice, mobile data), as it creates dependencies between the elements which are difficult to re-configure. Each re-configuration would need thorough testing before the network can go live. The 5G networks, as we discussed earlier, are geared to deliver numerous services with differing KPIs to a multitude of vertical sectors. These differing service requirements necessitate the agile creation of 'network slices' which will be like multiple networks supporting the exact KPIs for each of the services. Some of the slices need to be created and taken down dynamically to support the dynamic nature of some vertical services. Having a point-to-point architecture will not provide the necessary agility for the network to adjust quickly and at a lower cost to facilitate these changes [24].

The key to an SBA is to create well-defined, separate functions in the CN and interconnect them through service-based interfaces. In the current 5G core, the control plane functions do connect over these service-based interfaces (marked with a dot at the end of connections in Figure 5.26) and the user-plane functions connect through traditional point-to-point links. The service-based interfaces are commonly referred to as APIs (Application Programming Interfaces), which are standardised to enable easy programmability. With increased use of virtualisation (where the physical functional nodes can be located elsewhere, usually in the 'Cloud'), the use SBA gives the network operator flexibility of easily configuring, running and dismantling services [24]. We will discuss the topic of virtualisation in detail in Section 5.9.

The control plane functions in 5G core include the main AMF (Access control and Mobility management Function) which is similar to the MME in 4G LTE. The main tasks of AMF are denoted in Figure 5.27. The SMF (Session Management Function) sets up and manages the UE sessions, according to the network policy. Again, the specific details of its main functions are given in Figure 5.27. The specific modular control functions include the NSSF (Network Slice Selection Function), which, as the name suggests, helps to select the correct network slice for each UE session. The NRF (Network Repository Function) enables registration and discovery for network functions. The Unified Data Management (UDM) function stores subscriber profiles as the HSS operates in 4G. The Authentication Server Function (AUSF) executes user authentication as similar to the AAA functionality of 4G. The NEF (Network Exposure Function) is new to 5G, where external entities like application functions and service clients can view the 5G network service capabilities in a secure manner. The PCF (Policy Control Function) defines the policies to manage mobility, roaming and more uniquely to 5G, network slicing. This is a similar entity to PCRF in 4G. In the user plane, the UPF (User Plane Function) is responsible for packet data unit handling and connecting 5G core to external data networks. As noted above, this UPF is connected to the RAN nodes and other core entities through point-to-point links.

You will see that there is a clear separation between the control plane and user plane functions in the 5G core. This is another key feature of 5G core architecture, known as CUPS (Control and User Plane Separation). In 4G, the user plane and control plane functionalities were inter-twined in entities such as the S-GW and P-GW. In 5G, the separation has enabled clear evolution paths for each of the key entities in these two domains. Many of the control plane functions can be configured remotely (in the 'Cloud'), for example, and the interfaces can be created using standardised programming. The diverse control functions (detailed above) available in 5G core facilitates different selections of these functions to suit the requirements of a particular network slice. A detailed, in-depth reference for the developing 5G-core functionalities can be found in [25].

5.9 The Centralisation and Virtualisation of the Mobile Network

The application of centralised and virtualised components in the mobile networks has been a key research area over the past decade. This will bring fundamental changes to the way mobile networks are designed, built and operated. The basic thinking behind these concepts is that much of the processing functionalities involved in the cellular network can be removed from the physical entity and can be conducted in a centralised or a Cloud-hosted server. Virtualisation and Cloudification concepts are applicable to both RAN and CNs. In this section, we will look separately at RAN and CN changes that will happen through the application of these concepts, mainly in a technical sense. In Chapter 12, we will look at how the business and operational models of the mobile operators are likely to change with these fundamental shifts in the mobile network.

5.9.1   The Centralised RAN (C-RAN)

The centralisation is applicable to the RAN, where in traditional networks the gNodeBs in 5G or the eNodeBs in 4G would conduct all of the processing related to radio transmissions, radio resource control and management. This kind of network is now termed as having a distributed RAN (D-RAN) architecture and all deployed gNodeBs or eNodeBs need to have this processing capability. With a centralised C-RAN (C-RAN) functionality, the physical g/e NodeB locations can be reduced up to the basic radio transmission level and all related processing and RRC/RRM can be grouped together and conducted in a central data centre. The radical centralisation ideas became feasible with the increased computational power of data centres and capability to provide fast enough interconnections between these and the peripheral radio units.

The overarching benefit of C-RAN is a cost reduction and an efficiency improvement in the deployment and O&M (Operations and Maintenance) of the RAN. The potential cost reductions achievable with C-RAN were highlighted in Chapter 3. In common C-RAN terminology, the central processor is called the Base Band Unit (BBU). Having a centralised BBU enables deployment of general purpose computer units (GPU) and servers, which incurs a lower overall cost than deploying a few hundred specialised g/e NodeBs. The operations and maintenance of the centralised RAN also become much easier and more cost efficient. Also, scaling up of the RAN becomes easier with additional GPUs added at the centre with only the hardware radio units needing to be installed at the periphery. Any software upgrades, for example, need only be carried out at the central BBU.

In C-RAN architecture, connections between peripheral radio units (also called Distributed units –DU) and the central BBU is called the Fronthaul. Typically, these links can be optical fibre or even microwave, particularly in rural areas. These Fronthaul links are required to carry higher data rates than the traditional backhaul links we discussed in Chapter 6. The connection between the central BBU and the CN is called the Backhaul and its functionality is similar to the Backhaul links of the D-RAN networks. A recent standard called eCPRI (a follow on from CPRI – Common Public Radio Interface) was developed to standardise data transfer between the RF unit and the protocol layer 1 [26]. In many of the C-RAN systems, this CPRI interface is being used, as we will discuss below.

The split point between centralised functionality and distributed radio unit functionality can be varied across the protocol stack. In broad terms, there are two basic functional splits. A split at the lower layers (typically physical layer- L1) only enables the remote radio heads (RRH) to be deployed at the periphery which needs to do only the basic radio processing and transmission. This results in cheaper, less power hungry RRH units at the periphery of the network and a lot of processing conducted at the BBU. However, higher data rates need to be supported in Fronthaul, for example, this split needs to carry even the I-Q samples of modulation symbols up to the central BBU. In contrast, a higher layer split in C-RAN would enable some of the processing to be done at the peripheral units while protocol functions like RRC (in layer 3) are carried out at the central unit. This split requires more expensive distributed units but the amount of data that needs to be transferred in the Fronthaul networks (and hence the Fronthaul operational costs) are lower. This is an important network design consideration, as many operators now lease the Fronthaul from third parties and lease prices are based on capacity used. Figure 5.32 depicts C-RAN architecture with the different types of splits along with traditional D-RAN architecture.

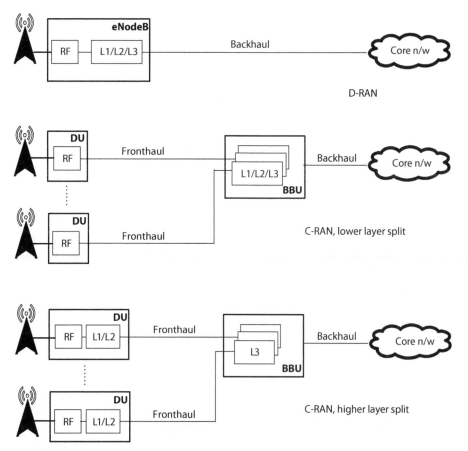

Figure 5.32 CRAN split options and comparison with D-RAN.

As noted above, the connection between the RF unit and Layer 1 can be configured with eCPRI or CPRI. In the case of the lower layer split shown in Figure 5.32(b), the total Fronthaul link can be configured through CPRI/eCPRI. Also, if the option is available, the central BBU can be located closer to the CN nodes, so the backhaul links and associated costs can be reduced.

3GPP has specified a number of possible C-RAN split points on the L1, L2, L3 protocol layers and sub-layers. These give flexibility for an operator to configure the C-RAN network to suit their individual needs, based on the broad trade-off points discussed above. These split points are illustrated in Figure 5.33.

The Cloud RAN can be seen as the next evolution of the C-RAN concept, when the central BBU functionality can be implemented in a Cloud data centre. This can be seen as virtualisation of the RAN, where RAN functions are now executed on virtual machines in the Cloud. This can make the network more flexible with additional temporary coverage and capacity (for example, for a large music event) can be easily configured and removed. One of the potential issues in Cloud RAN can be that the end-to-end latency levels can be too high for certain 5G vertical applications like V2X (which we will discuss in Section

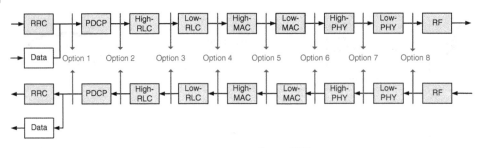

Figure 5.33 The 3GPP specified C-RAN split points. (*Source* [27]).

12.2.1), due to the large distances signals need to travel from the radio units to the Cloud BBU. A possible solution is to draw on the mobile edge computing (MEC) schemes to create an 'edge Cloud' and configure the RAN processing closer to the physical radio nodes and the CN components. We will also look at MEC at the end of the next section, in terms of an edge node to run some of the CN functionalities.

A more recent development in this sphere is the Open-RAN concept, which pushes for the ability to create a fully open, virtualised and inter-operable RAN also with the use of Artificial Intelligence (AI). Although cellular standards are meant to make the RAN open and inter-operable, in reality some components, especially the interfaces between different functional modules contain proprietary software which makes this almost impossible. The O-RAN alliance with a number of main operators and vendors [28], is at the forefront of Open-RAN design. They have developed an Open- RAN architecture, which contains a virtualisation layer, open interfaces (including Fronthaul), an open software reference design, open white box hardware (radio units) and an AI-driven RAN Intelligent Controller. We are likely to see a significant push for Open-RAN based 5G deployments in the near future. One such network, by Rakuten in Japan, is already taking shape and we will cover this in Section 12.1.

5.9.2 NFV (Virtualised Network Functions) and SDN (Software Defined Networking) Concepts

The NFV and SDN are now seen as key technology disruptors that will shape 5G and beyond cellular networks in the years ahead. NFV development is pioneered by the ETSI NFV Industry Specification Group (ISG) and aims to leverage the standard IT virtualisation technologies in mobile networks [29]. In basic terms, NFV means the ability to execute network functions in virtual machines (VM) anywhere (including in the Cloud) created on COTS platforms. Thus, NFV can replace the need for dedicated, specialised hardware running their own proprietary software in the network. The software-oriented programmable functions can be installed, operated, updated and deleted in any of the standard virtual machines. A detailed account of NFV and SDN, as applicable to 5G can be found in [30].

The SDN concept is complementary to NFV and concerns network control. A basic tenant of SDN is the separation of the network data forwarding plane from the control plane. This makes the network control functions directly programmable. SDN provides a logically centralised view of the network to the applications above. This gives the agility for the network administers to dynamically change network attributes to suit the application types and traffic flows. SDN is also based on open standards, and allows the option to build network control functions in a vendor neutral manner. The Open Networking Foundation (ONF) laid the foundations for SDN development and continues to develop new SDN solutions to suit the next generation of mobile and fixed networks [31]. In 2012, the ONF introduced OpenFlow specifications to enable SDN-based network control. With the developments of SDN, the specifications have also evolved, with ONF introducing CORD (Central Office Re-architected as a Data centre) specifications in 2017, which enables the building of agile data centres at the edge of the network (the 'edge' is detailed in a later paragraph) utilising NFV, SDN and Cloud technologies.

Through NFV, the network entities that carry out the control and data forwarding functions can be virtualised to run anywhere in the Cloud. With SDN, these dispersed virtual resources can be brought under a single policy or control framework and adapted to rapidly changing environments. There are a number of possible architectural implementations of the NFV and SDN currently discussed by ETSI and other related organisations. Figure 5.34 illustrates some of the possibilities for the placement of the SDN controller, in a generic NFV implementation.

The NFV architectural framework is built up from three main components. The Virtualised Network Functions (VNF), the NFV Infra-structure (NFVI) and the Management and Orchestration (MANO) system. The VNFs are the software implementations of the network functions which are deployed on the NFVI. The NFVI consists of all computing, storage and networking resources. Physically, the NFVI can span several

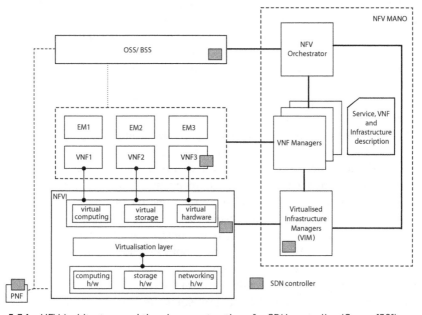

Figure 5.34 NFV Architecture and the placement options for SDN controller. (*Source* [32]).

locations. The virtualisation layer in the NFVI uses the hardware resources to generate the necessary virtual computing, storage and networking capabilities which, in turn, are mapped on to VNFs. The NFV MANO has the overall management and orchestration functionality. It consists of an NFV Orchestrator which communicates with the network's OSS and BSS (Operations and Business Support Systems) the VNF managers and VI (Virtualised Infrastructure) Managers. The OSS and BSS are not unique to NFV architecture, they are also present in traditional networks. The service, VNF and Infra-structure descriptions, for example, are fed to the MANO so it can configure the VNF and NFVI (through the respective managers) to achieve the required service levels.

Figure 5.34 also notes the possible options where the SDN controller can be placed in this architecture. The options include the SDN controller to be merged with the VIM in a way they are fused together, implement the SDN controller at the VNF, NFVI or the OSS. The SDN controller can even be installed separately as part of a Physical Network Function (PNF), although this option does not yield some of the benefits we will discuss below.

These two concepts combined together with the Cloud computing capabilities are defining the future directions for mobile CNs. The benefits to be achieved through NFV and SDN are highly significant. The ability to replace dedicated hardware with standard off-the shelf platforms running software-oriented functionality reduces the costs significantly. The networks can be scaled up or down very easily as the virtual machines (VM) can be added or removed effortlessly to function within the SDN control. The virtualised networks are also highly agile, with SDN enabling network administrators to adjust and manage the traffic flow through the network dynamically to meet the changing traffic patterns and demands. In traditional networks, as we will see in Chapter 6, the physical networks (both the RAN and the core) are always designed to meet the peak traffic demands, which will occur only for a fraction of the time in daily traffic patterns. This methodology of over-dimensioning the networks is wasteful for majority of the time, when the networks are not at their peak demands. The emerging standards framework for SDN will enable inter-operability between different functions, interfaces and devices without tying the operators to a single vendor with its own proprietary protocols. This will give more choice and flexibility in network design for the operators, driving down costs. We will have a more detailed look at this changing landscape in Chapter 12.

Looking back at Figures 5.16 and 5.27, you will recall that the 5G CN architecture is designed with specific functions and a separation of the user (data forwarding) and the control planes. This architecture renders itself very easily implementable with SDN and NFV. Perhaps the greatest benefit of such an implementation will be the ability to create end-to-end slices to support 5G verticals. As we will see in Chapter 12, each of these emerging 5G verticals have highly distinctive requirements in terms of the technical KPIs from the network. The network slices would share the same physical infra-structure but can logically be configured to support these distinct KPIs. Some of these logical networks will only be needed for a limited time period (for example, emergency networks in times of a serious event) and this NFV- and SDN-based architecture allows the creation, execution and dismantling of such network slices easily.

One of the potential issues in executing some of the network functions remotely in the Cloud would be the latencies it will add to processing times. For some applications in the URLLC domain requiring very low latencies (like vehicle-to-vehicle communications, which we will discuss in Section 12.2.1) and in mobile-gaming applications, this latency would adversely impact performance. So a related virtualisation concept was developed to enable some of the time critical processes to be executed nearer the physical edge of the network, called Multi-access Edge Computing (MEC), also formerly known as Mobile Edge Computing. The MEC servers offer the opportunity to deploy applications, services and content closer to the end users which can dramatically reduce the service response times. There is an on-going ETSI work on MEC, with the MEC ISG (Industry Specification Group) developing future standards for MEC. A useful white paper on how MEC will influence 5G can be found in [33].

5.10 Conclusions

In this chapter, we have carried out a detailed analysis of the mobile network and how it has evolved over cellular generations, from 2G to 5G. We looked at the main components of the mobile network, the RAN and the core, and their main generic functionalities. Also, the concept of the protocol stack was introduced with examples of how the stack is configured and the distinct functionalities of each of the layers. Across cellular generations, we have looked at how the protocol stack and the RAN and core features have changed to cater for increasing complexities brought on by evolving user requirements.

We have focussed on the 3GPP developed standards throughout this chapter, starting from 2G GSM to 5G-NR. 3GPP has been the most successful standardisation body by far and has cemented its dominance in 5G as the sole global 5G standard developer. In studying through the developments in each of the generations, some of the clues for this success of 3GPP can be understood. 3GPP standardisation is a continuing process, adapting to the needs of the cellular industry and the consumer. 3GPP has been able to strike a fine balance between providing incremental changes in releases within a certain generation and also, at the same time, doing the ground work for more substantial changes that will formulate the next generation. Certain features that have been studied and then introduced in releases of one generation have become indispensable in subsequent generations. A prime example is the dual connectivity feature introduced for support of small cells in 4G Release 12 to become indispensable in the NSA (non-stand-alone) configurations of 5G-NR.

The mostly evolutionary nature of the standards has many technical and commercial reasons. Technically, mobile networks perform vital functions in many spheres of modern life, and introducing any new functionality should not disturb the stability of the networks. Hence, new features are added after thorough testing and validation trials. Commercially, the mobile network operators underwrite huge investments in deploying these extensive infra-structure. They need to see good returns on the investments and any additional costs in modifying or replacing the networks should be based on sound business cases. At the end of the day, mobile users do not really care about the technology driving their mobile phones but the quality of their experience. So far, the incremental and evolutionary changes in the 3GPP technologies have changed the user experience positively and there is little reason to deviate from this tried and tested method.

A relevant question on the mobile network evolution is 'have some networks become redundant and can be taken down?' This question is more pertinent in terms of 2G and 3G.

In fact, some operators in technologically advanced countries have already taken down 2G and 3G networks. However, it is clear that many other countries still have a large subscriber pool (based on their older mobile phones and the level of subscription they are willing to pay) still relying on 2G and 3G networks. From a technology point of view, many of the older generations still serve a unique purpose such as 2G for remote area voice coverage and 4G for wider area mobile data and providing voice through VoIP. The 3G networks technologically seem to be doing a bit of both and hence can be seen as a likely candidate for redundancy. Commercially, each of these mobile generations have developed complex ecosystems around them from vendors to operators and different types of users. Some of the business users (like in telemetry use cases) have long-term contracts running for decades and will make certain networks (like 2G for telemetry) need to operate for some more years.

Turning our minds to the future, we will undoubtedly see a 6G with new and exciting features. A brief overview of what we believe as salient features of 6G is presented in the final chapter, Chapter 12. 6G will again capture the needs and demands of that particular time. Yet, some of the driving principles in managing change in the mobile networks in an evolutionary manner will no doubt remain.

References

[1] Sauter, M. (2017). *From GSM to LTE-Advanced Pro and 5G: An Introduction to Mobile Networks and Mobile Broadband*, 3e. Hoboken, NJ: John Wiley Publishers.

[2] Valdar, A. (2006). *Understanding Telecommunications Networks*. London: The Institution of Engineering and Technology.

[3] Zimmerman, H. (2007). OSI reference model: The ISO model of architecture for open systems interconnection. In: *The Best of the Best: Fifty Years of Communications and Networking Research* (ed. W.H. Tranter, D.P. Taylor, R.E. Ziemer et al.), 599–606. Ottowa, ON: John Wiley.

[4] Russell, A.L. (2013). OSI: The internet that wasn't. https://spectrum.ieee.org/tech-history/cyberspace/osi-the-internet-that-wasnt (accessed 16 June 2021).

[5] Korhonen, J. (2014). *Introduction to 4G mobile communications*. Boston, MA: Artech House Publishers.

[6] Perez, A. (2012). *Mobile Networks Architecture*. Hoboken, NJ: Wiley.

[7] Kukushkin, A. (2018). Global System Mobile, GSM, 2G. In: *Introduction to Mobile network engineering: GSM, 3G-WCDMA, LTE and the Road to 5G* (ed. A. Kukushkin), 59–102. Hoboken, NJ: Wiley.

[8] Halonen, T., Romero, J., and Melero, J. (eds.) (2003). *GSM, GPRS and EDGE Performance: Evolution Towards 3G/UMTS*, 2e. Hoboken, NJ: Wiley.

[9] ITU (2003). Deployment of IMT-2000 systems. https://www.itu.int/dms_pub/itu-r/opb/hdb/R-HDB-60-2003-PDF-E.pdf (accessed 16 June 2021).

[10] Toskala, A. (2004). Physical Layer. In: *WCDMA for UMTS; Radio Access for Third Generation Mobile Communications*, 3e (ed. H. Holma and A. Toskala), 99–148. Hoboken, NJ: Wiley.

[11] GSMA (2012). Remaining Relevant with Rich Communication Suite. https://www.gsma.com/futurenetworks/wp-content/uploads/2012/10/RemainingRelevantwithRCSInteropTechnologies-January2012.pdf (accessed 16 June 2021).

[12] Camarillo, G. and Garcia-Martin, M.A. (2008). *The 3G IP Multimedia Subsystem (IMS): Merging the Internet and the Cellular Worlds*, 3e. Hoboken, NJ: Wiley.

[13] Hunukumbure, M., Allen, B., and Beach, M.A. (2007). Code orthogonality for wideband CDMA systems with multiple transmit antennas. *IEEE Trans Veh Technol* 56(6): 3749–3756.

[14] Holma, H. and Toskala, A. (2011). *LTE for UMTS, Evolution to LTE-Advanced*, 2e. Hoboken, NJ: Wiley.

[15] ITU (2012). ITU global standard for international mobile telecommunications 'IMT-Advanced'. https://www.itu.int/en/ITU-R/study-groups/rsg5/rwp5d/imt-adv/Pages/default.aspx (acccessed 16 June 2021).

[16] Olsson, M., Rommer, S., Mulligan, C. et al. (2009). *SAE and the Evolved Packet Core: Driving the Mobile Broadband Revolution*. Cambridge, MA: Academic Press.

[17] Sesia, S., Toufik, I., and Baker, M. (2011). *LTE - the UMTS Long Term Evolution: From Theory to Practice, 2e*. Hoboken, NJ: Wiley.

[18] Nakamura, T., Nagata, S., Benjebbour, A. et. al. (2013). Trends in small cell enhancements in LTE advanced. *IEEE Commun Mag* 51(2): 98–105.

[19] ITU-R (2015). IMT Vision – Framework and overall objectives of the future development of IMT for 2020 and beyond. https://www.itu.int/dms_pubrec/itu-r/rec/m/R-REC-M.2083-0-201509-I!!PDF-E.pdf (accessed 16 June 2021).

[20] 3GPP (2017). First 5G-NR specs approved. https://www.3gpp.org/news-events/1929-nsa_nr_5g (accessed 16 June 2021).

[21] GSMA (2019). 5G implementation guidelines. https://www.gsma.com/futurenetworks/wp-content/uploads/2019/03/5G-Implementation-Guideline-v2.0-July-2019.pdf (accessed 16 June 2021).

[22] Ahmadi, S. (2019). *5G NR: Architecture, technology, implementation, and operation of 3GPP new radio standards*. Cambridge, MA: Academic Press.

[23] 3GPP (2017). Archive. https://www.3gpp.org/ftp/Specs/archive/38_series/38.300 (accessed 16 June 2021).

[24] Heavy Reading (2017). Service-Based architecture for 5G core networks. https://img.lightreading.com/downloads/Service-Based-Architecture-for-5G-Core-Networks.pdf (accessed 16 June 2021).

[25] Rommer, S., Hedman, P., Olsson, M. et al. (2020). *5G core networks: Powering digitisation*. Cambridge, MA: Academic Press.

[26] CPRI (2019). Industry leaders releasing the new eCPRI Specification for 5G – eCPRI V 2.0 with additional functionality for interworking. http://www.cpri.info/press.html (accessed 16 June 2021).

[27] 3GPP (2016). Archive. https://www.3gpp.org/ftp/Specs/archive/38_series/38.801 (accessed 16 June 2021).

[28] O-RAN (2019). Website. https://www.o-ran.org (accessed 16 June 2021).

[29] ETSI (2014). NFV one year later. https://www.etsi.org/newsroom/blogs/entry/nfv-one-year-later? (accessed 16 June 2021).

[30] Zhang, Y. (2018). *Network Function Virtualization: Concepts and Applicability in 5G Networks*. Hoboken, NJ: Wiley.

[31] Open Networking Foundation (2021). Software-Defined Networking (SDN) Definition. https://www.opennetworking.org/sdn-definition/?nab=1 (accessed 16 June 2021).

[32] ETSI (2015). Network functions virtualisation (NFV); Ecosystem; Report on SDN usage in NFV architectural framework. https://www.etsi.org/deliver/etsi_gs/NFV-EVE/001_099/005/01.01.01_60/gs_NFV-EVE005v010101p.pdf (accessed 16 June 2021).

[33] ETSI (2018). MEC in 5G Networks. https://www.etsi.org/images/files/ETSIWhitePapers/etsi_wp28_mec_in_5G_FINAL.pdf (accessed 16 June 2021).

6

Basics of Network Dimensioning and Planning

In this technical chapter, we will look at how a mobile communication network is dimensioned and planned, to provide services to a target population. Network dimensioning is the process where the number of cells needed to provide coverage and capacity to a certain area/population is estimated. The network planning process follows afterwards, where the exact placement of cell sites is determined on a geographic (and possibly a demographic) map of the area. In this chapter, the cellular fundamentals are more explained in the dimensioning process, whilst the planning capabilities in modern planning software tools are explained in the radio planning section. We will focus more on the dimensioning process, to understand how coverage and capacity is provisioned in radio access network (RAN) as well as in the backhaul (and fronthaul) transport network. A more detailed account of the cellular network planning process can be found in [1].

There are two basic aspects related to mobile network dimensioning/planning; one to provide signal coverage throughout the target area and the other to provide the aggregated data rates (or the capacity) demanded by subscribers in the target area. Depending on which aspect is prioritised, coverage-limited networks or capacity-limited networks can be generated. Typically, in urban areas, the priority is to support the high demand in capacity, giving rise to capacity-limited networks. In rural areas on the other hand, the capacity demand is not so high, yet users are sparsely spread over a wide area. This gives rise to coverage-limited networks as the priority is to provide coverage across the area.

Before moving onto the technicalities of network dimensioning and planning, we will first explore some of the basic principles governing radio signal propagation. There are three properties that dictate if the wanted message can be properly received at a given location in the network. These are the Signal strength, Noise power and Interference level. The signal strength is the received signal power of the wanted signal, which is confined to a radio resource in time, frequency and space (generally as per the serving cell of the user). The noise is an unwanted electrical signal generated by the thermal and electrical properties of the materials used in the radio circuits. The interference is also an unwanted signal, which comes mainly from the re-use of the same radio resource to support other users in other neighbour cells. Interference can also come from the use of neighbour frequencies, where unwanted out-of-band emissions can spill into the frequency band of concern. We

The Technology and Business of Mobile Communications: An Introduction, First Edition.
Mythri Hunukumbure, Justin P. Coon, Ben Allen, and Tony Vernon.
© 2022 John Wiley & Sons Ltd. Published 2022 by John Wiley & Sons Ltd.

will briefly look at the physical laws that govern the variability of these properties and then investigate how they are modelled in network dimensioning.

6.1 Properties of Signal Strength, Noise and Interference

As noted before, mobile radio signals propagate as any other electro-magnetic radiation. It does not need a medium to travel. For a simplistic analysis, we can assume we construct an omni-directional, point-source transmitter and allow free-space propagation. The high MHz and the low GHz frequencies (500 MHz to 2.6 GHz) where we are typically interested in for LTE mobile communication systems, there is no radio-frequency (RF) signal absorption effects by gases or water vapour in the air. Thus, for all practical reasons, open-air propagation can be treated as free-space propagation. For this hypothetical experiment, we transmit the signal through an omni-directional point source and detect the signal at another point, distance 'd' away from the transmitter. This is depicted in Figure 6-1.

If the signal is transmitted with power P_T at the point source, this propagates spherically in all directions from the source. This is called an isotropic antenna source, with unity antenna gain. At a distance 'd' the power P_T is spread (equally) across an area representing the surface area of a sphere with radius 'd', or an area of $4\pi d^2$. A receiver placed at distance 'd', would thus receive a power density P'_R of:

$$P'_R = \frac{P_T}{4\pi d^2} \tag{6.1}$$

In free space, the electro-magnetic power of the transmitter does not degrade but is simply spread over a spherical area. We also consider an isotropic antenna at the receiver, with unity antenna gain. As the receiver can only capture a fraction of this power, the power density degrades to the relationship of $P'_R \propto \frac{1}{d^2}$. This degradation is called path loss (PL)

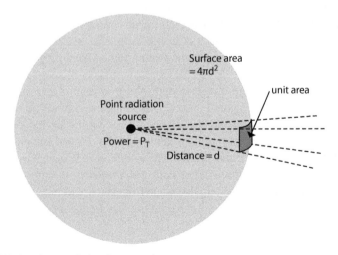

Figure 6.1 EM signal transmission from a point source.

and the power index on the distance term 'd' is called PL exponent (PLE) and is denoted by n. Thus, for free space, the PLE is n = 2. In real environments, such as urban or sub-urban areas, there are varying densities of buildings and other obstacles on the propagation paths, the n would be modelled as a markedly higher value. On the other hand, in scenarios like propagation through tunnels and street canyons, the radio signals are reflected and can concentrate back on to the receiver, making the effective PLE slightly less than two. In logarithm terms, the PL in free space can be expressed as $20.\log_{10} d$. You will see how this term is modified in real-world PL models later.

The frequency dependency of the free-space PL is explained with relation to the antenna aperture size of a received antenna. The aperture or the effective antenna area (A) is proportional to $\dfrac{\lambda^2}{4\pi}$, where lambda is the wavelength of the signal. Intuitively, this would imply that the higher the frequency, the smaller the aperture area of an antenna would be and as per the earlier relationship, the smaller fraction of the energy that would be captured. The frequency dependency on the received power can thus be illustrated as (with $\lambda = \dfrac{c}{f}$);

$$P_R = P_R'.A = P_R'\left(\frac{c^2}{4\pi f^2}\right) \tag{6.2}$$

Again the received power is degraded in an inverse square relationship, i.e. $P_R \propto \dfrac{1}{f^2}$. It should be emphasised that the reduction of received power due to higher frequency is **not** due to any frequency related attenuation effects in free space. This is simply because of the need to maintain unity gain in the antennas and higher frequencies imply smaller unity gain antennas to capture the energy. In logarithmic terms, the frequency dependency on free-space PL can be expressed as; $20.\log_{10} f$.

The received signal power, after combined free-space PL with distance and frequency effects, can be written as shown below. This also includes any antenna gains at the transmitter (G_T) and receiver ends (G_R). This is widely known as the Friis transmission formula [2].

$$P_R = G_T.G_R.P_R'.A = G_T.G_R.\left(\frac{P_T}{4\pi d^2}\right)\left(\frac{c^2}{4\pi f^2}\right) \tag{6.3}$$

The PL components in (3) can be written in logarithmic terms as:

$$FSPL_{dB} = 10.\log_{10}\left(\frac{4\pi df}{c}\right)^2$$

$$FSPL_{dB} = 20.\log_{10}\left(d_{km}\right) + 20.\log_{10}\left(f_{MHz}\right) + 32.45 \tag{6.4}$$

This expression is valid when the distance is measured in kilometres and the frequency is measured in mega-Hertz. The free-space PL (FSPL) is given in Decibels (dB), the common measure for calculations in network planning. We will later analyse how these values are changed when the real-world propagation conditions are considered in practical PL models.

Now, we can analyse the second of the parameters, the noise. If we assume a single transmission point operating at frequency f in free space, the signal strength at the receiver should degrade as per the PL equation above. At what point does it become undetectable? To work this out, we have to consider various noise components that add up in the system.

The fundamental noise component in all electrical systems is called Thermal noise. This is generated when electrons are randomly agitated in a conductor. As indicated by its name, thermal noise is proportional to the operating temperature. The generated noise power is also linearly proportional to the operating bandwidth. Thermal noise is called a type of 'white noise' because the noise level (or spectral density) across the frequencies remains the same, just as the colour white is made up of equal amounts of other colours (or frequencies in the visual EM spectrum). The noise power N can be expressed as:

$$N = k_B TB \tag{6.5}$$

The term k_B is the Boltzman constant $= 1.38 \times 10^{-23}$ Joules/Kelvin. The term T refers to the operating temperature, most commonly taken as 27^{0}C or 300 K (Kelvin). B refers to the operating bandwidth. At 1 Hz bandwidth, the Noise power works out to be N $= -174$ dBm. The unit dBm is Decibels with reference to 1 mW of power. This -174 dBm is typically considered as the noise floor of the system, onto which other noise components are added. It is easy to calculate the noise power at a given bandwidth with the following expression;

$$N_{dBm} = -174 + 10.\log_{10}(B_{Hz}) \tag{6.6}$$

For example, in a system with 10 MHz bandwidth, the thermal noise power would be N $= -174 + 70$ dBm $= -104$ dBm. Hence, the wider the bandwidth, the more thermal noise would be added to the system.

All receiver chains add some amount of noise into the wanted signal. The ratio of noise power at the output of a device to the input noise power (which can be attributed to thermal noise) is called the Noise Factor of the device. In Decibel terms, this is called the Noise Figure (NF). The particular noise level at the output of a device is thus given by the addition of NF onto the Thermal noise levels.

$$N_{dBm} = -174 + 10.\log_{10}(B_{Hz}) + NF \tag{6.7}$$

The devices of superior quality add little noise and thus have lower NFs. Most of the devices we consider are made up of chains of RF components (like mixers, feeders and amplifiers). The overall noise factor (OF) of such a chain is given by the Friis formula [3];

$$OF = F_1 + \frac{F_2 - 1}{G_1} + \frac{F_3 - 1}{G_1 G_2} + \ldots \tag{6.8}$$

Where F_1 is the NF of the first component the incoming signal encounters and G_1 is the signal gain (or amplification) by this component. Clearly, the overall NF of the chain is heavily dependent on the NF and the gain of the first component of the chain. Hence, it is customary to position a low-noise amplifier (LNA) with a very low-noise factor and high gain at the front end of the receiver RF chain. In the base station, a relatively new design concept is to place a tower-mount amplifier (TMA) as close as practically possible to the antennas. This can bring down the overall NF of the base station to around 2.5 dB. The mobile station components are not of as high quality (due to cost and size considerations) as the base station, hence the typical NFs of the mobile stations are around 7–8 dB.

The total noise power at the receiver is given by N_{dbm} in equation (7), but how does it relate to the signal level? For a particular type of modulation and coding for the data bits, a certain Signal to Noise ratio (SNR) value has to be maintained. This is usually given in Decibels. Thus, the minimum signal level needed at the receiver can be expressed as;

$$Smin_{dBm} = N_{dBm} + SNR_{dB} \qquad (6.9)$$

We will return to these SNR levels in the link budget calculations and also when dimensioning the capacity provision from a cell.

Interference is generated when the same time/frequency resources are utilised in a neighbour cell to support another user. We call this resource collision, especially for OFDM systems. Considering radio signal attenuation effects, interference is greatest at the cell boundaries, usually called cell edges. For a given user at the cell edge, the wanted signal is often weak, due to the high PL experienced when covering a longer distance. If there is interference from the neighbour cell, it could easily be as strong as the wanted signal. If a cell edge demarcates boundaries of multiple cells, the combined interference can be stronger than the wanted signal. Hence, it is not uncommon to see Signal to Interference levels of less than unity (SIR < 1), or in logarithm terms SIR < 0 dB.

The interference effects are usually added as a margin to the link budget, where the link budget is explained in the next section. What this means in effect is that the signal level at the receiver should be higher by this additional margin in order to counter for the effects of interference. The typical effect of adding a margin is to make the cell size smaller, hence the signal strength at the cell edge has more probability to meet the required level.

The interference margin is defined to be the 'noise rise' above the noise level. Thus. the interference margin I_m is defined as (in logarithm terms).

$$I_{m(dB)} = 10.log_{10}\left(\frac{N+I}{N}\right) \qquad (6.10)$$

This could also be expressed as the difference in SNR and SINR levels in Decibels.

$$I_{m(dB)} = SNR_{dB} - SINR_{dB} \qquad (6.11)$$

We have, so far, looked at the fundamental behaviour of signal, noise and interference power. Both the signal and interference powers vary upon the PL, which we showed to be a function of distance and frequency (for a unity gain antenna aperture size). The noise is modelled as an RF system generated property and does not depend on radio propagation.

6.2 The Link Budget and Coverage Dimensioning

The link budget is an estimation of the power gains and losses in provisioning radio links. The final aim of the link budget is to estimate the distances that can be covered by radio transmissions both in the downlink (from the base station to a user device) as well as in the uplink (vice-versa). The power gains come from the transmit power levels, the antenna directivity gains and specific gains related to each technology like LTE or UMTS that is being deployed. The losses are many; the main one being the PL from the transmitter to the receiver. Other losses include the various margins and additional losses we have to account for to counter for various radio impediments. We will study the link budget components in detail in this section and also demonstrate how to build a link budget for an LTE system.

6.2.1 The Transmit Power

The transmit power is measured at the output of the power amplifier, where the signal is then fed to the transmit antennas. This is usually depicted in dBm (Decibel levels w.r.t. a 1 mW of power). For larger macro-cell deployments, it is common to find power levels in the order of 40–46 dBm (or 10 W to 40 W in linear scale) at the base stations. This is the power level that drives the transmit signal for the down link. The handset or mobile device transmit powers are distinctly lower, around 23 dBm or 0.2 W, driving the uplink transmit signal.

6.2.2 The Antenna Gains

In the previous section, we looked at how isotropic antennas relate to PL. By definition, isotropic antennas radiate equally across all directions in the 3D space. With directional antennas, that energy is concentrated into a certain direction, giving more gain in that direction. Typical base station antennas are designed to cover only 120^0 in the azimuth plane. On the vertical plane, the directivity is even sharper, covering about 8^0 to 15^0. Thus, as a consequence of this directivity, the base station antennas have 17–18 dBi gain. The term dBi references to the isotropic antenna, i.e. how much antenna gain there is in relation to isotropic antenna. The mobile station antenna needs to be receptive at any angle, thus it is usually considered as omni directional. The antenna gain at the mobile device is thus taken as 0 dBi. The antenna gains are effective in both transmitting and receiving a signal, thus the base station antenna gains are applicable at both ends of a radio link.

The base station antenna directivity is useful, not just to extend the coverage footprint, but also to limit the interference. The interference mitigation originates from the way the antenna beam is pointed from the height of the antenna tower with a down-tilt, in the elevation plane. This is illustrated in Figure 6.2 below.

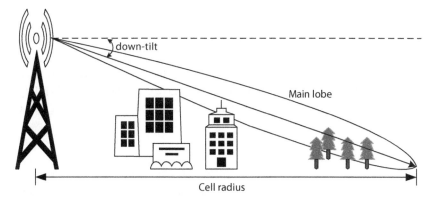

Figure 6.2 Cellular transmit tower antenna radiation pattern in the elevation plane.

The maximum gain direction (called the bore-sight) of the antenna in the vertical plane is directed towards the cell edge. The narrower elevation pattern of the antenna beam ensures that the effective signal (or interference to the neighbour cell) power is mitigated beyond the cell boundary by the combined effects of lower antenna gain and the distance-related PL. The antenna gain is similarly mitigated towards the inner areas of the cell. However, here, the distance-related PL decreases, hence the coverage is not negatively impacted as much as beyond the cell edge. In fact, this trade-off in the inner cell areas enables a smoother degradation of signal strength towards the cell edge from the cell centre.

6.2.3 Transmit and Receive Diversity Gains

Diversity schemes improve performance through replication or redundancy of data streams. We will study diversity schemes in detail in Chapter 9. In this link budget, we will include the generalised transmit and receive diversity gains for having two transmitters and two receivers. As noted above, this enables multiple signal paths across the radio channel, generating some gains which ultimately lower the SNR level which the system can operate on. To avoid double counting, the SNR values indicated in the link budget should thus be for a single antenna transmit and receive links (so called SISO), if the diversity gains are taken separately. Also, there is a limit on the total transmit power levels and the power per each transmit stream should be down scaled when considering transmit diversity.

6.2.4 The EIRP

The EIRP, or the Effective Isotropic Radiated Power is the combination of the transmit power and the antenna gains (for transmit antennas), minus any cable losses (from the power amplifier to the transmit antennas), if they are significant. EIRP represents the energy pumping out from the radio system in a desired direction. With the typical values, the EIRP for the base station stands around 60 dBm, whilst for the mobile station stands around 23 dBm. We will later examine how this disparity is compensated in the link budget.

6.2.5 Modelling the Path Loss

We discussed the free-space PL in the previous section. In real environments where cellular networks are deployed, conditions are much less benign than in the free space. The radio signals become attenuated, reflected or diffracted by various obstacles such as buildings or foliage. So, how can these be accommodated into a simple model? This is largely an empirical process, meaning fitting parameter co-efficients to best suit the observed PL values in a series of measurements.

6.2.5.1 COST –HATA model

A lot of field measurements and model fitting has been done over the years to derive simplistic PL models to cover a reasonable range of distances and frequencies. Perhaps the most utilised model is known as the COST-HATA model. This model has its origins in the Okumura model, proposed in 1968 on the basis of a series of field-trial measurements done in Tokyo. Then in 1980, Masaharu Hata simplified the considerations of the original Okumura model, to develop the Hata model [4]. These simplifications meant that the Hata model depended on only four basic parameters (the propagation distance, carrier frequency, base station height and mobile station height) and it could be applied only up to 1–10 km distance range and to a carrier frequency range of 150–1500 MHz. The COST 231 EU research project (1986–1996) further refined the Hata model to encompass a wider range of frequencies. The PL from the COST-HATA model can be expressed as follows [5]:

$$PL_{dB} = 46.3 + 33.9\log\left(f_{MHz}\right) - 13.82\log\left(hb_m\right) - a\left(hr\right) +$$
$$\left[44.9 - 6.55\log\left(hb_m\right)\right]\log\left(d_{km}\right) + C \qquad (6.12)$$

The notations f, hb, hr and d relate to parameters of carrier frequency, base station height, mobile station height and distance respectively, with their units noted in the subscripts. The term a(hr) is defined as;

$$a\left(hr\right) = \left(1.1\log\left(f_{MHz}\right) - 0.7\right)hr_m - \left(1.56\log\left(f_{MHz}\right) - 0.8\right) \qquad (6.13)$$

The term C is a correction factor depending upon the clutter type in the propagation environment. The typical values suggested in the model are:

$$C = \begin{cases} 3dB & - \textit{for metropolitan areas with high rise buidings} \\ 0dB & - \textit{for medium cities areas and suburban areas} \end{cases}$$

The COST-HATA model is recommended for application within the following value ranges [5].

Propagation distance – 1–20 km
Carrier frequency – 150–2000 MHz
Base station height – 30–200 m
Mobile station height – 1–10 m

Looking at the COST-HATA model, it is clear to see how the PL increments at a higher rate than the free-space PL for the factors of frequency and distance. The model also accounts for the fact that when the base station is higher, the less chance there is for the clutter (buildings and other obstacles) to block signal propagation to the receiver. This effect is captured as parameter a in the PL as a direct reduction as well as a co-efficient reduction for the distance-related term. It is thus crucial that the COST-HATA model is applied in situations where the base station antenna height is above the level of the clutter surrounding it. The height of the mobile station also has a negative impact on the PL, but this is much less significant than the impact of the base station height.

The popularity of the COST-HATA model comes from its flexibility to be applicable to a wide variety of parameter values (as depicted above) and to clutter environments. The clutter correction factor gives the additional flexibility of fitting this model to different environments with suitable adjustments to the factor. Although the model specifies an applicability limit at 2 GHz and minimum distance of 1 km, it is not uncommon to see the model applied to higher frequencies (up to around 3 GHz) and to lower cell radii.

6.2.5.2 SUI Model

The SUI (Stanford University Interim) model was a result of some extensive channel modelling work carried out by the Stanford University and the IEEE 802.16 WiMAX standardisation group. They focussed on developing channel and PL models for the Multi-point Microwave Distribution Systems (MMDS) in the 2.5–2.7 GHz range in the USA. The SUI PL models are generally applicable to mobile systems in suburban/rural environments up to 2 GHz carrier frequency.

The basic SUI PL model is given by the following

$$PL = A + 10\gamma \log\left(\frac{d}{d_0}\right) + s \tag{6.14}$$

The parameter d_0 is the reference distance, taken as 100 m. The constant A refers to the free-space PL at $d_0 = 100$ m, considering the distance and frequency related factors. Thus, A is defined as (where λ is the wavelength of the radio carrier);

$$A = 20log\left(\frac{4\pi d_0}{\lambda}\right) \tag{6.15}$$

The parameter γ can be regarded as an effective PL exponent (n) we discussed before. This value depends upon the propagation environment (terrain) and the base-station antenna height (h_b). The parameter γ is defined as

$$\gamma = a - bh_b + \frac{c}{h_b} \tag{6.16}$$

The co-efficients a, b and c have pre-defined values based upon the propagation environment or terrain type, which is categorised as A, B or C. The most challenging terrain is category A, defined as hilly with moderate to high tree densities and foliage. The most benign terrain is category C, which is flat terrain with light tree densities and foliage. Category B is

Table 6.1 The co-efficient values for the SUI PL model.

Co-efficient	Category A	Category B	Category C
a	4.6	4	3.6
b	0.0075	0.0065	0.005
c	12.6	17.1	20

terrain inbetween, either hilly terrain with light tree densities or flat terrain with moderate to dense tree densities. The co-efficients are defined as per the Table 6.1 above.

The parameter s in equation (14) is to compensate for shadow-fading effects. Shadow fading occurs when the signal path is blocked by buildings or trees/foliage, so the user experiences a long-term reduction in the received signal level. The specified value range for s is between 8.2–10.6 dB. We do look at margins in the next sub-section and the log normal fade margin (LNFM) is added to compensate for shadowing effects. Hence, if the SUI model is employed with the parameter s, the LNFM should not be added to the link budget, to avoid double counting of shadowing effects.

The specified value range for the BS antenna heights is 10–80 m and for the carrier frequencies is at or below 2 GHz. The UE antenna heights should be below 2 m. For carrier frequencies above 2 GHz and for UE antenna heights of 2–10 m, two correction factors are brought into the model. This generates the extended SUI model, which is not elaborated here. The reader is referred to [6] for a useful comparison on SUI and COST-HATA models.

6.2.5.3 The COST-Walfisch-Ikegami Model

When the transmitter and/or receiver or both are below the roof-top level, the dominant mode of propagation happens through multiple diffractions. The Walfisch-Ikegami model considers this fact and provides a PL model which considers building heights, street widths and building separation distances along the vertical plane linking the transmitter and receiver. The COST 231 project had generalised this model to be applicable to a wider range of parameter values [5]. The main difference from the COST-HATA model is that this model can be applied to base station antenna heights in the 4–50 m range, effectively covering below roof-top deployments. The model has two variants for LOS and NLOS conditions. The PL calculation for LOS is very simple, whilst for NLOS a detailed calculation is needed using the above building and street-related parameters. We will not present the full model here, but interested readers are referenced to [5] for the calculation formulae.

We have looked at three popular PL models, but there are many more being employed to suit different deployment conditions. Such a notable condition is indoor propagation, where more and more small cell (SC) networks are deployed. We will discuss SCs in detail in Chapter 10 and also cover some of the propagation aspects. As noted before, all models discussed are empirical models, where model parameters have been tuned to reflect the median PL values observed in a number of radio propagation measurements. When the radio transmitter is placed below the roof-top levels of local clutter, the geometry of the building (and other clutter) layout plays a significant part in determining the PL. The COST-Walfisch-Ikegami model gives some consideration to this building layout. With precise 3D models of indoor or outdoor layouts, ray tracing methods can be deployed to determine a 'heat map' of received signal power at any given point away from the transmitter. Ray tracing is a deterministic method, where the significant signal paths from the transmitter to receiver is 'traced' (or mapped) with the signal attenuations due to FSPL, reflections or diffractions along the way [7]. This can yield very accurate PL values at a multitude of receiver points of interest. However, the cost is the very high computational complexity. So Ray tracing techniques are generally applicable to only limited indoor and outdoor areas.

The PL is the main 'cost' item in the link budget. The other cost items are the various margins and losses we would add to ensure the signal is received across the cell, particularly at the cell edge. Additional margins and losses shrink the cell radii, ensuring better coverage for the reduced cell area.

6.2.6 Modelling the Log Normal Fade Margin

The LNFM accounts for the slow fading effects in cellular environments. The slow fading or shadowing effects are caused by signal blockage due to buildings and other obstacles in the environment. These shadowing effects are felt for longer time durations or distances and cannot simply be compensated through signal processing techniques. It is generally accepted that the logarithmic value of this slow fading follows a zero mean Gaussian distribution. In simpler terms, with this additional fading, the top of the 'bell' of the Gaussian curve lies around the median PL obtained by a PL model and more of the slow-fading components will be around this point. We can say that for 50% of the slow-fading occurrences, no slow-fading margin needs to be added to the PL and the received signal power will be at, or higher than the value given by the PL model. However, when we need a guarantee that for more than 50% of the instances the received signal power needs to be above a certain value, the LNFM needs to be added to the PL value and, accordingly, the received signal power should go down. This concept is illustrated in Figure 6.3.

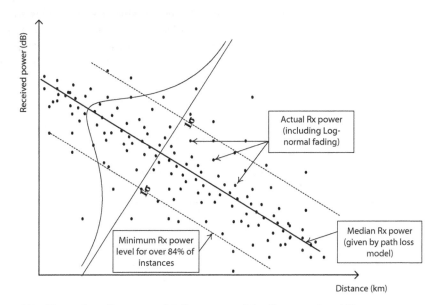

Figure 6.3 Illustration of Log-normal fading on top of the distance related PL.

Jake's equations [8] can be used to relate this slow-fade margin to the probability of guaranteeing a received signal level at a distance R from the base station (i.e. at cell edge for cell radius R). We will not go into the mathematics of these equations, but the essence of this seminal work is to relate the cell-edge coverage probability to the Gaussian probability distribution curve of LNFM. The area covered under the Gaussian curve relates directly to the cell-edge coverage probability.

Further mathematical analysis could relate the cell-edge coverage probability to the area coverage probability of the cell, which is usually greater. The standard deviation σ of the Gaussian (bell) curve determines the spread (or width) of the curve. σ is dependent upon the nature of clutter in the environment. For dense urban environments, where buildings are densely packed, it is customary to use higher σ values. For a lighter concentration of buildings in suburban or rural environments, a lower σ value is more suitable. In the absence of any specific clutter information, it is customary to use the value as $\sigma = 8$ dB.

Thus, for a given standard deviation (σ) of slow fading and a required cell-edge coverage probability R, the calculation of LNFM involves finding the inverse Normal value of the $N(0, \sigma)$ curve to satisfy probability R. Whilst this is seemingly a very detailed mathematical task, there are many software tools available to do this calculation. For example, using Microsoft Excel, the NORMINV(R, 0, σ) function would give the required probability value. The following Table 6.2 would give the LNFM values with a $\sigma = 8$ dB Gaussian distribution for a given set of cell-edge coverage probabilities. The corresponding cell-area coverage probabilities are also noted.

6.2.7   The FFM

The fast-fading effects are caused by the vector addition of various signal paths (known as multipath) from the transmitter to the receiver. As the receiver moves, this fast fading (as the name implies) causes a more rapid fluctuation of the signal strength. This effect (at the extreme) is also known as Rayleigh fading. The Fast Fade Margin (FFM) is an allowance to compensate for this type of fading effect. The use of diversity and interleaving techniques (as will be studied in Chapter 9), can combat the impacts of fast fading. If such techniques are in use, the FFM should be set accordingly. Typical values for FFM to be used in the link budget are around 3 to 5 dB.

Table 6.2   Cell edge and area coverage probabilities linked to LNFM.

Cell-edge coverage Probability	Cell-area coverage Probability	LNFM
50%	77%	0 dB
75%	90%	5.4 dB
85%	96%	8.3 dB
90%	97%	10.3 dB

6.2.8 Building Penetration Loss

Provision of in-building coverage is a main requirement for the cellular network. Research shows that up to 70% of voice and 80% of data connections originate indoors [9]. Although in-building radio coverage provision has developed as a specialised field with different options [9], we will look at the baseline scenario of outdoor cell sites having to provide this coverage. In this case, significant penetration losses can occur. Whilst the exact loss is dependent upon the wall thickness, building material etc., typically a common loss value (a type of margin) is added to the link budget to account for this. The common values used range from 15–20 dB. Similarly in-vehicle coverage also incurs a penetration loss, typically around 8 dB.

6.2.9 Building the Link Budget

Now that we have identified the essential components that will make up the link budget, it is time to build up an example one. We will consider dimensioning an LTE network in a dense urban area. It is customary to build the uplink and downlink link budgets side-by-side as most of the elements in the link budget effect both these links. The link budget is presented in Table 6.3 below.

Table 6.3 The Link Budget for and LTE Macro cell in dense urban clutter.

	Uplink	Downlink	Calculation
Transmitter			
Tx power	23	43	a
Antenna gains	0	18	b
Diversity gains	3	3	c
Cable/body losses	0	2	d
EIRP	26	62	e=a+b+c-d
Noise and signal level			
Thermal Noise	−174	−174	$f=10*\log(kT)$
System BW (MHz)	0.54	20	
Noise BW	57.32	73.01	$g=10*\log(BW_{hz})$
Rx Noise figure	2.5	7	h
Rx Noise Level (dBm)	−114.18	−93.99	i=f+g+h
SNR level	−7.5	−8.5	j
min signal level	−121.68	−102.49	k=i+j
Receiver			
Antenna gains	18	0	l
Cable/body losses	2	0	m

(Continued)

Table 6.3 (Cont.)

	Uplink	Downlink	Calculation
Rx diversity gains	3	3	n
Receiver gains	19	3	o=l-m+n
Margins			
Cell edge coverage pr.	85%	85%	
LNFM	8.3	8.3	p
Interference margin	2	2	q
FFM	3	3	r
Indoor penetration losses	18	18	s
Total Margins	31.3	31.3	t=p+q+r+s
MAPL	135.38	136.19	u=e+o+k-t
carrier freq (GHz)	2	2	
PL model	COST HATA	COST HATA	
BS antenna height (m)	30	30	
Coverage distance (m)	860	900	

The network bandwidth will be assumed to be 20 MHz. This will be the downlink bandwidth employed by the base station to serve multiple users in the radio frame. The uplink bandwidth employed by a single user, however, is much smaller. It is with the combination of multiple users that the uplink radio frame is built-up. We will assume three PRBs (Physical Resource Blocks) are utilised by an uplink user. With LTE numerology, a PRB consists of 12 sub-carriers spanning a total of 180 kHz. Thus, the three PRB allocation spans 0.54 MHz. As discussed above, the accumulated noise power corresponds to the used bandwidth. With the reduced bandwidth in the uplink this noise power is lower, and it counterbalances (to a certain extent) the reduced transmit power of the uplink.

Most of the components used in this link budget are discussed in a 3GPP technical report [10] in an LTE Release 11 context. Whilst we consider the uplink and downlink data channels in our example [10], lists maximum PL values for most of the physical channels in LTE.

The example link budget we have shown gives the cell radius for uplink and downlink in a dense urban environment. The smaller of the two will be the limiting cell radius and this should be adopted as the baseline cell radius in providing coverage for that environment. A typical deployment for dense urban and urban Macro cells is with three sectors, where a single-cell tower feeds sector antennas each covering 120° in azimuth. The coverage area (A) of a three-sector cell can be estimated as;

$$A = 1.95 * r^2$$

A simple division of the total area (of the dense urban environment type, for example) by this cell area A will give the number of sites required to provide coverage. If the cell tower is to provide omni-directional coverage (a single sector with 360° azimuth coverage), the regular area of a hexagon (A = 2.6*r²) should be employed.

We assume a dense urban environment with an area of 24 km² in our example, to be covered by three-sector cells. The total number of cell sites needed for coverage provision is thus

$$Number\ of\ sites\ for\ coverage = \frac{24km^2}{1.95*0.86^2km^2} \cong 17\ sites$$

The total area intended to be covered by the planned network usually consists of multiple types of propagation environments. If so, it is usual practice to develop separate link budgets for different environments. Usually, four propagation environments are studied, the Dense Urban (DU), Urban (U), Suburban (SU) and Rural (RU). In many planned deployments, these areas can be separated as contiguous segments. We saw how the COST-HATA and SUI PL models can be tuned to suit the propagation environments. Similarly, the standard deviation of slow fading and even the cell-edge coverage probability can be adjusted to suit the propagation environment and coverage requirements for a certain environment. This will yield a respective LNFM for the environment. The building (and general clutter) densities usually reduce in the order of DU, U, SU and RU. The indoor propagation loss values should also be adjusted accordingly in the respective link budgets. The separate link budget calculations will finally yield the respective site numbers required to cover given environments.

The provision of coverage for all four propagation environments is an option for Greenfield deployments, where a new entrant comes to the market and aims to provide an extensive newly deployed network. However, the majority of deployments will be by existing operators, where they would try to leverage their existing networks as much as possible. The usual deployment plan would be to provide the new technology (e.g.: LTE) at first to dense subscriber areas like DU and U and later expand to other environments. In such cases, limited coverage of the LTE cells need to be provided only to the targeted areas and effective hand-off procedures to legacy networks should be established. Also, many of the cell towers from legacy networks will be re-used to carry the antennas of new technology, limiting site and antenna mast rental costs. This will be a particular requirement in the radio planning stage but, for dimensioning, the same steps as above will be carried out.

6.3 Capacity Dimensioning

The basic idea of capacity dimensioning is the provision of a cellular network to satisfy the capacity demand of subscribers. Thus, there are two processes involved with capacity dimensioning. The first is to accurately identify capacity demand from subscribers. The second is to estimate the number of sites required to meet this capacity demand.

In this section, we will only look at the radio link (or air interface) capacity provision. The capacity demand also influences provision of backhaul capacity (which we will briefly look at in Section 6.4.2) and also the design of core network (CN) elements.

The estimation of capacity demand is quite a complex process and there is no single fixed method to achieve this. Different practitioners and literature have suggested slightly different procedures but all contemporary approaches encapsulate the following points:

- The long-term growth of subscribers/demand for services needs to be considered. Deploying a network is a major investment and returns do come over a longer period of time. The financial planners would have done calculations typically stretching over 5 years. Thus, the network needs to be designed to meet capacity demand at least over this period.
- Different subscriber profiles need to be considered, instead of looking at one average profile. Typically, two profiles are considered, that of business user and residential user. Business users generally demand more quality of service (QoS) and they are willing to pay more for this. The times when each type of subscriber is most active (called busy hours) are also different and this has a profound impact on capacity estimations.
- The network should be dimensioned/planned to meet closer to peak demand, not average demand. Usually in cellular networks, there is significant variation in demand during the span of a day. Only through dimensioning and planning networks to meet peak demand can operators ensure the QoS provision to subscribers at an acceptable level, when they are most active. Thus, the busy hours should be properly identified and capacity demand estimated based on these busy hours. However, designing networks to fulfil the absolute theoretical peak will be an overkill. We do have to take into account that not all projected subscribers will be active even at peak times. This will be accounted for in the over-booking factor (detailed later).
- More recent technologies like LTE offer different QoS classes with different guaranteed bit rate (GBR), packet loss and packet-delay parameters. User requirements should be mapped to these QoS classes and capacity demand estimation should be developed on this basis.

6.3.1   The Capacity Demand Estimation Process

We will demonstrate the capacity estimation process for a hypothetical new network deployed over a wide area covering the same dense urban (DU) region we evaluated in Section 6.2 (Table 6.4). The marketing teams would have estimated subscription growth over 5 years and the probable split of business and residential subscribers. There are several approaches to estimate probable subscriber numbers and we will take an approach involving subscriber growth as a percentage of population. The marketing teams also estimate the likely numbers of subscribers leaving the network (primarily using historical data) – this is called 'churn'. These two values help estimate the net subscriber growth on an annual basis.

The fact that all subscribers will not be active at the same time should be taken into account, through what is called the overbooking or activity factor. This is usually given as a fraction 1 in N, where the 1 specifies active subscribers over a given time over the total number N. It is customary to allocate a higher activity factor for business users than residential users, as business users demand a higher quality (and availability) of services.

Table 6.4 Capacity estimation for a Greenfield LTE network in dense urban clutter.

Capacity demand estimate-DU

Population	180000					
		Year 1	Year 2	Year 3	Year 4	Year 5
Subcriber penetration(%)		3.50%	5.00%	7%	8%	8.50%
Subcriber numbers		6300	9000	12600	14400	15300
Subscriber churn (%)		6%	6%	7%	6%	5%
Net subscriber numbers		5922	8460	11718	13536	14535
Business subscriber (at 70%)	70%	4145	5922	8203	9475	10175
Residential subscribers (at 30%)	30%	1777	2538	3515	4061	4360
Business overbooking factor	1/4					
Residential overbooking factor	1/8					
Active business subscribers		1036	1481	2051	2369	2544
Active residential subscribers		222	317	439	508	545

An example table is shown below for the DU region. Total population for the region is estimated at 180 000. This population value is not just based on the people living in the metropolitan area (which generates the main residential customer component) but also people who come into the city for work.

The next step in capacity estimation is to determine the services mix for the average residential and business subscribers and then convert this to average data rate requirement values. As noted above, the services in LTE are categorised as QoS classes. There are two basic service types defined in LTE, the GBR (Guaranteed Bit Rate) and non-GBR. The following Table 6.5 (from [11]) defines nine QoS classes and the associated services with them, under the broad GBR and non-GBR service types. Whilst we do not use the delay budget in our dimensioning process for the radio link, this delay budget involves end-to-end delay, typically from the UE to the Policy and Charging Enforcement Function (PCEF) in the CN for client/server applications (e.g.: ftp, e-mail, web) [11]. The radio link delay can be estimated by subtracting a pre-defined value for the delay between PCEF and radio base station. The packet-loss rates, on the other hand, are used to determine SNR levels at the receiver side needed to achieve these threshold BLER (Block Error Rate) values.

6.3.2 Capacity Demand Estimation – Worked Example

We will take up a simple hypothetical scenario where three of the example services are available for business and residential users, to enable us to derive capacity demand. These services will be VoIP (packet-based voice), real-time video streaming and a non-real time data service. We will assume these three services are used by both business and residential users, but in different flavours and compositions.

Table 6.5 The QoS classes in LTE and their features (Source [11]).

QCI	Resource Type	Priority	Packet Delay Budget	Packet Error LossRate	Example Services
1	GBR	2	100 ms	10^{-2}	Conversational Voice
2		4	150 ms	10^{-3}	Conversational Video (Live Streaming)
3		3	50 ms	10^{-3}	Real Time Gaming
4		5	300 ms	10^{-6}	Non-Conversational Video (Buffered Streaming)
5	Non-GBR	1	100 ms	10^{-6}	IMS Signalling
6		6	300 ms	10^{-6}	Video (Buffered Streaming) TCP-based (e.g. www, e-mail, chat, ftp, p2p file sharing, progressive video, etc.)
7		7	100 ms	10^{-3}	Voice, Video (Live Streaming) Interactive Gaming
8		8	300 ms	10^{-6}	Video (Buffered Streaming) TCP-based (e.g. www, e-mail, chat, ftp, p2p file
9		9			sharing, progressive video, etc.)

Table 6.6 Data rates for the selected LTE services.

Service	Packet size	Overhead	Base rate	Data rate
VoIP	20 Bytes	200%	12 kbps	36 kbps
RT video	1000 Bytes	-	2 Mbps	2 Mbps
Non RT data	1000 Bytes	-	500 kbps	500 kbps

The first step is to estimate the average data rates for all three services. As all three services are packet based, we will also consider the packet overheads such as headers and encryption. For VoIP, packet sizes are small and relative overhead for headers is very large (about twice the packet size). For packet-based data applications, this overhead is much smaller, about 2% in TCP for example, as the packet sizes are much larger. Hence, the header overhead will be neglected in packet-based data services. Table 6.5 gives data rates for each service. It should be noted that these data rates apply for early LTE networks and the user-experienced data rates have evolved significantly since then, but this does not impact the capacity dimensioning steps we will illustrate in this section.

The next step is to identify the busy hours for business and residential customers and profile their usage across the time of day. In this example, the time of day is profiled in two-hour slots. The busy hour for the business user falls in the 10 a.m. –12 noon slot, whilst the busy hour for the residential user falls in the 8–10 p.m. slot. An illustration of typical user activity distribution against time of day is shown in Figure 6.4 for both business and residential users, for this worked example.

We will show in this worked example, how the busy hour data demand is estimated. In a complete analysis, all time slots of the day should be considered for data demand. For the business user busy hour, it is estimated that, on average, 3.5 VoIP calls are received, each with an average duration of 3 minutes. We will estimate how much data demand is added by a single user over the two-hour (busy hour) period. For a single user, it does not seem to

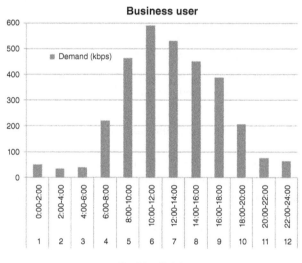

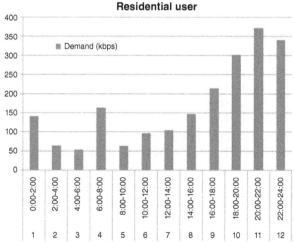

Figure 6.4 Traffic demands per user in business and residential segments.

make sense to 'dilute' the VoIP data rate over 2 hours. However, this is later expanded to all active users in that time slot, where the VoIP calls will be (roughly evenly) distributed across the two-hour period. So this is an initial step to come to that overall demand.

$$\text{The average VoIP demand per user} = \frac{36 * 3.5 * 3 * 60kb}{2 * 60 * 60s} = 3.15 \text{ kbps}$$

There will also be, on average, 6 RT video sessions each with 5-minute duration in the designated busy hour. The typical applications will be consumption of mobile video clips. This will mean that 30 minutes of the two-hour period will be utilised by the RT service for the average business user. We will again calculate the average data rate required to support this service for a given business subscriber.

$$\text{The average RT video demand per user} = \frac{2000 * 6 * 5 * 60kb}{2 * 60 * 60s} = 500 \text{ kbps}$$

Finally, there will be 5.2 non-RT data sessions during the busy hour, each with 4-minute duration. Typical business applications will be VPN e-mail and file transfer.

$$\text{The average non-RT data demand per user} = \frac{500 * 5.2 * 4 * 60kb}{2 * 60 * 60} = 86.67 \text{ kbps}$$

Thus, the total data rate requirement for the business busy hour would be 589.8 kbps, per business subscriber.

As we are looking at the total data rate demand during busy hours, the safer approach is to calculate both business and residential user activity at the two respective busy hours and look at the cumulative data rates. With the proliferation of office phones (work mobiles) and globally distributed business interactions, working out-of-office hours is fast becoming a norm for a growing percentage of the population. During the 8–10 p.m. residential peak slot, it is assumed that, on average, 1.2 VoIP calls, 0.6 Video call sessions and 1.4 non-RT data sessions will occur. For simplicity, we will also assume the durations of these sessions would be the same as in the business peak hour. Similar calculations to the above results in a total data rate of 74.4 kbps for business users in the 8–10 p.m. residential peak hour.

Next, we will look at the residential users' data consumption behaviours. It is generally known that residential users are more active in the evenings. In our example, we assume the 8–10 p.m. time slot as the busy hour for residential users. In pre-4G era, the usage was dominated by voice, but now the predominantly real-time and non-real time data services dominate the usage patterns. Popular examples of real-time data services for residential users include interactive gaming and video streaming. The non-real time data applications are dominated by web-browsing and peer-to-peer file sharing etc. The 'residential' category will also include users travelling home on public transport and consuming mobile data.

For simplicity, we will assume the average data rates for VoIP (which is accurate anyway), RT-data and non-RT data services remain the same but the activity times and numbers will vary. For VoIP, we will assume that, on average, 3.7 VoIP calls will be made, each lasting an average of 3 minutes. There will be 3.1 (on average) RT data sessions, each lasting 6 minutes. Finally, the average number of non-RT data sessions will be 2.8, each lasting 5 minutes.

$$\text{Average VoIP demand} = \frac{36 * 3.7 * 3 * 60kb}{2 * 60 * 60s} = 3.33 \text{ kbps}$$

$$\text{Average RT video demand} = \frac{2000 * 3.1 * 6 * 60kb}{2 * 60 * 60s} = 310 \text{ kbps}$$

$$\text{Average non-RT data demand} = \frac{500 * 2.8 * 5 * 60kb}{2 * 60 * 60s} = 58.3 \text{ kbps}$$

Thus, the total data rate requirement for the average residential user would be 371.7 kbps, for the busy hour of 8–10 p.m.

Similar to business users, we will look at residential data usage in the business user busy hour of 10 a.m.–12 noon. In the morning, office-hour residential activity is sparse. We will assume that 1.6 VoIP calls, 0.8 RT data sessions and 0.7 non-RT data sessions will take place on average, during this time, in residential user context. The average durations are assumed to be similar to peak hour times above, for the sake of simplicity. Similar calculations to the above will yield 1.44 kbps for VoIP, 80 kbps for RT data and 14.6 kbps for non-RT data. Total data rate requirement for the average residential user will be 96 kbps.

Now, with these average data rate requirements per user within the busy hours calculated, we can expand this to calculate the total data rate requirements for the total number of active subscribers. We will look at the subscriber numbers at the end of year 5, so this designed Greenfield network will be 'future proof' for the first 5 years. This calculation is shown in Table 6.7.

In a real dimensioning exercise, all time slots should be looked at for total demand and then the peak demand selected for radio resource provision. As you see in Figure 6.4 for busy hour illustrations, the sum of data demands can vary from the individual peaks. When considering other environments (U, SU, RU), usually the proportion of residential users will increment from DU to RU. This can create different busy hours for the combined traffic of business and residential users.

Table 6.7 Downlink data rate aggregate demand for the busy hour slots.

Time slot	....	10 a.m.–12noon	....	8 p.m.–10p.m.
Business user (BU) data rate (kbps)		589.8		74.42
Residential user (RU) data rate (kbps)		96		371.7
Active business users-Total	544			
Active residential users-Total	545			
Total data rate demand –BU (Mbps)		1500.45		189.27
Total data rate demand _RU (Mbps)		52.32		202.58
Maximum demand in peak hour (Mbps)		1552.77		391.85

6.3.3 Resource Provision – Worked Example

The second part of the capacity dimension example is the allocation of radio resources to meet the estimated demand. In network dimensioning and planning, we are in control of the spatial dimension of resource provision, i.e. how many cell sites should go in a design area. For this, we should first estimate the likely figure of data capacity a single LTE cell can provide. Then, assuming the users are uniformly distributed over the design area, we can estimate the number of cells needed.

The first step in capacity provision estimation is to determine what symbol rate an LTE data frame can support. In this example, we will assume an LTE system with 20 MHz bandwidth. In previous narrow-band generations like 2G GSM, the frequency resource was split and re-used according to a certain pattern amongst neighbouring cells. This avoided the use of the same frequencies in adjacent cells thus reducing interference. In the wideband generations of 3G and 4G, the whole spectrum is used by all cells, with advanced techniques for interference control. In this frequency reuse = 1 case, no explicit frequency planning is needed. The same LTE bandwidth is available for all cells in the network.

The LTE sub-frame (as we discussed in Section 5.8) is structured as in Figure 6.5, which shows the typical arrangement of resource elements allocated for data, control and reference signals.

An LTE sub-frame is made up of two slots in the time domain, each with seven symbols. In the frequency domain, a sub-frame contains 12 sub-carriers, and this time-frequency configuration is also called a physical resource block (PRB). Each of the time/frequency resource elements (168 in total) in the figure can be configured to carry data, control or reference signal symbols. In a typical arrangement, the first two columns (called LTE symbols) are dedicated for control signalling, and do not carry any data. Also, there are intermittent reference signals, which are used to estimate the channel at the receiver. So, out of

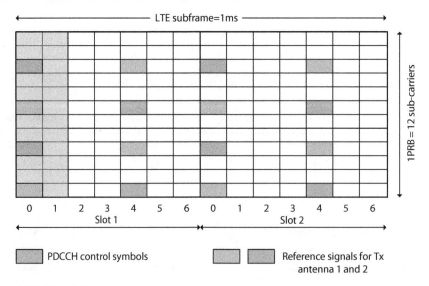

Figure 6.5 LTE sub-frame structure.

the 168 possible resource elements, 132 (168–12 x 3) elements can be used for data transmissions within a 1 ms sub-frame. In the 20 MHz BW LTE channel, there are 100 PRBs making up similar sub-frames, so the total number of resource elements is 13 200 (132 x 100) for the 1 ms sub-frame duration. As a symbol rate, this equals 13.2 Msymbols/s.

The actual bit rate a cell would support would depend on the modulation and coding scheme applied on the LTE resource element. As discussed earlier in Chapter 5, LTE (as well as other standards from 2.5G onwards) supports multiple adaptive modulation and coding (AMC). In LTE, 15 MCS (modulation and coding scheme) levels are supported [12], with the MCS levels allocated to each UE depending on the CQI (channel quality indicator) levels reported by the UE. The UE indications are based on measurements carried out by the UE on the downlink reference signals. Thus, the CQI's actually depend on both the SINR level experienced at a certain location and the quality of signal detection of each UE device.

However, in network capacity dimensioning, we assume a direct link between the SNR levels and the MCS (as the interference is separately added as a margin in the link budget). The higher the SNR, the higher the level of MCS can be applied, which carries more bits per symbol. For example, ½ rate QPSK would carry 1 data bit per symbol, ½ rate 16 QAM would carry 2 bits per symbol and ½ rate 64 QAM would carry 3 bits per symbol. In practice, the SNR level for a certain MCS will also take into account the QCI classes we discussed in Section 6.3.1.

The MCS levels used in this worked example are not the exact ones applied in LTE, but they will provide a simplified means of illustrating the capacity provision process. Whilst in dimensioning, we cannot ascertain all the MCS applicable across a cell, we can make some assumptions to determine a representative MCS for a region (or a band) of the cell with the average SNR values.

Remember that, in the link budget, we applied a certain SNR value to determine the cell radius. This would be the minimum SNR required to drive the lowest MCS level at the cell edge. To increase the MCS level, we have to ensure higher SNR values. This would, in turn, reduce the MAPL in the link budget and reduce cell radius. If we apply this process repeatedly, we would have co-centric rings in the cell, each defining a border beyond which a certain MCS will not have enough SINR to operate. The area (or the band) within two of these co-centric rings will be the region where a certain MCS will be supported. In this manner, we can work out the bands around a cell site, where a given MCS level will be supported.

In the link budget, we achieved a limiting cell radius of 860 m, limited by the uplink. Thus, in the downlink, a slightly higher level of MCS can be employed at the cell edge, to meet the higher link budget available. Table 6.8 demonstrates four example MCS levels (not the exact ones available in LTE), the relevant SNR levels, MAPL levels and the cell radii. We also include a MIMO-B version of ½ rate 64 QAM. MIMO-B is the spatial multiplexing version of the 2×2 multiple antenna systems considered here. When applied in MIMO-B mode, the data capacity per symbol can be doubled. Thus for ½ rate 64 QAM MIMO-B, 6 bits per symbol can be carried. However, the diversity gains associated with multiple receive and transmit antennas will no longer be applicable. This is enacted in the significantly lower MAPL for this option.

Table 6.8 Co-centric ring calculation (DL) for MCS levels.

MCS level	SNR (dB)	MAPL (db)	Cell radius (m)
⅛ QPSK	−7.69	135.38	860
½ QPSK	−0.5	128.19	540
½ 16 QAM	5.4	122.29	360
½ 64 QAM	11.3	116.39	250
½ 64 QAM − MIMO B	14.2	107.49	140

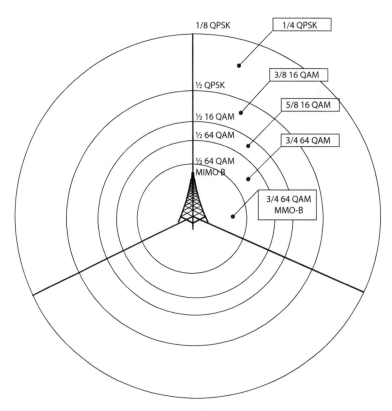

Figure 6.6 Co-centric rings and the applicable MCS levels.

In a complete dimensioning exercise with the MIMO-B option, it should be considered for all the MCS levels. Thus, Table 6.8 will also be inter-laced with MIMO-B related cell radii. We apply it only for the highest MCS here, for the simplification in illustrating the process.

The MCS levels shown here correspond to the MAPL levels at the edge of their respective co-centric bands. To find a representative MCS for the whole band, we consider the mean spectral efficiency (or the number of bits carried by a symbol) between the two band-edge values. Thus, a representative MCS corresponding to this mean value is allocated to the band. These co-centric bands with the allocated MCS levels are illustrated in Figure 6.6.

Table 6.9 The composite data rate calculation for MCS rings.

Symbol rate per cell (msym/s)	13.2		
Median MCS level	**Bits per symbol**	**% cell area**	**Data rates**
¼ QPSK	0.5	60.6	4
⅜ 16 QAM	1.5	21.9	4.34
⅝ 16 QAM	2.5	9.1	3
¾ 64 QAM	4.5	5.8	3.45
¾ 64 QAM – MIMO B	9	2.7	3.21
Total bit rate per sector (Mbps)			**18**

In a tri-sectored cell, we assume that each sector occupies one-third of the circular cell area, so the fraction of sector area occupied by each co-centric ring does not change.

We estimate the percentages of cell areas covered by each of the MCS options above, as a factor of the total cell area. Recall that we worked out a symbol rate of 13.2 Msym/s can be maintained per cell (or sector) for a 20 MHz LTE system. Now, assuming that users are distributed equally across the cell (or sector) area, we apportion the data rates that will be applicable at each of the co-centric areas. This apportioning is based on fractioning the total 13.2 Msym/s symbol rate as per the percentage areas of the co-centric rings. Table 6.9 illustrates the calculations and final data rate for the complete cell sector.

As we are considering tri-sectored cells in this example deployment (as is the case with most practical networks), the total cell capacity is 3 × 18 Mbps = 54 Mbps. Now, we can work out how many cells are needed for the dense urban area to satisfy capacity demand. Recall that we estimated a peak demand of 1552.77 Mbps for the business user busy hour. Simply dividing this peak demand with the capacity provided by one cell gives us the number of sites needed for capacity provision:

$$Number\ of\ cell\ sites\ for\ capacity = \frac{1552.77 Mbps}{54 Mbps} \cong 29\ sites$$

Comparing the number of cell sites to provide coverage and capacity, it is clear that the dimensioned network needs about 65% more sites to meet capacity demand. This will, in turn, reduce the cell radius than that planned with the link budget. The new cell radius can be calculated as

$$New\ cell\ radius = \left(\frac{24 km^2}{29 * 1.95} \right)^{1/2} = 0.65 km$$

When cell sizes reduce, MAPL reduces and a new set of higher MCS values will apply. This will, in turn, increase the new cell throughput. As the total capacity demand remains the same, the higher cell throughput will reduce the number of cells required to fulfil this demand. This will slightly increase the cell radius again. This is an iterative process and it converges to a certain number of cells, usually fairly quickly. As we are approximating (rounding up or down) the number of cell sites to the nearest integer, this iterative process can also result in the final site count oscillating between two successive integers. In this case, we select the higher number of cell sites as the final count.

As a rule of thumb, the mid-point between the number of sites given by the coverage and capacity analysis can be taken for the first iteration, to expedite the process. In this example, we will take 23 sites as that point and repeat the process shown in Tables 6.8 and 6.9 to work out what level of capacity 23 sites will provide.

With 23 tri-sector sites in the 24 km^2 coverage area, the cell radius (as per the above equation workings) would be 0.73 km. We will use this value to work out the co-centric rings and capacity partitions as per the previous tables. The resulting new values are depicted in Tables 6.10 and 6.11.

The throughput for the tri-sector cell will be 71.61 Mbps. Again, factoring in the 1552.77 Mbps capacity demand with this cell throughput would yield 22 cells as the requirement. The calculation will then oscillate between 22 and 23 sites. Thus, the conclusion will be to select 23, tri-sector sites to provide coverage and capacity to this dense urban area. It is always better to slightly over dimension rather than under dimension. In comparison, we can say that this network for the dense urban area will be capacity limited.

A similar capacity demand estimation and resource provision exercise should also be carried out for the uplink. Usually, capacity demand for the uplink is lower than for the downlink. Whilst the VoIP voice links would require the same capacity in both the uplink and downlink, the disparity comes from the data applications. Consumers are much more

Table 6.10 Co-centric ring calculation for MCS levels (2nd iteration).

MCS level	SNR (dB)	MAPL (dB)	
¼ QPSK	−5.24	132.93	730
½ QPSK	−0.5	128.19	540
½ 16 QAM	5.4	122.29	360
½ 64 QAM	11.3	116.39	250
½ 64 QAM – MIMO B	14.2	41.03	140

Table 6.11 The composite data rate calculation for MCS rings (2nd iteration).

Symbol rate per cell (msym/s)	13.2		

Median MCS level	Bits per symbol	% Cell area	Data rates
⅜ QPSK	0.75	45.3	4.48
⅜ 16 QAM	1.5	30.4	6.02
⅝ 16 QAM	2.5	12.6	4.16
¾ 64 QAM	4.5	8.1	4.81
¾ 64 QAM – MIMO B	9	3.7	4.4
Total bit rate sector (mbps)			23.87

likely to download data/content than to upload. The other factor to consider is that the capacity provision capability in the uplink radio frame is also limited. In LTE, the downlink utilises QPSK, 16 QAM and 64 QAM MCS schemes (as illustrated through our example above), but the uplink only utilises QPSK and 16 QAM due to the limited capabilities of the handsets. This significantly reduces capacity provision per radio frame in the uplink. The LTE uplink employs SC-FDMA compared to the OFDM in the downlink (please refer to Section 5.8), but the sub-frame structure (with 168 resource elements in a PRB with normal cyclic prefix) retains the same. Again, a complete assessment of the uplink demands and the capacity provision should be carried out to determine the number of cell sites required in the uplink. As in the downlink estimation, the higher number of sites should be selected as the limiting number of sites.

What we have shown here is a basic dimensioning process for LTE and there can be a number of limitations here. For example, we only look at the weekday traffic profiles and yet the weekend traffic profiles would be significantly different and may create unseen peaks in capacity demand. Incumbent operators can obtain more accurate forecasts of the traffic demand by analysing and extrapolating their current traffic data, particularly the Smartphone activity. We have studied a gradual capacity growth only with customer numbers yet, in real situations, there tends to be intermittent capacity demand surges, especially when new versions of Smartphone applications are released to the market. For example, when a new version of a popular video game or a new season of a popular video streaming TV series is released there tends to be capacity demand surges in mobile networks. However, this basic dimensioning example has explained the fundamental considerations in the link budget, capacity demand and capacity provision estimation in the RAN.

6.4 The Dimensioning of Backhaul Links

The backhaul links connect cell sites to the CN. It is essential to ensure the correct link coverage and capacity is provisioned in the backhaul, so that the data transfer from user devices to the CN is not impeded. The traditional backhaul methods all provide point-to-point links, from the base station (cell) site to the aggregation point in the RNC and then to the CN in 3G or directly to the CN entity in 4G LTE. Whilst the placement of this aggregation point has changed over the cellular generations, the basic requirements of the backhaul links are to provide the capacity, reliability and data security in connecting the base stations to the core, in what we term as the transport network.

Due to the wide variety of cell types in use in today's mobile networks, the backhaul connections also vary considerably. At one end, we have the sparsely located macro cells in rural areas, each with very large cell radii and large inter-site distances. These cells need backhaul links that cover tens or even hundreds of kilometres to the aggregation points. The microwave backhaul links are the choice for these types of cells, which are more cost effective in covering longer distances. Microwave links typically use frequencies in the range of 7 GHz to 40 GHz and the line of sight (LOS) link is made up of highly directive, very narrow beams. At the other end, are the dense deployments of SCs in urban and metropolitan areas. The backhaul links for these cells need to cope with very high capacity demands but the distances to cover are a few kilometres or even hundreds of metres, often with dense clutter and non-line of sight (NLOS) conditions. The fibreoptic backhaul links are the first choice for this kind of high capacity, shorter distance NLOS requirements. In some extreme cases, satellite links can be used to provide backhaul connectivity for far remote area cellular deployments but this also will add significant latencies.

As we move into the age of 5G, networks are gaining higher densification, with more SCs being packed into capacity hotspots. In these dense networks, the limiting factor can be provision of fronthaul (the connectivity for centralised networks, discussed in Section 5.9) and backhaul. New concepts of sharing radio spectrum for radio access and backhaul like self-backhaul and Integrated Access and Backhaul (IAB) are now discussed in the 5G domain. We will briefly touch on these new developments at the end of this section, but the main focus will be on backhaul link dimensioning for an LTE network, taking our worked example in Sections 6.2 and 6.3 further. A more in-depth and a detailed analysis of LTE backhaul planning can be found in [13].

6.4.1 LTE Backhaul Provision – General Aspects

LTE backhaul provision is vastly different to previous generations in a number of ways. LTE is the first all-IP network, so the backhaul should be designed for IP packet delivery. The flat architecture of LTE means that the eNodeBs are directly connected to the core through S1 links and neighbouring eNodeBs are also inter-connected through X2 links, so backhaul should be provisioned for both these links. Also through its subsequent releases, LTE has enabled a number of capacity enhancing features such as higher order MIMO and carrier aggregation. These have significant impacts on backhaul capacity, so links should be designed with future capacity increments in mind. Leasing dark-fibre capacity from third-party providers with the option to increment capacity has become a popular option for LTE networks. Dark fibre refers to unlit or currently unused fibre, which are laid underground in bulk as digging up ground incurs much more cost than the actual fibre.

The LTE backhaul transport network should provide both user plane and control plane connectivity in S1 and X2 links. The S1 user plane (S1-U) connects the eNodeB with the S-GW and P-GW entities of the CN. The S1 control plane (S1-C) connects the eNodeB with the MME of the CN. The X2 user plane and control plane connectivity is needed between neighbour eNodeBs, to execute handovers of users and inter-eNodeB coordination schemes like CoMP (Co-ordinated Multi-Point). These links are typically dimensioned to provide the five nines or 99.999% availability. This works out that the backhaul links are non-functional for less than 5 minutes per annum. Fibre backhaul link availability is enhanced by having redundant fibre available in case of a failure in operating fibre. For microwave links, redundant RF chains can be incorporated to operate, in case of a failure in the active link. A ring or mesh network topology can also be added on to the typical tree network topology (shown in Figure 6.7) to increase the reliability in case of a single link failure. In fact, recent advances in technology can even ensure six nines or 99.9999% availability now. This translates to less than 30 seconds of outage per annum. Compare this with the 95% of coverage guarantees we plan for the radio access link. Whilst the backhaul task is made easier with the point-to-point nature of the links, there are still many technical aspects to be considered in dimensioning network capacity. We will introduce some of these aspects in the sections below.

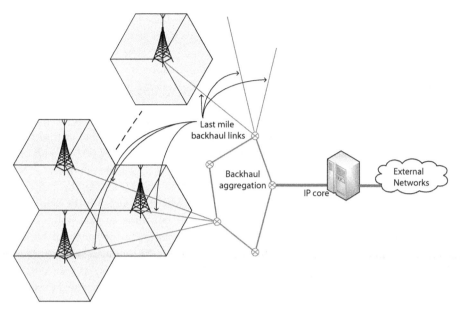

Figure 6.7 The 'tree and ring' network topology for the backhaul (transport) network.

6.4.2 LTE Backhaul Provision – Capacity Aspects

In considering capacity aspects, we will first look at how the backhaul links are accumulated in a typical LTE Macro cell network. We will follow the guidelines set by the NGMN alliance [14] in defining how the backhaul links are configured and also in estimating the backhaul capacities at each stage. We will consider both the traffic in the S1 link, linking eNodeBs with the S-GW and P-GW in the core, and the inter-eNodeB X2 traffic.

Figure 6.7 depicts how the typical LTE backhaul links are configured. Each of the Macro eNodeBs will have a dedicated backhaul link from a common aggregator point. This is called the last mile link. As detailed before, this is more likely to be a microwave link for larger rural cells or an optical fibre link for smaller urban cells. This last mile backhaul link should support the capacity demands of the Macro cell. The aggregation point accumulates the backhaul capacity of a number of macro cells. Physically, this aggregation can support a city wide network in urban areas or a region wide network in rural areas. The aggregation points are then accumulated into an IP core – linking to the EPC functions of the network. We illustrate how the backhaul network connects the eNodeBs with the core in a typical tree and ring network topology in Figure 6.7.

6.4.2.1 Derivation of Last Mile Backhaul Capacity

The backhaul capacity provision has to consider the competing factors of increasing cost to assign more bandwidth (or bit rate) on the backhaul link against the need to ensure good QoS. When each of the tri-sector cell sites is connected to the CN by a backhaul link, a simple calculation would suggest that 3 x sector capacity plus any overheads in the higher layers should be provisioned by the backhaul link. However, it is worth recalling that the radio access link capacity is calculated by looking at the busy hour capacity, when many users across the cell are active over the busy hour. This gives an averaging effect for SINR conditions users will experience and hence the data rates delivered to these users will also have a wide range, from high to low. The busy hour offered (or provided by the cell) traffic will represent these average conditions and this traffic demand will largely be consistent across the cells experiencing busy hour.

There is a paradox when considering traffic at quieter times. For most of the real-time data oriented traffic, a full buffer model is valid i.e. the user will consume or transmit as much data as is provided (or allowed) by the network. Imagine only one single user is active in the cell and the user is receiving very good SNR. This user can operate with MCS 64QAM MIMO B, giving spectral efficiencies of 12 bits/symbol. Under the full buffer model, this single user can consume all the access link radio resources in the cell. As per the calculations in previous sections, this user will consume a peak data rate of $13.2 \times 12 = 158.4$ Mbps. This peak data rate has to be supported by the backhaul, which is much higher than the data rates generated by the average traffic in the busy hours. Figure 6.8 illustrates this phenomenon. These rates are advertised by operators as their headline data rates, so it is important that the last mile backhaul capacity is dimensioned to support these peak data rates.

In our example, we see that the peak traffic demand can be as much as six times higher than the busy hour average demand for a single sector. In a tri-sector cell, it is highly unlikely that three single users from each of the three sectors, all in very good SINR conditions, will consume the peak data rates simultaneously. Thus, it is unlikely that the backhaul links (for a tri-sector site) having to support three times the maximum peak rate for a single user. However, the busy hour traffic follows similar patterns in all sectors and eNBs in the same area, so the busy hour traffic (mean traffic) needs to be multiplied by the number of sectors in an eNB.

In the capacity calculation for radio access links, we stripped out the control traffic, focussing only on user data demands. The control traffic is a necessary albeit a minor component in the contributors to backhaul traffic and hence we will consider it here. Looking back at the LTE sub-frame configuration in Figure 6.5, 36 symbols per resource block are

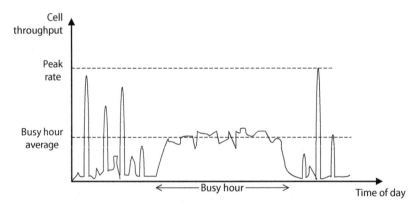

Figure 6.8 Illustration of busy hour traffic and quiet time peak rate.

utilised as control channel data. For the 100 resource blocks in the 20 MHz bandwidth, there would be 3600 symbols per millisecond, or 3.6 Msym/s. The control channel bits are transmitted in ½ QPSK for higher reliability and this results in a rate of 3.6 Mbps for the control information. For the busy hour mean traffic of the three sectors, the control overhead will add 10.8 Mbps, whilst for the single user peak traffic, we will assume an addition of 3.6 Mbps (although in practice this tends to be lower for a single user).

There is also the new X2 connection traffic to be considered in the last mile. This kind of direct inter-connection between neighbouring sectors and/or cells provides improvements in serving cell-edge users. If two eNBs serve the same user at cell edge for example, the potential interference in standard configurations now turns out to be useful signal for the users from multiple sources. The Co-ordinated Multi Point (CoMP) techniques have seen significant development in recent years and are part of the LTE-Advanced standards. For effective CoMP techniques, user data should be shared amongst participating eNBs or sectors through the X2 link. The NGMN white paper [14] assumes that 10% of backhaul traffic overhead will need to be supported in the X2 links, in cells with a large proportion of highly mobile users, also to facilitate handovers. As CoMP and handover techniques through X2 only benefit the cell-edge users, and as cell-edge users only contribute to the busy hour mean traffic, this overhead will be added only to the busy hour mean traffic.

Thus, the last mile backhaul traffic consists of two components, the peak traffic from a single user in very good SINR conditions, and the busy hour mean traffic from multiple sectors. The NGMN guidelines [14] state that the maximum from the peak user data rate in sector and the accumulation of three average busy hour data rates from the three sectors should be selected as the last mile backhaul capacity supporting a single eNB. With our calculated values, thus, the last mile backhaul capacity will be:

$$
\begin{aligned}
\textit{Last mile BH capacity} &= \max \begin{pmatrix} \textit{peak sector data and control capacity}, 3 \times 1.1 \\ \times \big(\textit{busy hour mean capacity} + \textit{control capacity}\big) \end{pmatrix} \\
&= \max\big(158.4 + 3.6, 3.3 \times 23.87 + 10.8\big) \\
&= max\big(162, 89.6\big)\,\text{Mbps} = 162\,\text{Mbps}
\end{aligned}
$$

6.4.2.2 Backhaul Aggregation

Different topologies can be used to inter-connect and aggregate backhaul links. Tree, ring and mesh network topologies are typically used. For the LTE Macro cell sites we are considering, the tree network topology will be considered, as in the NGMN white paper [14].

The last mile backhaul links are the topmost branches of this tree. They support the logical S1 links to the EPC (Evolved Packet Core of LTE) and the X2 links between the eNBs. We estimated the last mile backhaul capacity above, which ensures that peak sector capacity demand at quiet times (by one single user in very good SNR conditions) can be met. We assumed that peak capacities in the three sectors will not coincide, so only the maximum of the single peak or the multiple of three busy hour mean capacities have to be supported in the last mile of backhaul.

In estimating capacities for the aggregated backhaul links, we look at the large-scale trends and provision capacities accordingly. An essential feature here is that the aggregated links should be able to cater for unexpected surges in capacity. Basically, we are trying to find a happy medium between providing very large backhaul bandwidths to cater for these peak data surges in multiple sectors and limited bandwidths that only support the multiples of busy hour average demand. The solution here is to include buffers to store the data at the routers at each of the backhaul link aggregation points. However, there are performance implications with the size and frequent usage of buffers.

There are two simplistic approaches to apportion the size of the buffer and the bandwidth of the aggregated backhaul link. One approach is to include a large buffer and dimension the bandwidth of the backhaul link to fit the lower estimates of the sum capacity demands as noted above. Provisioning backhaul capacity is much more expensive than provisioning local buffer capacity. However, storing and retrieving data from buffers incur delay and overall performance would degrade. The other approach is to apportion backhaul bandwidth to more suit the highest possible demand and incorporate it with a small buffer. This would reduce delays and give very good performance, but the over-provisioned backhaul capacity would be underutilised for most of the time. This underutilised backhaul solution would not be cost effective in the long run. These two extremes are visualised in Figure 6.9.

The NGMN white paper [14] proposes a conservative estimate for the aggregated backhaul capacity, by assuming that when a quiet time peak capacity is demanded by one sector, the other sectors in the wider area are generating busy time mean capacities. Thus, when N sectors in the wider area are aggregated to a single point, the capacity requirement C_{ag} would be:

$$C_{ag} = quiet\ time\ peak + (N-1) \times busy\ time\ mean$$

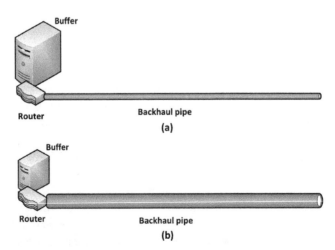

Figure 6.9 Two extremes on backhaul capacity provision. (a) Thin Backhaul Pipe with Large Buffer. (b) Thick Backhaul Pipe with Small Buffer.

To utilise this capacity lower bound in our example of 23 dense urban cells, the aggregated capacity that needs to be supported will be:

$$C_{ag} = 162 + (23 \times 3 - 1) \times \frac{89.6}{3} Mbps = 2193 Mbps$$

In a statistical sense, when the number of cells increase in size to large, the aggregated backhaul capacity that needs to be supported becomes closer to the lower bound of N times of the busy hour mean (per sector). This is illustrated in Figure 6.10 (from [14]).

It is worth noting that there are other approaches to estimate backhaul capacities at aggregation points. One approach is to define a statistical multiplexing factor (similar to the overbooking factor we discussed in access network capacity dimensioning) and use it to divide the extreme upper bound of total capacity demand. This upper bound would be when all sectors demand quiet time peak capacities. Typical statistical multiplexing factors or overbooking factors used in practice are between 3 and 5.

Another popular approach is to estimate the effective bandwidth requirements, using the self-similarity of backhaul traffic, i.e. with the Hurst parameter [15]. The feature of self-similarity means that a certain parameter will exhibit similar statistical properties when it is scaled up or down in time or spatial domains. For example, if there is a traffic burst for 10 minutes per one hour or within one sector for 20 cells, this pattern will be replicated in larger scales of time and space. In our test case scenario of backhaul traffic aggregation, the self-similarity applies mainly in the spatial domain. The normalised effective bandwidth calculation with the Hurst parameter generates a factor above 1, indicating by how many times the aggregated busy hour mean traffic needs to expand to account for the unexpected bursts. The Hurst parameter converges towards 1 when there are larger buffers and also

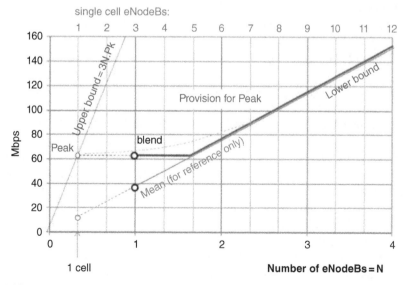

Figure 6.10 The estimation of aggregated backhaul capacity. (*Source* [14]).

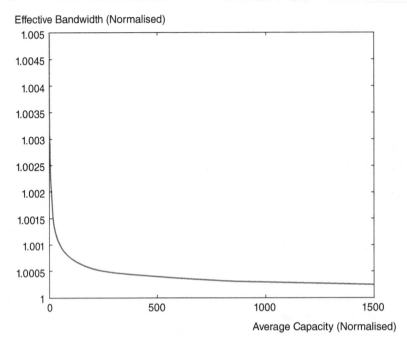

Figure 6.11 Effective bandwidth with a hypothetical Hurst parameter and a buffer size.

when the average traffic value grows bigger compared to the traffic burst. A typical normalised effective bandwidth variation is shown against the average traffic demand growth in Figure 6.11. The curve clearly indicates that when the average traffic demand is high, the effective bandwidth becomes closer to 1, i.e. when greater number of eNBs are involved, the quiet time surges do not have much impact. This can be compared to the result in Figure 6.10, where the aggregate capacity calculation with the NGMN proposed method also shows a similar result. An in-depth study on the use of self-similarity and Hurst parameter for LTE backhaul traffic estimation can be found in [16].

Sharing of eNodeB sites and transmit towers between operators is a preferred option and is even mandated by some regulators. In this case, the backhaul infrastructure (fibre or microwave) can also be shared effectively, which significantly reduces the transport network costs per operator. Even for Greenfield deployments who opt to utilise new eNodeB sites, the backhaul aggregation points and thicker backhaul links for aggregated capacity can be designed to make use of existing infrastructure. The backhaul capacities, particularly in IP backbone, can be leased from service providers, who cater for both wired and wireless networks.

Backhaul capacity is transported through 'tunnels' in the CN, as discussed in Chapter 5. This tunnelling allows user devices to retain the same IP addresses as it moves through eNBs and gateways [14]. However, this tunnelling incurs a packet overhead, typically estimated at 10% for shorter packet sizes. When backhaul capacity is leased by an operator, he may want to put in additional security for the traffic carried in the backhaul, from eNB to the EPC. Such additional protection to the user data can be granted by the use of IPSec

Encapsulated Security Payload (ESP) in the tunnel mode. This adds further overheads to the user data. A general overhead value of 14% is recommended by NGMN [14], when using the IPsec ESP. In total, the transport layer tunnelling and IP sec overhead adds a further 24% to the backhaul capacity. With the addition of these overheads, the last mile backhaul capacity per link would become:

$$Last\ mile\ BH\ capacity = 162Mbps*1.24 = 200.9\ Mbps$$

The overall aggregated capacities should also be upgraded accordingly, to account for this tunnelling and additional security.

As in the capacity dimensioning for LTE radio access, we have looked at quite a basic example of LTE in this backhaul dimensioning. However, the core principles explained would hold for even more advanced LTE/5G networks. A final note in this section is to re-iterate that the backhaul links should always be designed to absorb significant increases in capacity demand in future. Looking at LTE evolution from Release 8 to Release 14 (Section 5.8), the offered capacity in radio links have increased many fold, due to advances in technologies such as MIMO and carrier aggregation. This would also undoubtedly happen with 5G. So, any potential bottlenecks in backhaul links should be avoided from the initial link dimensioning stage.

6.4.3 New Developments in Backhaul/fronthaul Provision

A new development in RAN architecture sweeping the industry is termed as C-RAN (Centralised RAN), where some or most of the RAN functions can be performed by a centralised unit, leaving the eNodeB (or more appropriately named as a radio head) to carry out only the radio-transmission related functions. This centralised processing unit can even be located remotely in the Cloud (again termed as Cloud-RAN or C-RAN) architectures. We looked at C-RAN in some detail in Section 5.9, where the concepts of Fronthaul (to connect radio heads to the central processor) and backhaul (to connect the central processor to the core) were introduced.

With 5G, it is expected that this C-RAN architecture will be widely deployed, especially with densely packed SCs to support urban hotspots. These will open up new possibilities and challenges to the fronthaul/backhaul networks. The capacities needed in the fronthaul links will be several times greater than that of the backhaul links and we discussed how this ratio is impacted by the level of centralisation in Section 5.9. In these dense 5G SC networks, the main technical and financial challenges come with the design and deployment of fronthaul and backhaul networks [17].

A mixture of optical fibre and wireless techniques have to be used to meet the capacity, latency and QoS requirements of these dense networks, also impacted by the regulatory and landscape constraints in urban environments. 5G SCs are likely to rely heavily on millimetre wave (mmWave) frequency bands for achieving the required radio access capacity levels. The wide bandwidths available in this frequency range will also enable part of the band to be used for wireless fronthaul and backhaul provision. When LOS conditions are available, very narrow, pointed beams can achieve high SNR and hence high capacities. Wireless backhaul can also be used in NLOS conditions, but with lower SNR and hence lower capacities. There are a number of concepts being developed for the use of wireless backhaul, including the IAB topic in 3GPP [18]. In a dense SC deployment, the fronthaul/backhaul deployment topology is more likely to be a mesh network, so the presence of multiple routes will increase link reliability. With some nomadic and ad-hoc radio access points such as drones likely to become prevalent, the wireless fronthaul/backhaul will become a significant component in many 5G deployments.

Having both fibre and wireless backhaul options for 5G SCs can be highly beneficial in delivering the required levels of service quality and reliability for different vertical industry services that will become dominant (discussed in Chapter 12). Such vertical services will require network slices to be created with the required KPIs for each service. This will also include slicing of the fronthaul/backhaul networks, so the flexibility to configure and manage a certain set of KPIs (like latency, capacity and reliability) per slice will be a fundamental requirement for future fronthaul/backhaul networks. The EU funded 5G-Xhaul project gives some interesting research results and a test implementation of a futuristic 5G fronthaul/backhaul network [19].

6.5 The Network Planning Process

Whilst the coverage and capacity dimensioning processes give us an estimate of the number of sites needed for a particular network, it does not consider the specific coverage and capacity challenges and demands at individual locations within the area the network will be deployed. For this, we need to employ the network planning process with specific planning software. Within planning software, you can utilise 2D or 3D maps of the area where the network will be deployed, insert traffic demand maps on top of this, place the number of radio sites as determined by dimensioning, define antenna radiation patterns and analyse/optimise how the coverage and capacity targets are met. Each of these steps are described in the sub-sections below.

6.5.1 The Network Area Maps

The digital geographic maps of the area of network deployment is a basic requirement for the network planning process. Other layers of information can be added onto this map layer in the planning process, as we will discuss in the later sub-sections. These maps typically come in 30 m or 10 m resolution, with each pixel in the map representing a square area of relevant size. There are two basic spatial models used to define clutter features in a digital map and the associated RF attenuation values. A vector format is where the clutter features are defined as points, lines or polygons in the map, with differing lengths and sizes. A raster format is where each pixel is defined on the clutter type and maps are constructed as a collection of these pixels (of a given resolution). A typical raster model with different attenuation levels to different clutter types (shown by different shades) is depicted in Figure 6.12.

Working with 2D maps, cell-site placements and antenna patterns can provide quite accurate coverage predictions, where elevation features are not prominent, as in flat rural environments. Especially for rural network deployments with low demands on capacity,

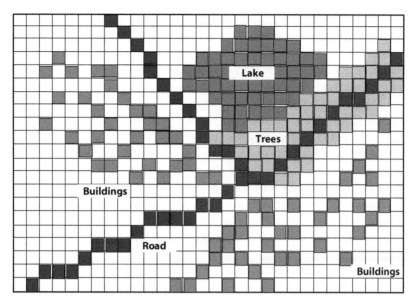

Figure 6.12 Raster 2D map with different clutter types.

the 2D map coverage based radio plan will be sufficient to give an accurate picture of the envisaged network.

3D maps, which also include information about features on the elevation dimension (such as heights of buildings and structures, hills and mountains) are useful for more detailed planning in urban environments, where often the networks are limited by capacity. 3D maps are increasingly used for LTE-Advanced and especially 5G network planning in higher frequencies. For millimetre-wave 5G networks, planar antenna arrays can be used which can provide 3D-beamforming patterns. Also, the intended traffic demand hotspots can be on different floors of high-rise buildings, shopping malls etc., which makes employing 3D maps imperative. Even with 2D maps, some planning tools offer the option to insert a layer of elevation as a DTM (Digital Terrain Model, including terrain features) or as a DEM (Digital Elevation Model, including terrain and building heights).

6.5.2 Site Placement and Antenna Radiation Patterns

Site placement is a key requirement in the RF planning process. There are several procedures being followed for this, depending on network type. If the operator's intention is to re-use the sites and antenna masts from a previous cellular generation or share sites with other operators, the planning team is constrained to use most of these existing sites. The RF plan will aim to introduce a minimum of new sites, as this would be the most cost-effective option. The RF designers have some flexibility with the site transmit powers (again, there are limits on the maximum power) and the down-tilt of the antennas to work within this constraint. If the network design is for a Greenfield deployment which does not use any existing sites, the aim of the RF design team will be to find locations with best coverage and capacity provision and also minimise overall site rental costs. It is very difficult to pinpoint a site to an exact location and then try to acquire this exact location, as it may not be available for rent. Hence, the RF designers provide an area for the potential location of the site as a search ring and once the design is finalised, the site acquisition team is tasked with securing a site for rental within this search ring.

It is an accepted procedure to utilise antenna heights which are more or less similar for all sites in the planned network. If there is a great variation in antenna heights, the taller sites can produce excessive interference to the neighbouring cells with shorter antenna masts. The antenna radiation patterns are also a critical element in determining coverage of the network. Some planning tools have a selection of digitised radiation patterns for common antenna models pre-stored in them. Some tools also offer the capability to define antenna patterns with main parameters like main beam half power beam-width, side lobe levels and positions of radiation nulls. As we discussed earlier, with 3D maps, the network designers have the option to use 3D-antenna radiation patterns. At the other end of the link, the receiver (UE) parameters can also be input, for example, the UE transmit power, receiver sensitivity and antenna gains and radiation patterns (usually this is an omni-directional pattern).

Once the antenna-related parameters are configured at both ends (including transmit powers and mast heights) and the sites are positioned in the digital map, the planning tool can generate coverage predictions. For this, the tools can use an established PL model or even a proprietary model. With 2D maps, the basic DL coverage predictions will involve estimating the distance-related PL from the serving cell transmitter, add-on a clutter-related loss at the reception point on the map and estimate received signal levels as per the nominal receiver (UE) properties. The transmissions from neighbour cells will be used as interference (as frequency re-use factor is 1 for LTE networks), if there is a need to generate SINR (Signal to Interference plus Noise ratio) maps at each of the pixels. With 3D maps, more advanced tools can use 3D-ray tracing methods to launch multiple signal rays from transmitters and geometrically evaluate their attenuations, reflections and diffractions, to derive the resultant signal strengths at each pixel. These advanced tools are more expensive and also need a large amount of processing power for their computations.

A basic downlink signal power estimation from multiple cells on a 2D map is depicted in Figure 6.13. This coverage map, developed for an urban area, clearly shows the street canyon effect, where the radio signal travels significantly longer distances along a street than in other directions.

6.5.3 Traffic Modelling and Capacity Provision Information

For capacity limited networks in dense urban and urban environments, it is particularly important to model how the traffic demand will be served by the planned network. The traffic demand can be interpreted in different ways. A basic method is to look at the subscriber densities in the planned network area and assign a traffic value proportionate to the subscriber density in a granular level, perhaps even at the pixel level of the map. This information can be inserted as another layer on top of the digital map to most of the planning

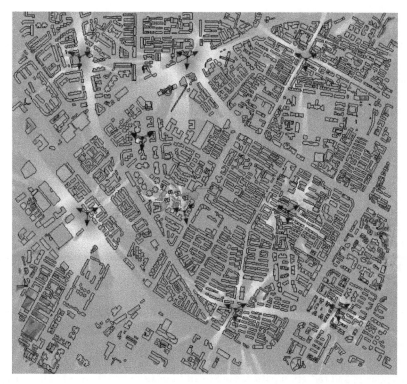

Figure 6.13 Basic SNR estimations from a planning tool with 2D map data. (*Source* [20]).

tools. Also, if the operator has historical or forecast information about the traffic patterns in the planned network area, these can be used as an information layer to the tool. Some tools have the ability to interpolate or extrapolate traffic data from a limited number of locations to build a picture across the map area. As noted before in capacity dimensioning, the busy hour traffic data should be used to map out the traffic demand across the planned network area.

Recall that in Figure 6.6 we illustrated how higher MCS is applicable and more capacity can be provisioned nearer to the transmit mast of a cell. Thus, placing sites closer to the traffic hotspots allow these sites to support more of the traffic demand. Also, increasing the number of sectors (from 3 to 6 for example) for sites in areas of high traffic demand enables them to support more traffic, at the expense of some increase in interference levels. A relatively new concept is to deploy heterogeneous networks or HetNets, where a layer of SCs (we discuss SCs in Chapter 10) on top of the Macro-cell network to absorb the high traffic demand from hotspots [21]. Many of the newer versions of planning tools allow this capability to add a SC layer on top of the Macro cells, either at the same operating frequency or another frequency, to plan these HetNets. In these cases, capacity demand is mainly absorbed by the SCs, whilst the Macro cells provide wider continuous coverage. A planning example of a HetNet with LTE Macro cells and SCs is shown in Section 10.7, with a resulting coverage plot in Figure 10.14.

6.5.4 Fine Tuning and Optimisation

Once the first run of the network planning is completed, it is important to establish the validity of the predictions. The PL models used, for example, only provide a generic behaviour and will need to be fine-tuned. One of the tried and tested methods for fine tuning the RF plan is by capturing measurements through drive tests. A test transmission can be set-up on the same frequency and measurements can be taken by a vehicle-based receiver on a number of drive routes. Then the measured PL values can be compared with the predicted values in the planning tool and necessary modifications to the PL model (usually to the clutter absorption loss values) would be carried out to bring it in-line with measurements.

The first run of the coverage and capacity predictions are likely to have a number of gaps, where the required service levels would not have been achieved. In the distant past, the network planners would have to manually tweak the transmit parameters in the tool, using a number of iterations until the desired results were achieved. The modern planning tools contain automated optimisation modules, where advanced algorithms can fine-tune the cell site positions, transmit powers, antenna down-tilts and other parameters to achieve the desired coverage and capacity level. A lot of the recent innovations in planning tools have come from the optimisation methods and features offered and now optimisation has evolved into an essential module that many network planners rely on.

A comparison of how optimisation can improve network coverage is provided below in Figure 6.14. The optimisation tool in the Ranplan Professional [21] is used to optimise the transmit powers and radiation patterns of SCs in this example. The results show a significant improvement in coverage (darker shaded regions) with the same number of SCs after optimisation.

The actual network optimisation efforts will continue even after the network is fully implemented on the ground. In the past, there were many teams of engineers recording measurements from various locations of the network, comparing them with expected

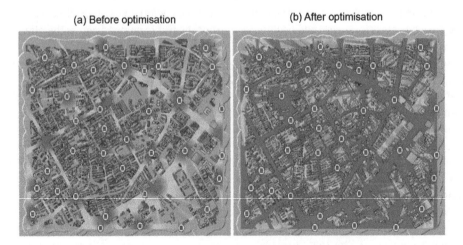

(a) Before optimisation   (b) After optimisation

Figure 6.14 Results from a network optimisation tool on an urban network segment. (*Source* [20]).

values and feeding any differences to optimisation teams. These optimisation teams would then fine-tune the network (for example to remove handover failure points in cell boundaries) over weeks and months. Nowadays, self-organisation network (SON) tools are commonplace in many 4G cellular networks. These SON tools can detect faults and fine-tune the network automatically, without the need for human intervention. We will look in some detail about these SON tools in Section 11.2.

6.6 A Look at 5G Networks

Throughout this chapter, we have mostly looked at dimensioning and planning for 4G networks, utilising well-established methods. We are now in the dawn of 5G network deployments and we covered some principles in fronthaul and backhaul provision for centralised 5G networks in Section 6.4.3. In this section, we look at the challenges and techniques for 5G radio access network dimensioning and planning more generally.

As we discussed in Section 5.9, 5G will be fundamentally different from previous generations, as it will aim to support many vertical industries, through the combined capabilities of eMBB, URLLC and mMTC. Future 5G networks will be capable of creating end-to-end network slices (or multiple virtual networks derived from the same physical resources) to provide the unique KPI (such as throughput and latency) needed for these verticals. However, the first 5G networks coming up are geared to support the wider consumer base with eMBB. Whilst the same fundamental principles of network dimensioning and planning will apply for these eMBB networks, there are some novel features of 5G which require especial attention.

The first 5G networks are more likely to come up as HetNets of dual connectivity, with LTE macro cells providing wider coverage and control plane connectivity and the 5G SCs providing higher capacity. The mmWave frequencies (above 6 GHz) are likely to be employed in these SCs, with planar antenna arrays capable of 3D beamforming. In order to determine the coverage properties of the millimetre-wave SCs with planar arrays, ray tracing techniques [7] can be employed, especially in indoor environments and confined outdoor spaces such as shopping arcades. However, due to smaller wavelengths and the wider bandwidths utilised, even minor features in the environment need to be considered for reflections or diffractions. Hence, these methods will involve a high computational cost and hence cannot be scaled out to wider areas or many numbers of cells.

For detailed simulations on the mmWave physical layer and for developing reliable link budgets, millimetre-wave channel models and PL models are essential. The 3GPP initiated developing the above 6 GHz channel model in 2015 [22] and this is now widely used for millimetre-wave related physical layer modelling. Equally important are the millimetre-wave PL models, that can reasonably capture the LOS and NLOS behaviours with millimetre-wave SCs. A PL model developed in [23] captures the probabilities of having LOS or NLOS links at a given distance from a millimetre-wave transmitter and also calculated an associated shadow-fading standard deviation value. The probable PL estimates from such a SC operating at 28 GHz, up to a 200 m transmission distance is shown in Figure 6.15. Due to this distinctly different behaviour in LOS and NLOS conditions, the PL profile shows a

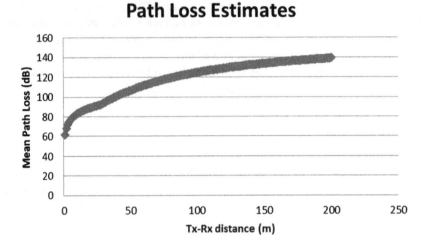

Figure 6.15 Path loss estimates.

non-linear curve with respect to distance, in contrast to the sub 6 GHz PL models we looked at in Section 6.2.5.

Due to the smaller wavelengths in the millimetre-wave frequency ranges, the foliage on trees also have a notable impact on their propagation. Experiments with fixed wireless access (at 28 GHz) systems have shown that in winter, when the leaves have fallen, the propagation conditions are more favourable than in summer. This aspect will also have to be taken into account in planning millimetre-wave mobile systems.

We looked at the fronthaul and backhaul provisioning (Transport network) options for 5G centralised RAN (C-RAN) in Section 6.4.3. This centralisation will also bring in novel challenges and opportunities to 5G radio access. With C-RAN, the remote radio heads (RRH) can be reduced to perform only the RF transmissions and the RAN processing can be centralised. In planning for the C-RAN network, due to the lower cost, size and power consumption of the RRHs, the site placement becomes simpler and more feasible to add sites as network demands increase over time. With millimetre-wave beamforming RRH, the beam patterns can be deployed, even in irregular patterns, to cover difficult urban locations like street canyons. More effort in planning will need to go into the positioning and capacity of the centralised processor, the BBU (Base Band Unit), which will need to have additional capacity to absorb any new RRHs that may be added in the future. The positioning of the BBU will influence the type of Fronthaul links and the capacity they can support and this may well be a limiting factor in 5G C-RAN design.

In designing the coverage and capacity from these 5G SCs (or even larger cells now operating at 3.5 GHz frequencies) new technologies like massive MIMO adaptations must be considered. Base stations (or gNodeBs – as they are called in 5G) with Massive MIMO capability will employ hundreds of antenna elements at the transmit end of the downlink, serving a number of single-antenna UEs. As with MIMO and MU-MIMO (discussed in Chapters 5 and 8), massive MIMO achieves significantly increased provided capacity (in line with the number of antenna elements), which must be taken into account in capacity dimensioning and planning. In addition, massive MIMO offers a feature called channel hardening, where variations in the channel gain (the instantaneous received radio signal profile per UE) are significantly reduced. Due to this property, the Fast Fade Margins (FFM) we considered in Section 6.2.7 for LTE systems, should be significantly reduced for such 5G systems. A practical estimation of this channel hardening with experimental results is presented in [24]. Due to improvements in RF design, signal processing and the likely higher SNR that can be achieved in 5G SCs, new higher layer modulation schemes at 256-QAM will be introduced in 5G. This will achieve four times higher spectral efficiencies than similar code rate schemes for 64 QAM in LTE, in the inner cell areas (refer to Figure 6.6), so the offered cell capacities in 5G can significantly increase, even for the same bandwidth.

As 5G networks evolve, we will begin to see networks supporting all eMBB, URLLC and mMTC applications with the required network slicing. Currently, as you would imagine, there is no widely accepted methodology on radio-network dimensioning or planning procedures to support URLLC and/or mMTC type traffic. Some of the underlying principles that can shape resource dimensioning for URLLC and mMTC traffic can be stated here. Increasing reliability whilst achieving low latency for URLLC would require increased use of radio resources in domains other than the time domain. For example, a URLLC packet can be replicated several times in the frequency domain (in frequency sub-bands for 5G) or transmitted with a higher power budget to ensure the necessary block error rate (BLER) levels at reception. For mMTC applications, it is vital to ensure the highest MAPL (maximum allowable PL) value between the transmitter and receiver as possible. This ensures large area coverage for mMTC base stations as well as increased battery life for sensor devices. One way to achieve this higher MAPL is through limiting mMTC transmissions to a very narrow frequency band, as then the noise bandwidth in signal reception reduces (recall the link budget in Table 6.2). This is how MAPL is expanded in current mMTC, as supported by 3GPP NB-IoT standards (Section 12.3) and other similar standards.

Although this chapter focussed on radio access and transport network dimensioning and planning, there are related methodologies and principles in the CN domain. In 5G, the use of NFV and SDN (detailed in Section 5.9) will bring in fundamental changes to CN dimensioning and planning. Many of the 5G CN elements can be virtualised to the cloud, yet their availability should be guaranteed with some level of redundancy in resources. Also, for some applications like URLLC, end-to-end latencies should be curtailed by placing resources closer to the physical network, through MEC (Mobile Edge Computing). It would be interesting to see what methodologies evolve in this segment as 5G networks migrate to a fully virtualised 5G core. The preferred option for early 5G networks is the NSA (non-stand-alone) mode, which utilises the LTE core, EPC. Further details on the evolving 5G network planning and optimisation concepts can be found in [25].

In this chapter, we have looked at the basic principles of radio network dimensioning and planning. The established methods and procedures were illustrated through a worked example in LTE, continuing through coverage, capacity and backhaul network dimensioning. Although the current focus in the industry is on 5G, no formal methods on dimensioning and planning for the diverse 5G networks have yet been made widely available. We have highlighted some key challenges with higher frequencies and the modifications that would be needed in 5G. It goes without saying that many of the methods (especially in capacity provisioning) used in LTE would still be useful for 5G, given that the 5G radio interface is based on OFDM, like LTE.

References

[1] De Alencar, M.S. and Carvalho Filho, D.D.M. (2017). *Cellular Network Planning.* Nordjylland, Denmark: River Publishers.

[2] Friis, H.T. (May 1946). A note on a simple transmission formula. In: *Proceedings of the I.R.E. and Waves and Electrons,* 254–256.

[3] Hill, G. (eds.) (2013). *The Cable and Telecommunications Professionals' Reference: Transport Networks.* 3rd ed., chapter 6 Focal Press.

[4] AWE communications note. http://www.awe-communications.com/Propagation/ .Rural/HO

[5] The COST 231 final report. Chapter 4, available at: http://www.lx.it.pt/cost231/final_ report.htm

[6] Abhayawardhana, V.S., Wassell, I.J., Crosby, D., Sellars, M.P., and Brown, M.G. (May 2005). Comparison of empirical propagation path loss models for fixed wireless access systems. In: *Proceedings of IEEE 61st VTC.* Spring.

[7] Geok, T.K. et. al. (April 2018). A comprehensive review of efficient ray-tracing techniques for wireless communication. *International Journal on Communications Antenna and Propagation* 8 (2).

[8] Jakes, W. (1993). *Microwave Mobile Communications.* New York: IEEE press, pp. 125–127.

[9] Tolstrup, M. (2015). *Indoor Radio Planning: A Practical Guide for 2G, 3G and 4G,* 3rd ed. Hoboken, NJ, USA: Wiley Publishers.

[10] 3GPP technical report, 3GPP TR 36.824. *LTE coverage enhancements (Release 11),* V11.0.0. Available at: https://www.3gpp.org/ftp/Specs/archive/36_series/36.824

[11] 3GPP technical specification, 3GPP TS 23.203 v12.0.0. *Policy and charging control architecture (Release 12),* Table 6.1.7. Available at: https://www.3gpp.org/DynaReport/23-series.htm

[12] Sesia, S., Toufik, I., and Baker, M. (2011). *LTE - the UMTS Long Term Evolution: From Theory to Practice.* Chapter 10, Hoboken, NJ, USA: Wiley Publishers.

[13] Metsälä, E. and Salmelin, J. (eds.) (2016). *LTE Backhaul: Planning and Optimization,* 1st ed. Hoboken, NJ, USA: Wiley Publishers.

[14] NGMN Alliance white paper. (July 2011). Guidelines for LTE backhaul traffic estimation. Available at: https://www.ngmn.org/publications/guidelines-for-lte-backhaul-traffic-estimation.html

[15] Vannithamby, R. and Talwar, S. (eds.) (2017). *Towards 5G: Applications, Requirements and Candidate Technologies.* Section 18.4, Hoboken, NJ, USA: Wiley Publishers.

[16] Polaganga, R.K.. (2017). Self-similarity and Modelling of LTE/LTE-A Data Traffic. MSc Thesis, University of Texas Arlington, available at: https://rc.library.uta.edu/uta-ir/handle/10106/26712

[17] GSMA Future Networks web article. (June 2019). Mobile Backhaul: An overview. Available at: https://www.gsma.com/futurenetworks/wiki/mobile-backhaul-an-overview /#8220e44fe960706ab834328617ff1e7e

[18] 3GPP Study Item description, RP-170821. (March 2017). Study on integrated access and Backhaul for NR. Available at: https://portal.3gpp.org/ngppapp/CreateTDoc.aspx?mode =view&contributionUid=RP-170831

[19] (2020). EU H2020 research project 5G-Xhaul. project website (viewed Feb.): https://www.5g-xhaul-project.eu/project.html

[20] Ranplan professional planning and optimisation tool. Web link: https://www.ranplanwireless.com/gb/products/professional

[21] Anpalagan, A., Bennis, M., and Vannitham, R. (eds.) (2016). *Design and Deployment of Small Cell Networks*. Cambridge, UK: Cambridge University Press.

[22] 3GPP TR 38.900. *Study on channel model for frequency spectrum above 6 GHz*. Available at: https://portal.3gpp.org/desktopmodules/Specifications/SpecificationDetails.aspx?specificationId=2991

[23] Sulyman, A.I., Nassar, A.T., Samimi, M.K., Maccartney, G.R., Rappaport, T.S., and Alsanie, A. (2014). Radio propagation path loss models for 5G cellular networks in the 28 GHz and 38 GHz millimeter-wave bands. *IEEE Communications Magazine* 52 (9).

[24] Gunnarsson, S., Flordelis, J., Van Der Perre, L., and Tufvesson, F. (2018). Channel hardening in massive MIMO - A measurement based analysis. In: *IEEE 19th International Workshop on Signal Processing Advances in Wireless Communications (SPAWC)*.

[25] Penttinen, J.T.J. (2019). *5G Explained: Security and Deployment of Advanced Mobile Communications*. Chapter 9. Hoboken, NJ, USA: Wiley Publishers.

7

Spectrum – The Life Blood of Radio Communications

7.1 Introduction

Any device that communicates wirelessly needs spectrum. Therefore, wireless communication cannot exist without spectrum. At the national level, innovative wireless solutions can greatly enhance a country's GDP. For citizens and consumers, radio-communication services such as mobile networks and wireless Local Area Networks, commonly known as Wi-Fi, have brought convenience to daily life. In this information age, demand for such services will continue to grow as staying connected and having ready access to information anytime, anywhere has become a part of life. Therefore, the radio spectrum is national wealth and provides a foundation for the wireless sector. The radio spectrum can also be a driver for economic development if carefully managed. Therefore, efficiently managing this scarce resource should be one of the important strategic objectives for all countries. In view of the increasing demand for wireless services and applications, frequency management policies and procedures need to be constantly modernised and revamped to accommodate new and innovative applications with little or no interference between users.

7.2 Spectrum Management and Its Objectives

Although there is ample amount of spectrum available, not all spectrum bands are desirable for radio communications. The part of the radiowave that is suitable to use for radio communications is limited. Since the demand for usable spectrum is higher than supply i.e. the availability of the spectrum desirable for radio communications, the spectrum is known as a scarce resource and needs to be managed efficiently. Radio spectrum is a national resource. Therefore, every country has a sovereign control over how it is used. The spectrum management responsibility lies on individual administrations. However, radiowaves do not stop at country borders. Therefore, administrations need to work together to coordinate spectrum usage to minimise interference.

The Technology and Business of Mobile Communications: An Introduction, First Edition.
Mythri Hunukumbure, Justin P. Coon, Ben Allen, and Tony Vernon.
© 2022 John Wiley & Sons Ltd. Published 2022 by John Wiley & Sons Ltd.

7.2.1 The Role of the ITU

The International Telecommunications Union (ITU) plays a key role in facilitating and coordination of spectrum usage to minimise interference. ITU facilitates the international connectivity in communications networks and plays a key role on managing radio spectrum at a global level [1]. For example, ITU facilitated the Geneva 2006 agreement (GE06) where national regulators across Europe agreed frequency plans to ensure digital TV services can operate within Europe without interfering with each other.

The guidelines related to the use of radio-frequency (RF) spectrum are available in the Radio Regulations, which contains the complete text adopted by the World Radio Conference (WRC). The Radio Regulations are covered by an international treaty governing use of the RF spectrum [2]. These Radio Regulations, originally adopted at the WRC-95, incorporates the decisions of the WRCs and are reviewed and, if necessary, revised at subsequent WRCs. The WRCs are held every 4 years. This mechanism provides an international framework for spectrum usage, which can be updated if and when needed.

The first step of the spectrum management process is to identify the spectrum for a specific service. This is known as frequency allocation. The objective is to use the same frequency band for the same service across different countries – an approach called 'harmonisation'. Harmonised spectrum usage results in lower equipment cost resulting from better economies of scale since the equipment manufacturers reach to the wider market. However, due to many reasons i.e. different requirements or differences in legacy usage in different countries, global level harmonisation may not always be possible. For the purpose of allocation of frequencies, the world has been divided into three Regions (Figure 7.1):

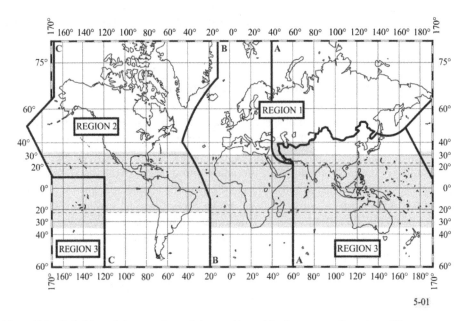

5-01

Figure 7.1 Definition of three regions of the world used for frequency allocations [3].

1. Region 1: Contains mainly Europe and Africa
2. Region 2: North and South American continents
3. Region 3: Asian continent including Australia.

The current approach is for administrations within each region to try to harmonise the spectrum allocations, at least at a regional level, to create benefits from economies of scale from any equipment needed for that particular spectrum usage.

7.2.2 Regional Bodies

Although spectrum management related to an individual country is the responsibility of an individual administration, spectrum matters are usually discussed and agreed at a regional level. This is because harmonisation of spectrum is a key enabler for the use of spectrum. Harmonisation (i.e. use of the same spectrum by different countries for the same service) enables spectrum users to receive the benefits of economies of scale. There are six main regional bodies:

1. Asia Pacific Telecommunity (APT): Covering the Asian and Pacific regions.
2. Arab Spectrum Management Group (ASMG): Covering the Arab states in the Middle East and North Africa.
3. African Telecommunications Union (ATU): Covering all of the administrations on the African continent.
4. European Conference of Postal and Telecommunications Administrations (CEPT): Comprising 48 member administrations including all of the EU countries, Russia and Turkey.
5. Inter-American Telecommunication Commission (IATC): Covering the Americas and the Caribbean.
6. Regional Commonwealth in the Field of Communications (RCC): Covering the Russian Commonwealth of Independent States, as well as the Baltic states which have observer status.

These regional bodies have a pivotal role on harmonising the regional activities which have a wider remit beyond spectrum i.e. cooperation on commercial, operational, regulatory and technical standardisation issues. In addition to international regulations, these regional bodies significantly influence national spectrum allocations. For instance, the European Conference of Postal and Telecommunications Administrations or CEPT has 48 members and covers almost the entire geographic area of Europe [4] extending from Ireland to Russia. It primarily focuses on policy making rather than operational aspects. CEPT facilitates discussions to promote further European harmonisation on regulatory issues related to post and telecommunications. These common positions developed by CEPT help strategic planning and influence decisions at ITU level to achieve European goals. CEPT has also coordinated frequency usage and cooperated in the ITU policy-making process. Although decisions of CEPT are not binding, the UK has committed to most CEPT decisions and it is hard for the UK to deviate from a consensus CEPT position in WRCs of the ITU.

7.2.3 National Regulators and Their Roles

The main objectives of the national regulators include efficient use of spectrum whilst promoting competition and encouraging investments. National regulators are usually independent bodies and have special powers to carry out these duties set to them. For instance, The Office of Communications or Ofcom is the national regulator in the UK which regulates communications services. Ofcom is an independent authority whose duties are defined by the Parliament [5]. Ofcom's principal duty is 'to further the interests of citizens in relation to communications matters, and to further the interests of consumers in relevant markets, where appropriate by promoting competition' [6].

Regulators engage with each other at regional level to harmonise spectrum use within their countries. At European level, harmonisation activities are usually initiated as a result of a mandate from the European Commission. This also includes development of band plans and technical license conditions. However, it is not mandatory for WRC decisions on frequency allocations to be adopted. The national administrations consider ITU decisions and regional requirements appropriately to develop national spectrum decisions.

Regulators have the difficult job of balancing conflicting requirements of different users:

- Balance spectrum requirements for different users
- Dedicate spectrum for specific use to provide certainty
- Manage cross-border interference and harmonised use
- Have the right framework to choose what the market needs.

Encouraging innovation is an important element of the duty of a regulator. Spectrum should be viewed as an enabler to make things happen and foster innovation. For this, spectrum should be available to users when they need to use it and at the locations they need. A sensible approach to regulation involves consulting the industry and spectrum users to understand the market requirements before making decisions. The core principles of effective spectrum policy are:

- to put in place procedures and processes to maximise the efficient use of spectrum
- spectrum is made available to those who most value the resource
- minimise the harmful interference between spectrum users.

Spectrum management tasks involve frequency allocation, allotments, and assignments wherein, the entire radio spectrum is divided into bands of frequencies established for a particular type of service (allocations) such as land mobile telecommunications which encompasses further divisions such as cellular radio, business radio and many others (allotments). The final subdivision of the radio spectrum takes place via the assignment processes during which an applicant is provided with an 'assignment', authorisation or license to operate a radio transmitter on a specific channel or group of channels at a particular geographic location or locations under specific conditions.

As a part of the spectrum management process, the national regulator adopts spectrum allocation identified at the WRCs in the form of a frequency allocation table (FAT). It contains entries of agreed spectrum allocations used for frequency assignment. The international frequency allocation table is published and maintained by the ITU whilst national frequency allocation tables are published and maintained by individual administrations. National administrations generally follow the international Frequency Table although there can be significant variations, particularly where the international regulations allow for a number of radio services in a specific band and the national administration restricts the number. In the National Table of Frequency Allocations, the spectrum available for civil, government or common uses are generally defined clearly and publicly. Some elements of the Frequency Allocation Table are as follows:

- Frequency band classification:
  - Exclusive bands
  - Shared bands
  - Receive-only bands
  - License-free bands (not by ITU decision).
- Radiocommunication services
  - 40 radiocommunication services are defined in ITU-R Radio Regulations Article 1
  - About 30 radiocommunication services appear in the Frequency Allocation Table [7]
  - More services could be defined by regulators.
- Radiocommunication service category:
  - Primary
  - Secondary.

7.2.4 The Spectrum Management Process

The spectrum management process is presented graphically in Figure 7.2

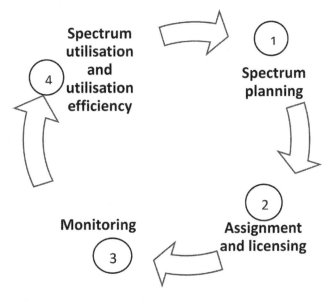

Figure 7.2 Example spectrum management process.

The tasks inherent in the spectrum management process are sequentially and logically connected to each other in the spectrum lifecycle as explained below:

1. **Spectrum planning:** Spectrum planning is the first step of this process. It involves the allocation of portions of the spectrum to specified uses in accordance with international agreements, technical characteristics and potential use of different parts of the spectrum, as well as national priorities and policies. The development of the spectrum planning process firstly requires understanding existing uses of the spectrum under consideration which can be achieved through a spectrum inventory. The conduct of spectrum planning is an abstraction of the national spectrum framework which outlines the associations of specific frequency bands with specific services. Band planning is about determining how bands will be used, how radio services operating in the given frequency band would be implemented, as well as which technologies would be accepted based on market or administrative processes.

2. **Assignment and licensing:** Once the spectrum planning steps are completed and allocations are identified, spectrum can be granted to the users by means of assignment and licensing. The licences usually contain technology and usage restrictions necessary for spectrum management reasons to manage the risk of harmful interference and to ensure compliance with statutory duties and international obligations. The regulator issues licenses to authorise the use of radio spectrum allocated for civilian use with the aim of securing optimal spectral efficiency. A licence typically contains clauses on technology and usage restrictions to ensure that harmful interference is minimised, if not eliminated, to ensure compliance with statutory duties and international obligations. Formal (and informal) coordination procedures should be established to deal with assignments in shared spectrum, interference problems and to review requirements on a regular basis.

3. **Monitoring:** Spectrum monitoring and compliance activities help users to avoid incompatible frequency usage through identification of sources of harmful interference. Monitoring can enhance the effectiveness of frequency assignment, assess spectrum occupancy and compliance with national rules and regulations, compliance with commercial, administrative and technical conditions enshrined in the license. Spectrum monitoring supports the overall spectrum management effort by providing general channel/band measurements, including channel availability statistics of a technical and operational nature, thereby giving a measure of spectrum occupancy. Monitoring is also a useful planning and operational tool as it can assist spectrum managers to ascertain the level of spectrum use compared to the assignments registered on paper or in data files. In some instances, a monitoring and measurement system can help where a solution to a problem requires more than knowledge of authorised or designed characteristics of radio systems. Monitoring and measuring systems also obtain information on the operation of individual stations, for regulatory, enforcement, and compliance purposes; hence can be used to establish location of interfering sources. In general terms, monitoring gives feedback to spectrum management on whether the practical use of the spectrum matches national policy and can also inform future spectrum use trends and/ or requirements by feeding forward information to managers.

4. **Review of spectrum utilisation and utilisation efficiency:** Since spectrum is a scarce and valuable resource, knowledge of the status of spectrum utilisation helps spectrum managers to plan and use frequencies efficiently, avoid incompatible usage,

and identify areas of spectrum scarcity. It is essential to monitor spectrum utilisation and efficiency of usage to maximise efficient use of spectrum. Furthermore, spectrum use planning and resolution of spectrum scarcity issues can be accomplished through study and analysis of spectrum occupancy data. Interventions may be necessary to clear existing users in order to allocate spectrum to another user who provides more socio-economic benefits to the country.

7.3 Spectrum Allocations

As mentioned earlier, spectrum management starts with dividing radio spectrum into ranges of frequencies known as frequency bands and allocating each band to one or more service. Radio regulations provide a number of different radio services:

- Terrestrial and space services
- Broadcasting services
- Fixed service
- Amateur services
- Radio astronomy service
- Radio services related to Earth observation.

Spectrum allocation is the process of distribution of the RF spectrum amongst different radio services. The role of the national spectrum management authority is to allocate specific frequency bands to a given area, without introducing harmful interference to existing users. Most of the time, spectrum allocation requires consideration of trade-offs between different potential uses. This is because spectrum is a scarce resource where demand outstrips supply.

Spectrum planning involves the allocation of portions of frequency spectrum to specified uses, in specific locations and in accordance with international agreements, technical characteristics and potential use of different parts of the spectrum, and national priorities and policies.

Once a frequency band has been allocated to a service, it is necessary to make provision for systems and users to access the frequencies in an orderly manner. The *Band Plan* is a plan for distributing or allocating bandwidth within a frequency band. The most commonly used method is by dividing the spectrum available into a number of channels. The bandwidth of the channels depends on the technology to be used and the required traffic capacity of the systems that will use the channel. Channel plans are defined e.g. by ITU-R Recommendations to maintain the necessary spectral separation between the transmitter and receiver. Band plans determine how bands are to be used, how radio services are implemented and, in some cases, which technologies will be accepted or if the market alone will establish what technologies prevail.

7.4 Spectrum Assignment

Spectrum assignment is the authorisation given by an administration for a radio station to use a RF or RF channel under specified conditions. In its simplest form, it is the assignment of operating frequencies to radio equipment in accordance with national plans and policies for frequency allocation. It involves thorough analysis of technical parameters of the radio equipment used for the communication without causing harmful interference to other radios users. Spectrum assignment strategies are designed to achieve optimum use of spectrum. It involves facilitation of access to, and re-use of, specific frequencies whilst ensuring that spectrum rights are given only to those users who can maximise the economic and social benefit from the spectrum and that users know their rights and obligations with sufficient precision to allow them to make plans and avoid harmful interference. There are different mechanisms used by national regulators to assign spectrum which are explained in the next sub-sections.

7.4.1 Administrative Assignments

Traditionally, spectrum was granted to various organisations or users by administrative action, i.e. directly by the regulator or by collective inter-governmental agreement. In other words, it was a command and control mechanism where the regulator defined who could use the spectrum and under which conditions. Most of these assignments were based on the First-come-first-served (FCFS) principle where licences were assigned to applicants in the order of their application. This is one of the simplest assignment mechanisms wherein applicants were awarded spectrum usage rights in the order that they request them. These assignments were purely based on the demand. i.e. economic value of spectrum was not considered. The administrative assignments were appropriate when demand for spectrum did not exceed supply. Most administrative assignments were carried without charging any fees or with a relatively small annual licence fee. Although the administrative-based assignments are rare in the current climate, assignments made on this basis still dominate in the national frequency table. In most cases, the usage rights of the incumbent spectrum users are protected when considering new entry. This has resulted in situations where the incumbent users lack incentives to efficiently use spectrum or give up on the spectrum for other usage.

7.4.2 Market-Based Mechanisms

As the demand for spectrum increased, many regulators moved away from the administrative assignments in favour of market-based mechanisms to assign spectrum. Other factors could include difficulties in finding new allocations; perceptions of spectrum hoarding and inefficient use; realisation of the economic value of spectrum; etc. New Legislations have enabled regulators to impose licence fees reflecting the economic and market value of spectrum rather than the administrative cost of issuing licences.

The general policy considered by national regulators is to release spectrum through a competitive process. Market-based mechanisms are used when the demand for spectrum is higher than the quantity of spectrum available. This is usually the case for mobile spectrum where the demand for spectrum is significantly higher than spectrum supply. Two most commonly used market-based mechanisms to assign spectrum are the beauty contest and auctions.

7.4.2.1 Beauty Contests

In this case, the licences are assigned to applicants that present the most compelling application, in the regulator's judgement, based on a set of pre-determined quantitative and/or qualitative criteria. Beauty contest mechanisms are likely to yield better usage outcomes than first-come-first-served mechanisms and can also allow the spectrum manager to include certain policy initiatives that can maximise benefits for consumers. Therefore, this approach is appropriate in cases where spectrum is being assigned for a specific end use.

However, there are significant challenges associated with setting up beauty contests. Two biggest drawbacks are; firstly, they are more prone to subsequent challenges by the applicants. The process of a beauty contest has to be transparent to minimise the challenges in a court of law which can create delays in the award of licences. The regulator in this case has to set very clear criteria on which selection between bidders is made. Even so, it is likely to select applicants with an element of judgement resulting in the introduction of risk during the selection process making it discriminatory and less transparent. The second biggest drawback is that it is difficult to enforce. For instance, if the recipient is unable to fulfil the conditions of the license a few years down the line, it is too late to relinquish spectrum and assign to another user. The winner would have made investments by this time.

7.4.3.1 Spectrum Auctions

Auctions are found to be the most common mechanisms to assign spectrum where demand exceeds supply. In this case, a bidding process is used and licences are awarded to the bidder who is prepared to pay the most money. The auction process is much more transparent compared to the beauty contest; therefore the outcome is less likely to be challenged. In auctions, the sum of money has to be paid by the winner at the end of the auction. It is expected that a user who is willing to invest a significant amount of money upfront, has a viable business case and is most likely to use the spectrum to recover the cost of using the spectrum. It is also expected that the user who values the spectrum the most is likely to produce the most valuable outcome to society.

However, the success of the auction largely depends on the auction design. The outcomes of poorly designed auctions can be very damaging to any country. An auction solely based on bidders 'willingness to pay' may not capture the impact of any social externalities, so may not be fully efficient. Auctions tend to yield better outcomes when there is sufficient competition in the market. One way to incentivise bidders is to have sensible reserve prices. Other policy objectives may also influence the design of the auction and qualifications for bidding. In the UK, 800 MHz and 2.6 GHz auction completed in 2013 [8], Ofcom's approach was being driven by its overriding duty to facilitate competition, and the assumption that the spectrum would be used for public mobile services based on advanced mobile technologies such as LTE. An auction can also provide significant revenues for governments, although, this should not be the objective of the auction. There are a variety of auction models used by regulators to achieve the desired outcome. This is because every spectrum auction is unique in terms of the competition, objectives, market conditions etc.

7.5 Spectrum Licensing

All transmitting stations that use spectrum require a license to operate unless specifically exempted. Licensing is an administrative process that grants permission for a spectrum user or organisation to use specific frequency(s). The licensing is usually in accordance with the Telecommunications Law in the country. Exemptions are generally made for many low-power applications such as Wi-Fi and Bluetooth devices and receive-only equipment.

It is essential that the spectrum use to be recognised and usage rights including the technical license conditions are defined such that existing users are protected. Examples include radio astronomy and satellite receive-only stations (which do not transmit but are very sensitive to interference). Note that the licensing condition in place are known as the 'least restrictive licence conditions' which are essential to protect the existing users in adjacent spectrum bands.

7.5.1 Spectrum for Mobile Services

To overcome the drawbacks of the first generation of mobile systems, the second generation of the mobile communications systems (known as 2G systems) was launched in the early 90s. The primary spectrum band used for 2G was 900 MHz spectrum. The world's first 2G call was made in March 1991 in Radiolinja's network (currently known as Elisa) in Finland [9]. Subsequently, the 1800 MHz band was introduced to enhance the competition [10]. The GSM1800 band was a UK initiative which was followed by other European countries, even although no other country had an immediate need for this at that time. The goal of 3G systems, progressed through the ITU's International Mobile Telecommunications (IMT) 2000 initiative, was to consolidate the diverse and incompatible mobile environments into a seamless radio and network infrastructure capable of offering a broad range of services on a global scale [11]. The aim was also to have globally harmonised spectrum band for 3G systems. The first band identified specifically for 3G, around 2 GHz band, achieved this objective as far as it could. The debate of identifying the need for more spectrum for advanced mobile services has continued since then and is still a major debate at the international fora. The challenge is that the desirable spectrum for mobile communications has already been allocated for different services. So how do we find spectrum for mobile services? There are three main options available for a new user to find spectrum:

1. Repurposing/re-farming existing spectrum
2. Acquiring new spectrum
3. Sharing the spectrum with other users.

These options are explained in the next sub-sections.

7.5.1.1 Repurposing/Re-farming Existing Spectrum Bands

Liberalisation or removal of restrictions of spectrum allows MNOs to use the spectrum used for legacy technologies to be used with more efficient technologies. For instance, due to the significant increase in data usage, many operators re-farm the spectrum used for GSM with more spectrally efficient technologies such as UMTS and LTE. This happens to a number of bands around the world. A good example of this is repurposing the 900 MHz band used for GSM for UMTS services.

Although GSM was one of the widely used mobile technology standards around the world, the data speeds supported by the evolutions of GSM was limited to few hundred kbits/second, compared to 10s of Mbits/second supported by 3G and 4G technologies. As data usage increased, MNOs started re-farming 900 MHz band initially to deploy UMTS (known as U900 carriers) around 2006 mainly to obtain benefits from the HSDPA introduced by 3GPP Release 5. The lower propagation loss in 900 MHz band results in more than two times additional coverage compared to 2100 MHz band, which is the primary based used for UMTS deployments around the world. Operator interest to re-farm 900 MHz band grew rapidly putting pressure on regulators to allow deployment of UMTS in this band.

Electronic Communications Committee (ECC) within the European Conference of Postal and Telecommunications Administrations (CEPT) has completed compatibility studies between the systems (i.e. mainly GSM and UMTS) operating in 900 and 1800 MHz bands in 2007, paving the way for regulators to liberalise these bands [12]. The compatibility studies usually take into account the systems operating in adjacent bands to ensure deployment of UMTS in the 900 MHz band does not cause harmful interference into those systems operating in adjacent bands. The study suggested remedies such as spatial geographic separation between deployments, installing external filters at the base stations to restrict the transmission envelope or introduction of the guard bands used to mitigate interference between different systems depending on the deployment scenario.

Understandably, MNOs who did not have 900 MHz band in their spectrum portfolio (i.e. MNOs deployed GSM in 1800 MHz band) resisted this development. However, that didn't stop deployment of UMTS in 900 MHz band. Elisa was the first MNOs to launch UMTS in 900 MHz band in Europe in late 2007 [13]. Subsequently, other MNOs such as Vodafone Portugal, Orange France and many other MNOs followed suit.

Similarly, 1800 MHz band used for GSM was repurposed to deploy LTE by many mobile operators. As such, 1800 MHz band became one of the most popular spectrum bands used for LTE deployments.

Both 900 MHz and 1800 MHz bands were subsequently used for land mobile services repurposed for the deployment of more spectrally efficient technologies such as LTE. A good example of repurposing spectrum for different services is the use of 800 MHz band for mobile use. Originally, 800 MHz was used for analogue TV transmission. However, the introduction of digital TV (known as digital TV switchover) enabled this valuable spectrum to be used for mobile services. To make the 800 MHz spectrum band available for mobile use, a national change programme was required making everyone get digital TV as the old analogue TV signal was switched off. In the UK, the Digital TV switchover program was completed over a four-year period from 2008 to 2012.

7.5.1.2 Acquiring New Spectrum Bands

Since most of the spectrum desirable for radio communications is allocated for different services, particularly within the spectrum sweet spot ranging from 300 MHz to 3 GHz, the industry began to look for alternative spectrum bands above 3 GHz. The evolution of the mobile RF chip design and advancements of technology features such as multiple antenna systems (known as multiple input multiple output or MIMO pave the way for spectrum bands above 3 GHz to be used for mobile services. The shorter wavelengths of higher frequency bands make it easier to develop antennas with multiple antenna elements (i.e. manageable size). Further advancements such as beamforming (i.e. a technique that focuses the signal towards a specific receiver, rather than spreading the signal into wider areas from the antenna) can compensate the propagation losses in higher frequencies, making them potentially suitable for achieving sensible cell ranges to deploy mobile services. These advancements make high frequency millimetre-wave bands such as 26 GHz bands widely considered for the deployment of 5G mobile services. Note that each spectrum band has a role to play in the network. Since the coverage is already achieved using lower frequency bands, these higher frequency bands which also have higher bandwidths provide the necessary capacity uplift required at the traffic hotspot areas. Inherent characteristics of higher propagation losses makes these high-frequency bands reusable more widely compared to the lower frequency bands. Table 7.1 shows characteristics of different spectrum bands and their role in mobile networks.

Currently, many regulators are working to make new spectrum bands available for mobile services. For instance, within Europe, the European Commission adopted an amending Implementing Decision to harmonise the radio spectrum in the 3.4–3.8 GHz (or 3.6 GHz) band for future use with 5G services. This decision required Member States to allow use of the 3.6 GHz band for 5G systems by 31 December 2020 in line with the European Electronic Communications Code [14]. MNOs can therefore follow the mechanism used by the national regulator, mostly market-based mechanisms such as auctions, to acquire these new spectrum bands.

Table 7.1 Characteristics of different spectrum bands and their role in mobile networks.

Spectrum band	Characteristics
Sub 1 GHz band (700–900 MHz)	Wide area and deep indoor coverage Limited quantity 800 MHz band: one of the most popular bands for 4G equipment 700 MHz band: Pioneer 5G band identified within EU
Low band (1.4–2.6 GHz)	Medium coverage, Popular with equipment 1.8 and 2.6 MHz bands: most popular bands for 4G equipment 2.6 GHz band: Local coverage, opportunity for high spectral efficiency
Mid band (3.4–3.8 GHz)	Considered for local coverage in cities Opportunity for high spectral efficiency Availability of wider bandwidths per MNO (~80–100 per MNO)
mmWave bands	Short range but compensated with massive MIMO Yet to be proven for area coverage Ample quantity in the excess of 400 MHz wide channel per MNO Very high spectral efficiency

The other option of acquiring spectrum is from an existing holder of spectrum through spectrum trading.

Spectrum trading is a process that allows the holders of certain spectrum licences to transfer or lease the rights to use spectrum to another party [15]. Spectrum trading is available for certain categories of licences and for certain frequency bands. Spectrum trading was introduced by the regulators as a means of encouraging more flexible use of spectrum, discouraging spectrum hoarding, and to meet growing demand. They are usually associated with company take-overs, mergers, etc. Depending on the regulation in the country, the licence may not be transferred but the rights and obligations attached to the licence may be transferred to another party [15]. Spectrum trading transactions may be subject to competition assessment to ensure there is no harm done to the market conditions after the trading.

7.5.1.3 Sharing Spectrum with Other Users

A dramatic rise in mobile broadband services has increased pressure on regulators to release additional spectrum needed for capacity expansion. Since most spectrum bands are already in use, regulators are struggling to identify new spectrum bands to meet demand. This has forced industry to explore spectrum management approaches beyond the commonly used approaches. Spectrum sharing, a collective use of spectrum for two or more services, is gaining popularity as a viable approach. Although most of the radio spectrum is shared between different services, spectrum sharing was not considered as a viable approach for mobile services until recently. Technological advancements enable effective management and minimise the harmful interference paving the way for spectrum users to viably share spectrum. This also creates a potential avenue for regulators to enhance efficient use of spectrum. There is, however, a significant amount of uncertainty and complexity as to how such sharing might work in practice and the value of such spectrum to the marketplace.

Spectrum sharing can be implemented in different ways depending on the type of spectrum users and their rights of use i.e. relative priorities and whether sharing is on a geographic or time basis. From a regulatory perspective, this sharing can be authorised under license or licence-exempt frameworks:

- Licensed Shared Access is a regulated sharing approach in which sharing is permitted between multiple spectrum users in a coordinated and managed way. This can be implemented in two different ways.
  - ○ Allowing sharing partners to have equal rights with some restrictions on the transmission power or where the spectrum is used i.e. geography to minimise interference or have different access rights.
  - ○ Allowing incumbent licensed users to sub-license spectrum to other spectrum users, thus enabling access to licensed spectrum. This is usually implemented with predefined access rights amongst the parties involved. In this case, the incumbent user or primary user usually has the highest priority. The secondary user who shares spectrum with the primary user, can use the unused spectrum channels by the incumbent. The European framework which has been introduced to foster a more efficient usage of spectrum within the band from 2.3 GHz to 2.4 GHz is a good example for LSA [16].

- Licence Exempt Access is a largely unregulated approach by which all parties use a band of spectrum as a common and shared resource without need for a licence. Typically, parties are subject to regulatory-defined mandatory constraints such as radiated power and to protocols that serve as 'politeness rules' for the commons. There is no hierarchy of use and spectrum availability is best effort.

We can organise sharing parties within 'user classes' [17]. These classes are a means to describe spectrum users who are subject to a particular set of sharing conditions and enjoy certain rights of protection against harmful interference from other users. They can be listed as 'Primary', "Secondary", and 'Other' user classes:

- Primary user: The incumbent protected from all other usage within the shared spectrum band.
- Secondary user: Protected from tertiary usage, but subject to restrictions arising from primary usage.
- Other user: Not protected against any other usage in the shared spectrum band.

Spectrum sharing is usually done by the incumbent entering into a sharing agreement with the sharing partner. It is the incumbent's role to provide spectrum to be shared with the sharing partner and to abide by the sharing commitments. To enable sharing, the incumbent needs to transition from an exclusively licensed spectrum environment to a shared environment, and thus inherently lose flexibility and freedom of spectrum use. There are considerable real and potential costs, complexities and risks associated with making this transition. Incentives can take a variety of forms to assist incumbents in realising the benefits outlined above. One option is the introduction of a 'Spectrum Currency' that rewards agencies 'that move quickly to promote more effective spectrum use by making some of their spectrum available for sharing with other users' [18]. Similarly, in Europe, forms of Administrative Incentive Pricing (AIP) provide a means to reward incumbents for making more efficient use of spectrum they hold by making unused spectrum available to commercial providers.

Sharing has the potential to bring significant costs and risks to the incumbent:

- Smaller geographic spectrum footprint and/or restrictions on day or time of use.
- Reduced freedom of choice and flexibility as to how spectrum can be used.
- Increased operational costs and complexity to coordinate and manage use within a shared spectrum environment.
- Potential to be locked into legacy technology or added complexity to introduce new technology.
- Risk of degradation of incumbent services and capabilities if sharing arrangements do not conform to the regulated or negotiated performance levels.

There are four key stakeholders involved in spectrum sharing for MNO use [19]: the incumbent non-MNO spectrum users; the regulatory bodies and policy makers that oversee and influence spectrum sharing approaches and outcomes; the MNOs and surrounding mobile ecosystem; and the mobile end users. Each party has an important role to play in making spectrum sharing a successful reality for the industry. For instance, the roles and motivations of the parties involved in mobile spectrum sharing can be summarised below.

1. **The incumbent**: provides access to spectrum to be shared. Sharing usually decreases the economic value of spectrum and adds additional restrictions on the use of spectrum. Incumbents may be required to be compensated depending on the situation.
2. **The sharing licensee:** benefits from sharing as it gains access to new spectrum which otherwise would not be available. There may be some conditions that the sharing partner has to abide by to minimise the harmful interference.
3. **The regulator:** facilitates the sharing framework such that all parties involved benefits from sharing. The regulator has to introduce clear steps and rules such that all parties can operate in an interference-free environment.

7.5.2 Dimensions of Spectrum Sharing

Spectrum sharing can improve spectrum utilisation in three dimensions: frequency, location and time. It enables a network to opportunistically use any available frequency channels at points in time and space when and where they are not in use. Since most technologies do not use RF channels continuously and in all locations, it is possible to share the same frequency by multiple services. There are multiple ways that spectrum can be shared. The simplest way is to carve out geographic exclusion zones so that no two systems operate simultaneously in the same area. Usually, primary users have higher priority rights on usage. This means that secondary users are required to protect primary users from interference in the shared spectrum bands. The secondary user must do this by restricting operation of their transmission so that its transmission does not cause harmful interference within their radii of operation. The restricted operation for the secondary user may not be attractive depending on the use cases. However, this is commonly used as a mitigation technique to minimise interference in the shared bands. It requires coordination between the spectrum users in the band. Use of geo-location databases is recently being trialled and implemented in some countries as a way to manage coordination between different users. A more complex implementation of this is dynamic spectrum sharing where the primary licensees can share spectrum with secondary users on a moment by moment basis.

7.5.2.1 The Role of Regulation for Spectrum Sharing in the Mobile Industry

The primary objective of most regulators is to increase the efficient use of spectrum whilst enabling interference-free operation. If this is possible, the economic value of spectrum can be increased. If sharing is introduced, regulators must ensure existing licensed incumbents are equitably treated and that incumbent services, subjected to sharing, are adequately protected from interference or interruption. Therefore, the use of spectrum sharing is highly dependent on each sharing situation between an incumbent and sharing partner. Because of this inherent need for customised solutions, regulators are somewhat reluctant to generalise the spectrum sharing approaches for all spectrum bands. However, it is important to clearly define the sharing terms that can be applicable to each sharing scenario. Hence, the regulator can only provide a light regulation in terms of defining the terms to serve the purpose. In many sharing circumstances, sharing partners would have to (i.e. incumbents and MNOs) negotiate and define customised terms that enable both parties to maximise utility of the shared band. However, the regulator can play an important role in enabling the sharing process:

- Identifying and encouraging sharing opportunities, particularly incentivising incumbents to participate.
- Developing frameworks to make sharing happen.
- Facilitating band harmonisation and standardisation within the regions or globally.
- Developing the licensing regime to facilitate sharing i.e. allow sharing as an interim step until the incumbent clears its operation or enabling incumbents to sub-licence spectrum to MNOs.
- Enabling opportunities for MNOs to acquire shared spectrum based on the market mechanisms.
- Facilitating and managing the negotiation process.

7.5.2.2 Spectrum-Sharing Examples

Some spectrum-sharing examples in use today can be presented below.

White spaces are the name given to parts of spectrum that are unused in a particular location and at a particular time. TV white spaces exist between airwaves primarily used for digital terrestrial TV broadcasting (470 MHz to 790 MHz). In the UK, Ofcom has published consultation proposals for authorising certain types of white space devices (WSD) on a licensed basis [20]. This follows publication of a statement outlining Ofcom's policy decisions on implementing TV white spaces and a draft Interface Requirement and draft statutory instrument for device licence exemption. The statement follows completion of coexistence and framework testing, policy development and industry trials (some ongoing). Ofcom was aiming for the framework authorising commercial use of white space technology to be in place in the autumn of 2015 to enable use of new wireless applications.

In the US, the TV white spaces framework follows a Dynamic Spectrum Access (DSA) approach based on geo-location databases. The FCC issued final rules to allow low power unlicensed devices to operate on unused channels in the TV broadcast bands in September 2010, following a six-year process that began in May 2004 with the first Notice of Proposed Rulemaking (NPRM) in 2004 [21]. FCC has developed the technical capabilities of sensing to reduce risk of interference, mandated a database-driven approach for transmissions including high power and fixed services [22]. The IEEE has developed two standards for communications in the TV white spaces, IEEE 802.22 and IEEE 802.11af. FCC has authorised few TV white space database systems for operation, and manufacturers are only now beginning to obtain FCC certifications for this category of devices [23].

Recently, a few regulators have initiated frameworks to share spectrum as a possible solution to mitigate the scarcity of spectrum and address the increasing wireless capacity demand. Some of them can be listed as follows:

- UK regulator Ofcom published a "framework for enabling shared use of spectrum for" four spectrum bands namely 1800 MHz band (i.e. 1781.7 to 1785 MHz paired with 1876.7

to 1880 MHz), 2300 MHz band (i.e. 2390 to 2400 MHz), 3800 to 4200 MHz band and 24.25-26.5 GHz (only available for indoor use) [24].

- Swedish regulator 'PTS' is promoting spectrum sharing between different spectrum usage in the long term, and even creating international spectrum sharing harmonisation [25].
- The European Union (EU) is strongly advocating the use of spectrum sharing to cope with spectrum shortage. The EU is also funding research activities in the use of shared spectrum for a range of applications, including wireless broadband [26].
- In the US, 3.5 GHz is under consideration for spectrum sharing. It is also recommended that spectrum sharing can 'double' the spectrum available.

7.6 Spectrum Bands Considered for 5G

The current trend is to release spectrum based on the technology neutral principal i.e. spectrum is not constrained with a specific technology. However, in practice, these conditions are derived on the basis of the most likely usage patterns in the future. Although regulators follow a technology neutral approach to assign spectrum, industry has considered some spectrum bands specifically for 5G mobile services. Within Europe, Radio Spectrum Policy Group (RSPG) formed an opinion identifying the following spectrum bands as pioneer bands for the strategic roadmap towards 5G in Europe [27]:

- 700 MHz band: Similar to 800 MHz band, 700 MHz band was used for TV transmission and cleared for mobile use. Propagation characteristics of the 700 MHz band enables achievement of nationwide coverage and deep indoor coverage.
- 3.4 to 3.8 GHz band: The RSPG considered the 3400–3800 MHz band to be the primary band suitable for the introduction of 5G services in Europe. This band is already harmonised for mobile networks and consists of wide bandwidths of up to 400 MHz of continuous spectrum enabling deployment of wide channel bandwidths.
- 26 GHz band (i.e. 24.25–27.5 GHz band): This band was proposed by Europe at WRC-15 in order to strengthen the global harmonisation for IMT services. The IMT-2020 (the name used in ITU for the standards of 5G) services are expected to support ultra-low latency and very high bit-rate applications which, in turn, require larger contiguous blocks of spectrum than those available in frequency bands that had previously been identified for IMT services. Although spectrum re-farming provides additional spectrum, this will not provide sufficient amounts of spectrum to allow MNOs to support new services envisaged for 5G. Millimetre wave (mmWave) spectrum is viewed as a key band considered amongst the new spectrum bands for potential mobile services. Although the propagation characteristics of mmWave is significantly different to sub 1 GHz, around 2 GHz bands commonly used for mobile services, the availability of a large quantity of spectrum allows support of much wider bandwidth which, in turn, provides increased mobile broadband speeds or ultra-high capacity. By definition, mmWave band is the spectrum between 30 GHz and 300 GHz. This is because wavelengths at these frequencies are between 1 mm and 1 cm long. However, mmWave term is commonly used to refer to frequencies above 24 GHz [28]. The 26 GHz band (24.25 GHz–27.5 GHz) was identified as the 'pioneer' mmWave band in Europe for 5G [27].

- In addition, there are a number of other frequency bands above 24 GHz (37–43.5 GHz, 45.5–47 GHz, 47.2–48.2 and 66–71 GHz [29]) which were identified for IMT services enabling deployment of 5G networks on a worldwide basis at the World Radio Conference in 2019.

7.6.1 Example Illustration of Spectrum Deployment Strategy for MNOs

It should be noted that a sensible approach for MNOs is to use a combination of these three 5G spectrum options. Figure 7.3 illustrates a high level technology and spectrum deployment strategy for a typical MNO. Table 7.2 shows a spectrum usage and re-farming example applicable for a typical MNO.

- **700 MHz band:** if acquired from the auction in 2020, it would be sensible to deploy for early release of 5G since the timing would work out just about right. Furthermore, this band is recognised as one of the pioneer spectrum bands for 5G in Europe; hence, it can benefit from economies of scale.
- **800 and 1800 MHz band**: LTE is already deployed in these bands in addition to some GSM operation in 1800 MHz band. It would be sensible to continue with no change for some time as it would also benefit from high handset penetration.

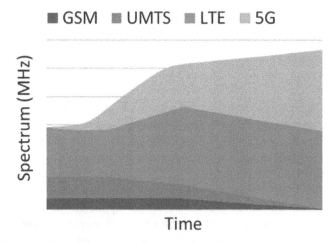

Figure 7.3   Illustration of spectrum re-farming and usage with time.

Table 7.2   Illustration of spectrum re-farming and usage for different bands.

Band (MHz)	700	800	900		1800		2100		2300		3400
Duplexing	FDD	FDD	FDD	FDD	FDD	FDD	FDD	FDD	TDD	TDD	TDD
Today		LTE	GSM	UMTS	GSM	LTE	UMTS	LTE	LTE		
Short/ med. term		LTE	GSM	UMTS	GSM	LTE	LTE	LTE	LTE	LTE	5G
Long term	5G	LTE	5G	5G	GSM	LTE	LTE	LTE	LTE	LTE	5G

- **900 MHz band**: MNOs have currently deployed HSDPA in this band. It would make sense to continue with that deployment in the short- and medium-term as investments have already been carried out. But, in the long term, it would make sense to migrate both technologies to 5G when the time is right i.e. more 5G capacity is required. It is also costly to run parallel networks i.e. 2G, 3G, 4G and 5G. At some points, 2G and 3G will be switched off to make cost savings and exploit the benefits from the latest technologies.
- **2100 MHz band**: This is the band with the highest UE penetration in the market for most operators today. Migrating to LTE would provide the most benefits immediately due to this high UE penetration.
- **2300 MHz band**: Most handsets also support this band as this is adjacent to band 1 i.e. 2100 MHz band. Again, to exploit the immediate capacity benefits, it would make sense to deploy LTE in this band.
- **3400 MHz band:** This is another band considered as Pioneer bands for 5G. It would make sense to deploy 5G when it has matured to gain benefits from economies of scale.

To develop a spectrum and deployment strategy, it is essential to understand the MNO's business priorities. This way we can develop customised technology, site and spectrum deployment strategies that best suit for MNOs. It also requires a deep understanding of these matters and understanding the big picture, together with the other influential factors, can help the wireless industry and users developing a balanced and realistic strategies.

7.6.2 Local Access Spectrum

Usually, in most countries, MNOs are assigned spectrum at the national level. The situation may be different in large countries such as the USA, where spectrum is assigned at regional level. However, the characteristics of mobile networks show that use of spectrum is not uniform across the country. This is because demand for mobile services is not uniform. Hence, demand for use of spectrum, is usually concentrated around where people live, commute or work. For instance, in the UK, 45% of mobile data traffic is generated in 1% of the geography [30]. This means that spectrum is not used optimally in some geographic parts of the country. Further, acquiring spectrum suitable to deploy mobile networks and operating a mobile network is a capital intensive operation where, typically, up to 5 MNOs are operating in a country. However, there may be other smaller spectrum users who may want to deploy small networks to serve local demand. To enable opportunity for such users to acquire spectrum, regulators have started releasing spectrum for local use. Below are some examples where national regulators have made spectrum available for local use at the time of writing this book:

- German regulator Bundesnetzagentur (or BNetzA) has set aside 3.7–3.8 GHz spectrum band for local use [31] in 2019. Applicants can pay a licence fee that depends on the assigned bandwidth (varies from 10 MHz to 100 MHz), the assignment term (in years or fractions of a year based on each month commenced), and the surface area covered by the assignment in square kilometres.
- UK regulator Ofcom sets out a licensing scheme known as Local Access licence that gives a way for other users to access spectrum that has already been licensed to the

MNOs, in locations where an MNO is not using their spectrum. The licence is initially available for a duration of 3 months for a single location or area but can be extended with the agreement with the MNO at a one-off cost of £950 per licence. It is available for all mobile bands currently licensed to MNOs i.e. 800, 900, 1400, 1800, 900, 2100, 2300, 2600 and 3400 MHz bands.

New users can apply to Ofcom for these licences provided they meet the conditions set by the Ofcom.

- The Swedish Post and Telecom Authority (PTS) is planning to make the 3.5 to 3.8 GHz bands available in two different ways [32] 1) Predefined geographic areas with high-population density and/or high demand; and 2) Smaller geographic areas located outside the predefined areas.
- The national regulator in France, ARCEP, has opened the 2.6 GHz TDD band (i.e. 2570–2620 MHz frequency band) for enterprises in May 2019 [33]. ARCEP made this spectrum available for local professional mobile radio (PMR) networks that satisfy the specific connectivity needs of certain companies and organisations that belong to 'vertical' markets. ARCEP has also provided an opportunity for enterprises wanting to test a technology or an innovative solution with a regulatory sandbox without necessarily having to comply with the entire regulatory framework that normally applies.
- In Finland, the Transport and Communications Agency Traficom, made the frequency band 2300–2320 MHz available for private local radio networks such as Private LTE Networks, in June 2020 [34]. These networks could be deployed to serve a limited user group based on mobile communications technology. Deployment of larger regional networks is not allowed using this frequency band. Furthermore, the networks must not be used for public telecommunications to a non-restricted user group.

Note that this information is up to date at the time of writing this book. Other regulators are expected to follow this trend in years to come making local use of widely available spectrum for new spectrum users.

References

[1] ITU article. https://www.itu.int/en/about/Pages/default.aspx
[2] World Radiocommunication Conferences (WRC). https://www.itu.int/en/ITU-R/conferences/wrc/Pages/default.aspx
[3] Article 5, ITU radio regulations articles, edition of 2020. Available at: http://www.itu.int/pub/R-REG-RR-2020
[4] CEPT web page. https://www.cept.org/cept/background
[5] Ofcom web page. https://www.ofcom.org.uk/about-ofcom/what-is-ofcom
[6] Ofcom report. https://www.ofcom.org.uk/__data/assets/pdf_file/0024/156156/annual-report-18-19.pdf
[7] ITU report. http://www.itu.int/pub/T-REC
[8] Ofcom report. https://www.ofcom.org.uk/spectrum/spectrum-management/spectrum-awards/awards-archive/800mhz-2.6ghz

[9] Radiolinja's history. https://web.archive.org/web/20061023214602/http://www.elisa.com/english/index.cfm?t=6&o=6532.50&did=10101

[10] Goddard, M. (2018). IET Berkshire prestige mobile lecture - where is the spectrum coming from? 16 July.

[11] Background of IMT-2000. https://www.itu.int/en/ITU-T/imt-2000/Pages/background.aspx

[12] CEPT report on COMPATIBILITY BETWEEN UMTS 900/1800 AND SYSTEMS, OPERATING IN ADJACENT BANDS, Krakow. March, https://www.ecodocdb.dk/download/815bc8b3-2614/ECCREP096.PDF

[13] Elisa web page. https://halberdbastion.com/intelligence/mobile-networks/elisa-finland

[14] (2019). News article on "Commission decides to harmonise radio spectrum for the future 5G". 24 January. https://ec.europa.eu/digital-single-market/en/news/commission-decides-harmonise-radio-spectrum-future-5g

[15] Ofcom report. https://www.ofcom.org.uk/__data/assets/pdf_file/0029/88337/Trading-guidance-notes.pdf

[16] ECC Report 205. LSA (LSA). Feb. 2014.

[17] http://www.gsma.com/spectrum/wp-content/uploads/2014/02/The-Impacts-of-Licensed-Shared-Use-of-Spectrum.-Deloitte.-Feb-20142.pdf

[18] President's Council of Advisors on Science and Technology. (July 2012). Realizing the full potential of government held spectrum to spur economic growth. p. ix

[19] Report on the impact of licensed shared use of spectrum, A report prepared in collaboration with Deloitte and real wireless for the GSM association. (23 January 2014).

[20] Ofcom report http://stakeholders.ofcom.org.uk/spectrum/tv-white-spaces

[21] (2011). Report on Unlicensed Operation in the TV Broadcast Bands, Second Memorandum Opinion and Order, 25 FCC Rcd 18661 (2010); Unlicensed Operation in the TV Broadcast Bands, Notice of Proposed Rulemaking, 19 FCC Rcd 10018 (2004); IEEE Communications. Emerging cognitive radio applications: A survey. March.

[22] (2008). Report on unlicensed operation in the Tv broad. Bands, 23 FCC Rcd 16807, 16386 71-73.

[23] Report on innovation in the broadcast television bands: Allocations, channel sharing and improvements to VHF. ET Docket No. 10-235, Report and Order, FCC 12-45 6 (rel. Apr. 27, 2012).

[24] Ofcom Shared access licences, https://www.ofcom.org.uk/manage-your-licence/radiocommunication-licences/shared-access

[25] report. http://pts.se/upload/Remisser/2014/PTS-Spectrum-Strategy-draft-summary-140224.pdf

[26] Report published by plum consultancy. http://www.plumconsulting.co.uk/pdfs/Plum_Jan2014_harmonised_spectrum_for_mobile_asean_south_asia.pdf

[27] RSPG opinion paper. https://rspg-spectrum.eu/wp-content/uploads/2013/05/RPSG16-032-Opinion_5G.pdf

[28] Ofcom report. https://www.ofcom.org.uk/__data/assets/pdf_file/0021/97023/5G-update-08022017.pdf

[29] ITU web article. https://www.itu.int/en/itunews/Documents/2019/2019-06/2019_ITUNews06-en.pdf

[30] 26 GHz - the opportunity for a fresh approach to licensing in higher frequencies, A report from real wireless for the UK spectrum policy forum. (January 2021). https://www. real-wireless.com/news/publications

[31] BNetZa web article. https://www.bundesnetzagentur.de/SharedDocs/Pressemitteilungen/ EN/2019/20191031_LokalesBreitband.html

[32] Preliminary study prior to future assignment of frequencies for 5G (3.4 – 3.8 GHz and 24.25 – 27.5 GHz). https://www.pts.se/globalassets/startpage/dokument/icke-legala-dokument/rapporter/2018/radio/preliminary-study-frequencies-5g-pts-er-2018-4.pdf

[33] Web page on ARCEP. https://en.arcep.fr/news/press-releases/p/n/businesses-digital-transformation-1.html

[34] Webpage on Traficom. https://www.traficom.fi/en/news/new-leaf-turns-history-telecommunications-fortums-loviisa-power-plant-gets-its-own-mob

8

Fundamentals of Digital Communication

In previous chapters, we approached modern communication systems from a macroscopic perspective. Yet, the operation of networks and management of multiple users in a system is built upon fundamental principles and processes that facilitate point-to-point communication, which each transmitter and receiver must carry out proficiently. In this chapter, we introduce the fundamental components of digital communication systems. The main aim of the chapter is not to give a very detailed technical exposition of the subject of digital communications, but to place the crowning achievements of communications engineers of the last century into the context of modern systems. Topics covered in this chapter include encoding and decoding, modulation and demodulation, the communication channel, system performance analysis, and system impairments, which engineers must cope with when designing and deploying systems in practice. The treatment of such a breadth of topics can only really be done qualitatively in a single chapter and, even then, it is impossible to cover every important topic. Hence, a section devoted to discussing other texts that readers might wish to consult for more detail is included at the end of the chapter. We begin the chapter by looking at the basic concept behind this magnificent, elegant technology.

8.1 Basic Digital Communication System Overview

A point-to-point digital communication system has a very simple structure. It is comprised of a transmitter, a receiver, and a channel. The channel is the medium that links the transmitter to the receiver; it carries the transmitted message from the former to the latter. The transmitter can be decomposed into several parts: an information source, various encoders, and a modulator (Figure 8.1). The receiver contains the inverse processes for each of these parts: a demodulator, a decoder, and an interpreter or sink.

Digital communication is completely analogous to human communication. In humans, an idea is first formed in the brain (information source) and is then structured into a sentence using a particular language (encoder). The sentence is spoken by creating sound waves using pulmonary pressure delivered by the lungs, which is shaped by the larynx, vocal tract, mouth, tongue, jaw etc. (modulator). These waves travel through the air (channel) to reach the intended recipient. The sound waves cause the eardrums of the recipient

The Technology and Business of Mobile Communications: An Introduction, First Edition.
Mythri Hunukumbure, Justin P. Coon, Ben Allen, and Tony Vernon.
© 2022 John Wiley & Sons Ltd. Published 2022 by John Wiley & Sons Ltd.

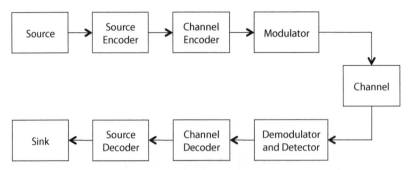

Figure 8.1 Basic digital communication system model.

to vibrate, and these vibrations are transferred through the middle ear to the inner ear (demodulation), where they stimulate the vestibulocochlear nerve (decoding). The message is then transmitted via this nerve to the temporal lobe of the brain, where it is interpreted.

In most human communication, information is represented in the form of words and sentences, which are built from an alphabet. The English alphabet has 26 characters, and words and sentences generally conform to a number of rules that have been developed over hundreds of years. Digital communication is actually much simpler, but again, it is analogous. The alphabet in this case is comprised of only two characters known as *bits*, which are typically taken to be 0 and 1. Words can be formed from different patterns of 0s and 1s. The lengths of these words can be fixed or variable. In fact, the idea of *source coding* relates to this concept of representing complicated words with sequences of bits. We will touch on this in more detail in Section 8.2.2. For the moment, let us assume that we form a word out of m bits. It is apparent that, without any constraint on the formation of the word (i.e. any combination of 0s and 1s can be used), we have 2^m possible words of length m.

In the most basic digital communication system, each of these words can be mapped to, or represented by, an electrical signal that is a function of time. This mapping is particularly useful for conveying information to a distant receiver, since electrical, or more generally electromagnetic, signals can be propagated efficiently over long distances using optical fibre, radiowaves, and various other mechanisms and media. The particular combination of bits that constitute a given word can be encoded in the amplitude, phase, frequency, or other structural aspects of the signal. When transmitting wirelessly or in the presence of other electromagnetic signals (interference), these signals can be used to *modulate* a carrier waveform, which is usually a sinusoid at a higher frequency. At the receiver, the signal is first *demodulated*, i.e. the information-bearing characteristics of the signal are extracted from the received waveform in preparation for further processing. The signal is then mapped back to a corresponding sequence of bits, and the transmitted message can be recovered.

In general, errors can arise when decoding the transmitted message at the receiver. Typical physical processes that lead to such errors will be covered later in the chapter. At a high level, it is intuitive that one may wish to put mechanisms in place during the encoding phase of transmission that would mitigate the effects of these errors. One way of achieving this aim is to append *redundant* bits to each word. In our example, where m bits are parsed to form a word, suppose for each bit of information (out of m bits), we introduce two more bits that are exactly the same. This is known as *repetition coding*, and is a simple way to improve the resilience of a communication system to possible errors that might occur during transmission of the word. The total number of bits in each *codeword*[1] is now $3m$. However, there are still only 2^m possible codewords since we started with m-bit sequences. The result of this encoding operation is that the receiver is able to recover the transmitted word even if some of the bits were altered during transmission (i.e. flipped from 0 to 1 or vice versa). This ability to correct errors introduced in the communication channel is what has propelled communication systems forward in the last 70 years. We will discuss this in more detail in the next section.

This very simple introduction to how a digital communication system works provides a high-level view of the main processes needed to convey information digitally between two devices. Of course, specific implementations of such systems will require careful design of the different components illustrated in Figure 8.1, and these designs will differ amongst implementations. For now, let us move on with a more in-depth discussion of the encoding operations undertaken in all digital systems.

8.2 Encoding Information

Referring to Figure 8.1, two encoding operations are performed at the transmitter of a digital communication system following generation of information at source: *source coding* and *channel coding*. We will provide details of how these operations are carried out in this section. First, we begin with a fundamental discussion of how we might convert data into a discrete form that can be naturally communicated using digital means.

8.2.1 Sampling

Communication signals (as with most real-world signals) are physically analogue by nature (e.g. a voltage). However, modern integrated digital circuitry is built to cope with signals that take on discrete values (i.e. digital signals). In an information theoretic sense, transforming signals from the analogue (continuous) to the digital (discrete) world has some fairly noteworthy advantages, as we will see below. The transformation from the analogue domain to the digital domain is known as *sampling*. Effectively, this process involves observing a continuous signal and taking snapshots at regularly spaced intervals (the *sampling interval*). This is illustrated in Figure 8.2.

The question is: how often should we sample a continuous signal such that we can perfectly reconstruct the original signal from the sampled data points? The answer lies with Nyquist's Sampling Theorem.[2] This theorem states that, for a band-limited signal $X(f)$ with $X(f) = 0$ for $|f| > W$, the sampling interval T_s must be no greater than $1/2W$ seconds for perfect reconstruction to be possible. Equivalently, the *sampling frequency* $f_s = 1/T_s$ must be no less than $2W$ Hz for perfect reconstruction to be possible.

The reasoning behind this fundamental result is somewhat intuitive given our basic knowledge of time-frequency analysis. The sampling operation is mathematically equivalent to

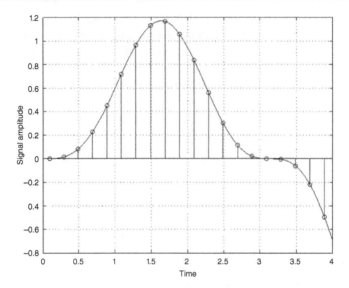

Figure 8.2 Illustration of sampling. The solid line represents the original signal, whereas the points denoted by markers correspond to the samples taken at regular intervals.

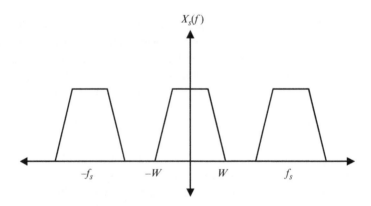

Figure 8.3 Sampled spectrum for $X(f)$, which is denoted here by $X_s(f)$. The sampling frequency f_s in this example is greater than $2W$.

multiplying the continuous signal by a sequence of impulses spaced apart by T_s seconds. In the frequency domain, this is equivalent to the convolution of the signal spectrum and a sequence of impulses spaced apart by f_s Hz, which basically replicates the signal spectrum every f_s Hz (Figure 8.3). Hence, if $f_s < 2W$, the signal spectrum will overlap at the edges, a condition known as *aliasing*. No amount of filtering can be done to remove the effects of aliasing once it has occurred, so we must ensure that $f_s \geq 2W$ for perfect reconstruction.

Once we have a discrete version of a signal, we can proceed with digitisation. This process involves mapping a discrete sample to a sequence of bits, as briefly discussed above.

A huge amount of research has been conducted to devise optimal methods of performing this task of *quantisation*. We will not concern ourselves with the nuances of these methods here. Instead, we will assume that an appropriate mapping operation is carried out, after which we will be left with a sequence of bits.

Before moving on, it is important to note that sampling is not necessary in all digital communication systems. Indeed, if we wish to transmit the works of Shakespeare, we only need a method of encoding the letters and symbols that constitute the English language (which is a discrete set of characters) into sequences of bits. From such a starting point, we can move forward with a discussion of *source coding*.

8.2.2 Source Coding

Source coding is the process by which a set of symbols, or samples, used to represent information is mapped to a sequence of binary digits. In general, the a priori probabilities of the letters in the alphabet are different. Consider, as an example, the English language. Some studies have shown that the letter 'a' appears in English text about twice as often as the letter 'd'. A good source encoder will take this information into account when mapping symbols to bit sequences. Here, we will discuss two methods of source coding: fixed-length coding and variable-length coding.

8.2.2.1 Fixed-length Coding

For discrete sources without memory,[3] mapping letters or symbols to sequences of bits is relatively straightforward. Suppose the alphabet in question has M letters and our source produces a letter every T seconds. Then we need $R = \log_2 M$ bits to represent each symbol (assuming M is a power of 2), and the *rate* of encoding the source message is R bits per symbol, or R/T bits per second. The rate R is actually greater than this if M is not a power of two. Mapping symbols to bit sequences of a constant length is known as *fixed-length source coding*.

This approach to source coding is simple, but can be inefficient in the sense that bits can be wasted in the encoding process. For example, if we have an alphabet with three symbols and we encode each symbol with two bits, we only use three of four possible bit sequences. This inefficiency is particularly common when the size of the alphabet is small. In such a case, it is beneficial to group symbols and treat each group as a *supersymbol* in the encoding process. Mapping a group of J symbols to a sequence of N bits results in a rate of $R = N/J$ bits per symbol. Letting J grow large results in a very efficient encoder.

The source-coding approach discussed here is *distortionless* since the mapping from source symbols to bit sequences is one-to-one. However, we can reduce the rate by constructing a many-to-one mapping, which can be particularly useful if we know a priori that some source symbol sequences will occur with greater probability than others. In this case, we can map those uniquely (one-to-one), whilst creating a many-to-one mapping for the remaining low-probability source sequences. But this approach introduces distortion, or an error, each time the many-to-one mapping is invoked in practice. Claude Shannon[4] was able to show that the probability that distortion is introduced can be made arbitrarily small by using the aforementioned grouping method. Shannon's result was a mathematical theorem now referred to as the *source coding theorem* [2].

8.2.2.2 Variable-length Coding

Another way to improve efficiency of a source code is to allow the encoding operation to map symbols to bit sequences of different lengths. Let us explore this idea by, once again, considering the English language. This language is constructed from an alphabet with 26 symbols. But, as we noted before, some letters are used more often than others. Does it make sense to encode each letter using the same number of bits (in this case, five bits per symbol would be required)? From the point of view of maximizing efficiency, or minimizing the encoding rate, it certainly does not make sense. Intuitively, it would be best to use fewer bits to represent frequently used letters. This idea was recognised quite some time ago, and forms the basis of Morse code.[5]

In general, the approach described here is known as *variable-length source coding*. This method of source coding is best described by way of example. Consider an alphabet of three symbols, a, b and c, where $P(a) = 1/2$ and $P(b) = P(c) = 1/4$. The lowest possible rate of the source code is 3/2 bits. If we use fixed-length codewords to represent these three symbols, we would need two bits, and the efficiency would be 3/4, which is quite poor. However, we could implement the following mapping, based on our a priori knowledge of each source symbol:

$$a \mapsto 0$$
$$b \mapsto 10$$
$$c \mapsto 11.$$

The average rate of this code turns out to be 3/2, and thus the code is maximally efficient. It is important that variable-length codes are *uniquely* and *instantaneously* decodable. That is, we want to be able to map a bit sequence back to the original source symbol uniquely (the mapping should be one-to-one). And, we want to be able to do this quickly! It is possible to design a mapping that is one-to-one, but where the decoder must wait a very long time before the inverse mapping back to the source alphabet can take place, this is clearly not a good way to design a practical communication system. An ingenious and simple algorithm devised by David Huffman in 1952 (whilst a PhD student at MIT) yields codewords that are uniquely and instantaneously decodable whilst being maximally efficient. This technique is now referred to as *Huffman coding*, and the basic principle is used in many applications today, including multimedia codecs and data compression.

8.2.3 Channel Coding

In the early twentieth century, Ralph Hartley, a scientist at Bell Labs, developed a way to quantify the rate at which information could be sent over a channel with limited bandwidth. Specifically, he suggested that one pulse in a set of M distinct electrical pulses is capable of representing $\log_2 M$ bits of information. Combining this idea with Nyquist's results on sampling in band-limited channels (see Section 8.2.1), Hartley postulated that the maximum rate of communication, measured in bits per second (bps), could be mathematically expressed as

$$R = 2W \log_2 M$$

where W denotes the bandwidth of the channel in units of Hz.

Hartley's Law, as it is known, was modelled on systems where noise did not corrupt transmitted signals. This was clearly a weakness of the theory, which did not go unnoticed. In the years that followed Hartley's initial work, Shannon proved that *reliable* (i.e. zero error-rate) communication could be achieved even in noisy channels as long as the rate of transmission is less than a positive upper bound, which is known as the *capacity* of the channel. Shannon's result is a mathematical theorem known as the *channel coding theorem* (see Box 8.1). He published his work in 1948, and this contribution is credited as laying the foundation for what became the field of mathematical engineering known as *information theory*.

The channel coding theorem describes the conditions under which error-free communication is possible, and it makes reference to the existence of *codes* that achieve this condition. But it does not tell us how to construct these codes. Shannon envisaged a simple communication model (basically, the same one shown in Figure 8.1) where information that has been mapped to a sequence of bits via a source encoder could then be encoded further to provide robustness against noise. This encoding operation is the *channel encoder* shown in Figure 8.1. There is a corresponding *channel decoder* at the receiver.

Encoding to provide robustness of this nature can broadly be classified into two categories: automatic request-for-repeat (ARQ) coding and forward-error correction (FEC) coding. The goal of the former is to detect error events and inform the receiver when an error has occurred. Thus, ARQ codes are typically known as *error detection* codes. Once an error has been detected, the receiver requests a re-transmission. This is basically analogous to asking someone to repeat what they have just said to you if you did not hear it the first time. There are many flavours of ARQ, and it is a complex subject that brings in many different aspects of the communication system, from the physical encoding and decoding of binary sequences to network level functions.

In the spirit of focusing on the digital transceiver model depicted in Figure 8.1, we will focus on FEC for the remainder of this section. FEC codes are typically capable of detecting and correcting errors at the receiver without any additional re-transmissions. Hence, they are sometimes called *error correcting codes* (ECC). FEC codes can be categorised as *linear codes* and *nonlinear block codes*. The latter is rarely, if ever, used in practice; hence, we will only touch on the former. Linear codes are labelled as such because the linear combination of any set of codewords is, itself, a codeword. The class of linear codes can be further divided into *linear block codes* and *trellis codes*.

Box 8.1   The channel coding theorem

For a discrete memoryless channel, all rates below capacity C are achievable. Specifically, for every rate less than C, there exists a sequence of codes with this rate that yield a probability of error tending to zero.

Conversely, any sequence of codes with probability of error tending to zero must have rate less than C.

8.2.3.1 Linear Block Codes

A linear block code is constructed by mapping a sequence of, say, k bits to a sequence of $n > k$ bits known as the *codeword*. A code constructed in this manner is said to have rate $r = k/n$. Hence, low-rate codes correspond to the case where a significant amount of redundancy is added to the original information sequence, whereas high-rate codes have little redundancy included, and are thus more susceptible to errors introduced by the channel. It is often desirable for a linear block code to have the form illustrated in Figure 8.4. A linear code with this structure is called a *linear systematic code*. Such codes are useful in applications where the message part is combined with a hash function that facilitates a quick verification of the correctness of the message part. In this case, the decoder can simply observe the message part without going through the potentially lengthy and computationally expensive process of fully decoding the codeword.

For any linear block code, encoding can be achieved by arranging the information bits into a (row) vector $\mathbf{b} = (b_0, \dots, b_{k-1})$ and multiplying this vector by a $k \times n$ *generator matrix* $\mathbf{G}$, resulting in the codeword $\mathbf{c} = \mathbf{b}\mathbf{G}$. The generator matrix uniquely determines the code structure. Furthermore, for each generator matrix, there exists a *parity check matrix* $\mathbf{H}$ such that $\mathbf{G}\mathbf{H}^T$ is the $k \times (n - k)$ matrix of zeros. In other words, the generator matrix is orthogonal to its corresponding parity check matrix. This is a key property of linear block codes. Geometrically, one can think of the encoding operation as a mapping of the information sequence to a point in a higher dimensional subspace (Figure 8.5). The generator matrix provides this mapping and, in fact, the rows of this matrix span the higher dimensional subspace. The parity check matrix, on the other hand, lies in the *null space* of this higher dimensional subspace, i.e. it corresponds to all the codewords that are orthogonal to each

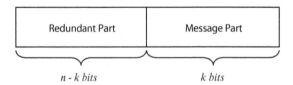

Redundant Part	Message Part
n - k bits	k bits

Figure 8.4 Illustration of a systematic codeword.

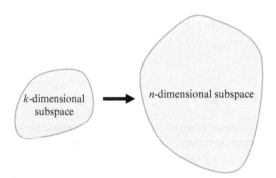

Figure 8.5 Illustration of the mapping operation from a subspace of low dimension to one of high dimension that a channel encoder performs.

valid codeword generated by **G**. This geometric separation is what gives the code its error detecting and correcting capability.

At the receiver, the codeword may be received with certain bits in error (i.e. flipped). A plethora of decoding algorithms exist for different linear block codes, which can be used to recover the transmitted codeword even in the presence of errors. A simple method that can be used for small codes is known as *syndrome decoding*. Without going into detail, suffice to say that the key to correcting errors introduced in the channel lies in the relationship between the generator matrix and the parity check matrix. By multiplying the received codeword by the parity check matrix, the decoder can compute the *syndrome*, which gives locations of the erroneous bits (provided there are not too many errors). With this information, the correct message can be recovered. It is important to note that it may be possible for a decoder to *detect* that errors in the received codeword are present, but not possible for it to *correct* these errors. Indeed, different codes have different error detecting and correcting capabilities, and a huge amount of effort has been expended over the past 70 years to obtain particular codes that exhibit good error detection/correction properties.

To understand the basic properties of linear block codes a little better, it is useful to consider a few basic codes that one might encounter in practice. Let us start with the *repetition code*, for which a single bit is repeated n times to achieve a rate $1/n$ code. This is a low-rate code with reasonably strong error detecting and correcting capabilities: up to $n-1$ errors can be detected, and roughly $n/2$ errors can be corrected. Contrast this example with the *single-parity-check code*, where a modulo-2 sum of the k information bits is appended to this word to create a codeword of length $n = k + 1$. This is a high-rate code ($r = (n-1)/n$), but the error detecting/correcting capability of this code is very low: a single bit error can be detected, and the code is not capable of correcting any errors.

Arguably, the first sophisticated codes developed can be attributed to Richard Hamming (a mathematician at Bell Labs in the middle of the twentieth century). *Hamming codes* are a class of linear block codes of length $n = 2^r - 1$ derived from $k = 2^r - r - 1$ information bits, where r is an integer greater than one. For example, take $r = 3$, which yields a rate 4/7 code that can detect up to two errors per codeword and correct a single error. For comparison, the repetition code that yields the same error detection/correction performance has length $n = 3$ and rate 1/3.

Many other more powerful codes exist, and it is impossible to list all of the most important examples here. However, to conclude this discussion on linear block codes, one more class of codes must be named due to their significance in modern communication systems: *low-density parity check (LDPC) codes*. LDPC codes were originally discovered by Robert Gallager in the 1960s [3], forgotten for 35 years, and then rediscovered by David MacKay and others in the mid-1990s [4]. These codes are particularly special for the following reasons:

1) Their decoding performance is within a fraction of a decibel of channel capacity;
2) They can be decoded using practical circuits on state-of-the-art integrated chips;
3) Because they were discovered many years ago, there is no fundamental intellectual property (IP) protection in place that would prohibit commercialisation of systems using LDPC codes.

For these reasons, LDPC codes have received attention in many fields for many practical applications, including digital video communication by satellite and terrestrial means as well as in non-volatile memory devices (e.g. NAND flash memory).

8.2.3.2 Trellis Codes

The term *trellis codes* is used to describe a general class of codes for which encoding is often performed in a sequential manner rather than block-by-block as is done for linear block codes. The result is that the encoder maintains a state that changes with each new input bit. The evolution of the states can be mapped through a structure that resembles a garden trellis (Figure 8.6). This trellis structure can be used to simplify the decoding procedure to achieve near optimal performance. We will not delve into the details of trellis codes apart from to mention two very important types: *convolutional codes* and *turbo codes*.

Convolutional codes derive their name from the process used to encode a stream of bits. Typically, a sequence of bits is entered one at a time into a sort of binary filter. Figure 8.7 shows the archetype convolutional encoder. For every bit that is entered, two coded bits are output. Thus, this is a half-rate code. In the figure, the two blocks store the input bits. Thus, we see that this 'filter' has a memory order of two bits. Hence, the trellis shown in Figure 8.6 describes the encoder state. The optimal decoder will find the most likely path through the trellis given the received bit stream.

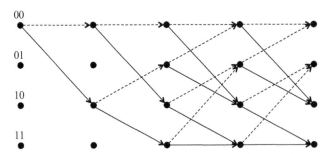

Figure 8.6 An example of a four-state system modelled dynamically as a trellis. At each stage moving left to right, a single bit is appended to the front (left) of the state (represented by two bits), and the right-most bit is dropped. This yields a new state.

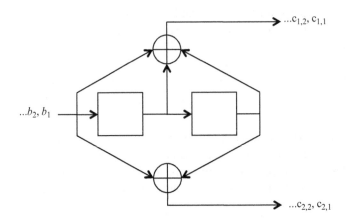

Figure 8.7 Archetype half-rate convolutional encoder.

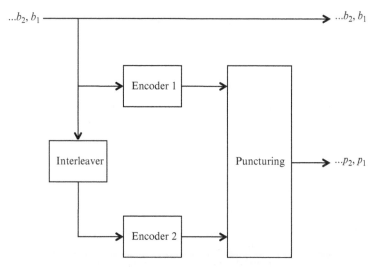

Figure 8.8 The basic structure of a parallel, concatenated convolutional encoder.

The era of capacity-approaching codes with iterative decoding truly began in 1993 with the report of the discovery of *turbo codes* at the International Communication Conference in Geneva. These codes are built from constituent convolutional encoders running in parallel, with a *bit interleaver* employed between them, which permutes the bits over a large block size to create independence between the two encoded streams. This independence ensures that burst errors, which may affect contiguous segments of the coded bit stream, can be more easily corrected. At the outputs of the encoders, *puncturing* – i.e. erasing bits according to a specified pattern – is often employed to increase the code rate. To better understand how puncturing works, consider a coded bit stream with rate 1/3 (for every information bit, three bits are output from the encoder). By removing (puncturing) every sixth bit, the code rate is increased to 2/5, which arises from a 1:3 ratio of information to coded bits multiplied by a 6:5 ratio of coded bits to punctured bits.[6] Finally, the encoded bits are appended to the original bit sequence to form a systematic code. An illustration of this *parallel, concatenated convolutional encoder* is shown in Figure 8.8.

The name *turbo code* is actually derived from the decoding operation. The inventors of turbo codes intuitively felt that an iterative decoding procedure could improve the performance of the code as long as the information fed back into the decoding process with each iteration was effectively unbiased, i.e. it did not contain information about the original received message, only new information gleaned from the decoding operation. This feedback resembled a sort of *turbo principle*. Many modern decoders, including those used to decode LDPC codes, operate in an iterative manner.

8.3 Signal Representation and Modulation

As mentioned above, once digital information is encoded, groups of bits are mapped to waveforms in preparation for transmission through the channel to the receiver. In this context, a communication signal can be represented as a function of time, say $s(t)$, which occupies some nonzero bandwidth and is centred at a frequency f_c. Such a signal is termed a *band-pass signal*. Systems engineers find it useful to employ a *low-pass equivalent* model to represent communication signals and systems for the purpose of analysis and design. This enables them to abstract the actual carrier frequency (within reason) and focus on other design elements that can be influenced at *baseband*, which is the low frequency part of a communications circuit (from DC to some nominal maximum frequency that loosely defines the rate of communication). The relationship between the low-pass and band-pass signals is given by

$$s(t) = \Re\left\{s_l(t)e^{i2\pi f_c t}\right\}$$

where $s_l(t)$ is the low-pass signal, and $\Re\{\cdot\}$ denotes the real part of the argument. Notice that the multiplication of the low-pass signal and the exponential in the time domain corresponds to the convolution of the spectra of these two signals in the frequency domain. This result is known as the *modulation property* of Fourier transforms and, thus, this process of attaining a band-pass signal from a low-pass information-bearing signal is called *modulation*. For purposes of illustration, an example of the signal spectrum of the band-pass signal is shown in Figure 8.9.

Notice that the mathematical representation of the band-pass signal given above suggests that, in general, the low-pass signal is complex-valued, i.e.

$$s_l(t) = x_s(t) + iy_s(t).$$

In fact, this seemingly unphysical representation of the signal has a well-defined origin in the use of orthogonal properties of sinusoids. Without going into detail, it is sufficient to interpret the low-pass signal $x_s(t)$ as the signal that modulates a *cosine* waveform and the signal $y_s(t)$ as the one that modulates a *sine* waveform at the same carrier frequency

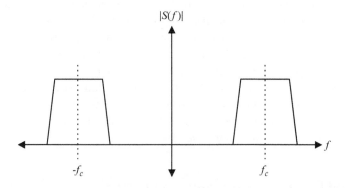

Figure 8.9 Illustration of the magnitude of the spectrum of $s(t)$, denoted here by $|S(f)|$. Each component of this spectrum resembles the original spectrum of the low-pass signal $s_l(t)$, but shifted from the origin by $\pm f_c$.

f_c. Hence, the standard terms used to refer to these two signals are the *in-phase* and *quadrature* components of $s_l(t)$. It is these components that carry the information of the signal.

8.3.1 Mapping Bits to Signals

Typically, a group of bits is represented by a particular pair of low-pass waveforms $x_s(t)$ and $y_s(t)$. At baseband, the model discussed above reduces to a much simpler representation by noting that $f_c = 0$ and, thus, the quadrature component $y_s(t)$ is not used. The signal is then given by $s(t) = x_s(t)$. Baseband modulation schemes are the simplest to understand. Bits are usually encoded in a rectangular pulse of period T seconds. A common and intuitive approach to mapping a single bit to a pulse is the *non-return-to-zero* (NRZ) scheme, whereby a 1 is represented by a positive level of a given amplitude for the duration of the pulse, and a 0 is represented by a negative level (Figure 8.10). Various other similar schemes exist for encoding information in a binary manner. Some of these facilitate self-clocking by forcing the waveform to have a transition at the beginning or middle of each pulse (e.g. *return-to-zero*, *Manchester*), whilst others are energy efficient or enable communication without an accurate clock (e.g. *inverted* schemes). Because there is no carrier waveform at baseband, these modulation schemes are typically used in wired systems, and are thus often referred to as *line codes*.

In the cellular access network, telecommunication systems operate at band-pass. Hence, both in-phase and quadrature carrier waveforms are modulated. Information is encoded by again mapping bit sequences to pulses $x_s(t)$ and $y_s(t)$, each of T seconds in length. The shape of each pulse can be optimised to satisfy design criteria. Typically, the amplitudes or frequencies of these pulses provide the degrees of freedom required to convey information. For example, in a simple *amplitude shift keying* (ASK) transmission, the low-pass signal can be represented as

$$s_{l,k}(t) = a_k g(t)$$

where $g(t)$ is the basic pulse and a_k signifies the amplitude (positive or negative) of the kth signal. In the simplest binary case, the amplitude can be zero or one, which yields the aptly named *on-off keying* (OOK) scheme. Figure 8.11 illustrates an example of band-pass OOK where the symbol period T is 0.2 seconds and the carrier frequency is $f_c = 40$ Hz.

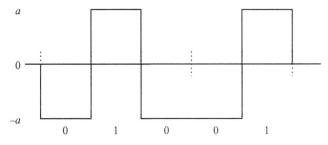

Figure 8.10 An NRZ line code.

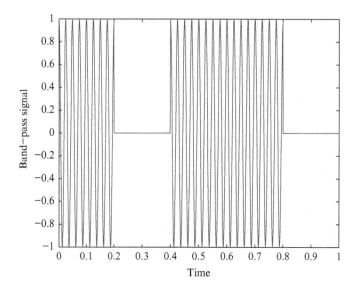

Figure 8.11 Illustration of OOK with a symbol period of 0.2 seconds and a carrier frequency of 40 Hz.

As an alternative to encoding information in the signal amplitude, the *phase* of the band-pass signal can also be used. This is known as *phase shift keying* (PSK). For example, suppose we have $m = \log_2 M$ bits that we wish to transmit.[7] Using phase only, we can let

$$x_s(t) = g(t)\cos\left(\frac{2\pi}{M}(k-1)\right)$$

$$y_s(t) = g(t)\sin\left(\frac{2\pi}{M}(k-1)\right)$$

with k denoting the index of the signal transmitted. For $M = 4$, the resulting band-pass signal may look something like the waveform shown in Figure 8.12. This example is known as *quadrature phase-shift keying* (QPSK) since four phases are used to convey information. Notice that the amplitude of the signal is constant; only the phase changes (every 0.25 seconds).

When working with complex low-pass signals, it is convenient to represent the possible signal set using a *constellation diagram*. For the QPSK example shown in Figure 8.12, the constellation diagram has four points in the complex plane, as illustrated in Figure 8.13. When encoding information in both amplitude and phase, the constellation diagram is more complicated. An example can be seen in Figure 8.14, which depicts the constellation of a *quadrature amplitude modulation* (QAM) scheme with $M = 16$ possible signals. Many variants on these basic signal sets exist, including schemes that exploit variations in frequency to encode information. However, most modern telecommunication systems utilise some form of amplitude and phase modulation, along with slightly more complicated multi-plexing techniques (see below), to convey information.

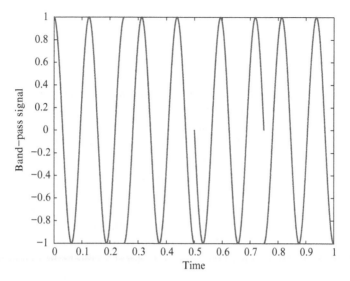

Figure 8.12 Band-pass QPSK signal.

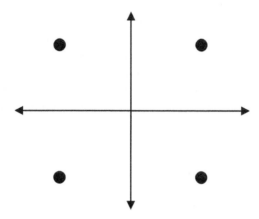

Figure 8.13 Constellation diagram for QPSK.

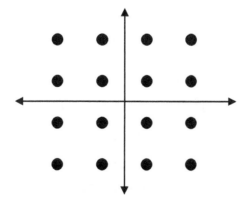

Figure 8.14 Constellation diagram for 16-QAM.

8.3.2 Signal Spectrum

So far in this brief introduction to modulation, we have focused on the time-domain representation of the signals. We briefly noted that a basic pulse – represented by the function $g(t)$ – forms the template of the modulated signal. In fact, this pulse defines the spectral characteristics of the signal, which are very important when considering how close in frequency different transmissions (from different users or devices) can occur. Modulated signals can be viewed as random processes since one does not know a priori what a transmitted message will be. Thus, to study the spectral properties of these signals, we must consider the power spectral density of the process.

In the case of ASK, PSK, and QAM signals that have zero mean – e.g. the constellation diagrams exhibit symmetry and all constellation symbols are transmitted with equal probability – it turns out the power spectral density of the low-pass signal is proportional to the square of the spectral representation of the pulse $g(t)$. If the pulse is rectangular, its Fourier transform is a sinc function $(\sin(\pi fT)/(\pi fT))$. Hence, using the modulation property of Fourier transforms, the spectrum of the band-pass signal is simply the low-pass spectrum shifted up and down in frequency by f_c. An example of the power spectral density of a QAM signal is shown in Figure 8.15.

It is worth noting that rectangular pulses yield power spectra that decay like f^{-2}. Here, we are referring to the *side lobes* of the spectrum, which are the minor peaks in the spectrum that can be seen on either side of the main peak in Figure 8.15. In applications where signals from different users must be frequency multi-plexed with limited bandwidth, such a slow decay may lead to interference problems amongst adjacent transmissions. Of course, other pulse functions exist, some of which exhibit power spectral decay of f^{-6} (e.g. the *raised cosine pulse*). From our discussion of time-frequency analysis, we

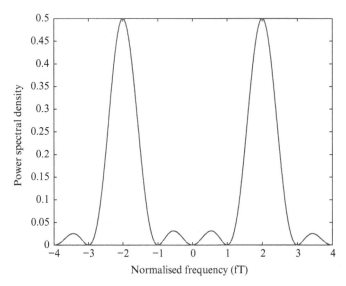

Figure 8.15 Illustration of the power spectral density for zero-mean, memoryless modulation schemes.

recall that localisation in one domain yields spreading in the other. Hence, any increase in the rate of spectral decay necessarily results in a longer pulse duration for a fixed bandwidth.

8.4 Signal Demodulation and Detection

Once a bit sequence is encoded, mapped to signals, and a carrier waveform is modulated, the band-pass signal can be transmitted over the channel to the receiver (see Figure 8.1). In this section, we first develop a very basic model of such a transmission and show how we can begin to recover the transmitted message at the receiver.

8.4.1 System Model and Sources of Noise

In almost all communication systems, the channel between the source and destination distorts the transmitted signal. The most basic model that describes this distortion is the *linear time-invariant* model, whereby the receiver observes an attenuated signal in the presence of additive noise. Mathematically, the simplest linear time-invariant model can be described by the equation

$$r(t) = hs(t) + n(t)$$

where $r(t)$ is the received signal, h denotes an attenuation factor, and $n(t)$ is additive noise.

Typically, the noise has zero mean and corresponds to a Gaussian stochastic process[8] with power spectral density $N_0/2$ W/Hz, with N_0 being a constant. This statistical model for the noise corresponds to so-called *thermal noise*, which manifests as small, random voltage fluctuations in electronic components located in the receiver. These voltage fluctuations result from spontaneous changes in electron energy states, which occur at temperatures above absolute zero regardless of whether or not an external power source is present. Thermal noise is typically modelled as being *white*, i.e. all frequency components are equally present in the noise signal. This is a good approximation for practical communication systems since they are band-limited, but, in fact, thermal noise decays quickly at very high frequencies due to quantum effects.[9] As a result of the white noise approximation, the basic channel model given above is known as the *additive white Gaussian noise* (AWGN) channel.

It should be noted that other sources of noise exist in communication systems. *Shot noise* can affect signal fidelity in low-current portions of the network, particularly in sensitive parts of the optical backhaul network. This type of noise results from the discrete nature of charge and quantum mechanical effects, such as electron tunnelling in semiconductors. In low-frequency parts of the network, *flicker noise* (a.k.a. $1/f$ *noise*) can cause problems. This type of noise is generally thought to result from electron emission processes in circuit components, but there is no single unifying theory that can be used to predict flicker noise. From a macroscopic perspective, the most important general observation is that noise at low frequencies increases approximately like $1/f$ as f gets small.

8.4.2 Demodulation

We characterised the properties of the signal $s(t)$ in Section 8.3. Here, we are concerned with recovering this signal at the receiver in the presence of AWGN. Signal recovery is split into *demodulation* and *detection*. Demodulation is the process of getting the received signal into a form that can be used to make a decision about what was transmitted. Detection is the process of making that decision.

Let us focus on demodulation for linear, memoryless modulation schemes (e.g. ASK, PSK). Throughout this discussion, we will assume $h = 1$ in our mathematical model stated above. Demodulation is typically performed in one of two ways, known as *correlation demodulation* and *matched-filter demodulation*. Both operate in a similar manner by applying a number of linear operations (averaging processes) to the received signal.

The principle of the correlation demodulator is based on the fact that the total set of all possible signals that could have been transmitted lie in a subspace that can be characterised by a set of *orthogonal basis functions*. Essentially, this means that the kth signal $s_k(t)$ can be represented as a weighted sum of these basis functions, where each weight is determined by integrating the product of the signal and the basis function over a period. If the basis functions are known a priori (by design), then the weights completely describe the transmitted signal. Referring to the basic AWGN system model described above, this suggests that it is possible to obtain an estimate of the transmitted message at the receiver by integrating the product of the *received signal* and each basis function over a symbol period. Such an operation would result in the transmitted signal weights corrupted by noise samples. The process of integration amounts to *correlating* the received signal with the basis functions, which effectively projects the received signal onto the corresponding function. This process gives the demodulator its name. Note that correlation must be performed for each basis function. Thus, correlation demodulators are typically implemented in a parallel manner as shown in Figure 8.16.

For some line codes or ASK, the signal is one-dimensional. Consequently, there is a single basis function proportional to the transmitted pulse $g(t)$ (see Section 8.3). For PSK and other two-dimensional (complex) constellations, two basis functions must be used. There are many ways to define the basis functions for complex signals and, in fact, many receiver designs turn out to be equivalent because of this fact. One valid approach is to consider the band-pass signal and choose the basis functions to satisfy orthogonality conditions. In practice, however, the carrier frequency f_c can be very large, and correlation of such high-frequency signals can be difficult. A solution to this problem is to *down-convert* the received signal to an *intermediate frequency* where correlation can be performed using much more conventional electronic circuitry. Another approach is to down-convert the signal to baseband, i.e. shift the frequency of the received signal by f_c and consider the real and imaginary components of the signal separately.

Instead of performing a set of correlations, we can pass the received signal through a set of linear filters and sample at the end of each symbol period (Figure 8.17). This amounts to convolving the received signal and the filter impulse response over a symbol period. The key to designing this demodulator is to specify the filter responses correctly. It can be shown that if one chooses each filter response to be proportional to one of the possible transmitted signals, the received signal-to-noise ratio (SNR) at the output of the

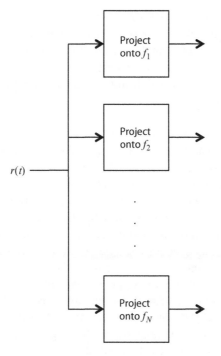

Figure 8.16 Correlation demodulator. The functions $f_1, \ldots, f_N$ are the orthogonal basis functions that span the signal set.

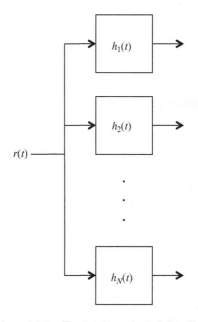

Figure 8.17 Matched-filter demodulator. The functions $h_1, \ldots, h_N$ are the filter responses.

matched-filter demodulator is maximised. Hence, this demodulator is optimal in this sense. The *matching* of the filter response to the transmitted signal gives the demodulator its name. In practice, it turns out that the *matched filter* demodulator is identical to the correlation demodulator when each filter response is proportional to the corresponding time-reversed basis function in the correlator shifted by a symbol period, i.e. $h_k(t) \propto f_k(T - t)$.

8.4.3 Detection

Correlation demodulation and matched-filter demodulation often lead to a very similar, if not the same, set of outputs that can be used for detection. These outputs are discrete since the demodulation operation naturally samples the continuous received signal. The outputs are actually *statistics* since their properties are unknown a priori and they are affected by random noise. In fact, they are *sufficient statistics* that can be used to decide what signal was transmitted, i.e. no relevant information has been thrown away during the demodulation process, and thus the statistics can be used to detect the transmitted signal.

Detection is performed optimally by determining the most probable transmitted signal given the received statistics. This is known as *maximum a posteriori (MAP) detection*. When the a priori probabilities of the transmitted signals are all equal, the MAP detector reduces to the *maximum likelihood (ML)* detector, which determines the signal that, given it were transmitted, would yield the most probable received signal. Note the subtle difference in the MAP and ML schemes. In a sense, they only differ in the hypothesis that is chosen, i.e. either the received signal or the transmitted signal is taken to be known for the probability calculation. When the transmit signals are not equally probable (and this information is known at the receiver), the MAP scheme yields the correct outcome more often than the ML scheme. In an AWGN channel with equally probable transmitted signals, the ML detector chooses the signal that minimises the Euclidean distance between the received statistics and the signal weights. This is known as the *minimum distance criterion* and is a standard approach to detector design in practice.

8.5 Performance Analysis

Now that we understand the basics of how to convey information digitally from a transmitter to an intended receiver, we turn our attention to two fundamental performance metrics that are used in practice to aid the analysis and design of communication systems. These metrics ultimately relate to our physical understanding of the channel coding theorem, which links rate to reliability. The first metric we will consider is the *capacity* of the system which, by definition, quantifies the maximum rate we can achieve. The second metric is the *probability of error* or *error rate*, which quantifies reliability in practical scenarios.

8.5.1 Capacity

Shannon's channel coding theorem promises that reliable communication is possible as long as transmission takes place at a rate that is less than channel capacity. Practically, we may wish to determine what the channel capacity is. Here, we focus on the most important channel model for telecommunications: the band-limited additive white Gaussian noise (AWGN) channel described by the system equation

$$r(t) = s(t) + n(t).$$

It is possible to calculate the capacity of this channel under the constraint that the transmit power cannot exceed P watts. Assuming the noise power spectral density is $N_0/2$ W/Hz and the bandwidth is W/Hz, it is possible to show that the capacity of this channel is

$$C = W \log_2 \left(1 + \frac{P}{WN_0}\right) \text{ bits per second}$$

This equation is possibly the most referenced equation in information and communication theory, and forms the benchmark for many practical systems. Given the trend to ever-greater bandwidth in some communication systems, the observant reader will pose the question: what is the capacity as W grows greater whilst maintaining a fixed total transmit power constraint of P watts? A Maclaurin series expansion shows that the limiting capacity for a large bandwidth system is

$$\frac{P}{N_0} \log_2 e \text{ bits per second}$$

From these two expressions, we see that, in the finite bandwidth case, an increase in transmit power results in a logarithmic increase in the achievable rate of the system. Yet, when the bandwidth of the channel is large, we can achieve a *linear* increase in capacity by turning up the transmit power. This observation somewhat motivates the use of larger bandwidth in single-user systems, and is the key prediction that led to the idea of *ultrawideband* communication for personal-area networks. The problem, of course, is that most wireless systems must cope with increasing amounts of interference from other users. Hence, research into ultrawideband technology is typically limited to specialist applications at present.

We can use the fundamental capacity equation given above to establish an important bound on the energy required to transmit each bit of information. To do this, we note that the rate of reliable communication must be less than the capacity C. Let this rate be denoted by R, such that $R < C$. Furthermore, it is easy to deduce that the power constraint P can be expressed in terms of the rate of communication (in bits per second) and the *energy per bit* E_b as follows

$$P = E_b R.$$

Substituting into the capacity expression for the band-limited AWGN channel and performing some algebra yields the required relation

$$\frac{E_b}{N_0} > \frac{2^{R/W} - 1}{R/W}.$$

The ratio R/W is known as the *spectral efficiency* of the system, and it has units of bits per second per Hz. As a result of this inequality, we can deduce that the ratio E_b/N_0 must be greater than $\ln 2$ or -1.59 dB in order for the spectral efficiency to be positive, i.e. so that reliable communication can take place at a positive rate. This is an important limit that can be used to analyse the performance of FEC schemes in AWGN channels. Effectively, designers of codes aim to achieve a very low probability of bit error for E_b/N_0 ratios close to this limit. So-called *capacity-approaching codes* achieve very low error rates at E_b/N_0 ratios only a fraction of a dB above the lower bound.

8.5.2 Bit-error Rate and Symbol-error Rate

Another key performance metric in communication systems operating in AWGN channels is the probability that a bit is detected erroneously. This is usually referred to as the *probability of bit error* or the *bit-error rate* (BER). Theoretically, the BER can be calculated for a given modulation scheme (without accounting for FEC) by determining the probability that the MAP or ML detector decides in favour of an erroneous transmitted signal. As an example, consider a simple binary PSK scheme (BPSK) operating in an AWGN channel, whereby a 1 is signified by a positive amplitude and a 0 is signified by a negative amplitude. Gaussian noise corrupts each transmitted signal, so that the distribution of the received signal looks something like that shown in Figure 8.18. In this case, the ML detector decides upon the transmitted signal by observing the sign of the received signal since, as can be deduced from the figure, this corresponds to the most likely outcome. Thus, the probability that a signal was erroneously detected is equal to the Gaussian tail probability obtained by integrating the curve from zero to positive or negative infinity, depending on which signal was actually transmitted. In our example, if a $+1$ were transmitted, the probability of error would amount to the area under the right-hand Gaussian curve in Figure 8.18 that lies to the left of the vertical axis. The width of each Gaussian curve in the figure corresponds to the standard deviation of the noise. Hence, for systems with little noise, most of the area of each curve will be confined to one side of the vertical axis, meaning that the probability of error will be low.

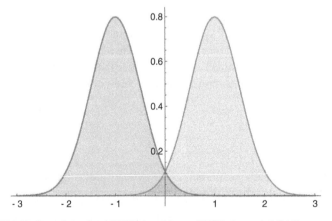

Figure 8.18 Distribution of received BPSK signal in an AWGN channel. A 0 bit corresponds to a -1 having been transmitted; a 1 bit corresponds to a +1 having been transmitted. At the receiver, a mixture of Gaussian signals is observed, one centred at -1 and the other centred at +1.

These calculations differ for each modulation scheme, and a tremendous amount of research has gone into characterising the BER performance of popular schemes, including ASK, PSK, QAM, and more advanced methods such as differential modulation. Some of these calculations are very involved, and approximations must be made. It is often more straightforward to calculate a slightly more macroscopic performance metric: the *symbol-error rate* (SER). The method used to do this is effectively the same as was outlined above in the BPSK example. Here, we forego a discussion of the mathematics and instead focus on comparing the SER performance of two different modulation techniques. Through such a comparison, we will be able to make several important observations.

Let us begin with ASK. In this modulation scheme, we only use signal amplitude, which is a real-valued parameter, to convey information. We can add any number of amplitude levels that we like to encode more information; however, these amplitude levels will become closely packed as their number increases if we constrain the average power that can be used to transmit the message. This manifests in worsening SER performance as the number of levels increases, as can be observed in Figure 8.19.

Now, let us turn our attention to the performance of PSK in an AWGN channel. In this case, we encode information using two real degrees of freedom – the amplitudes of the real and imaginary components of the signal – under the constraint that the resulting signal has constant amplitude. Hence, as we increase the number of constellation points, we pack these more closely along the unit circle in the complex plane, but we do not increase the required transmit power. This is a more efficient use of resources, and the resulting SER for a given E_b / N_0 is lower than the corresponding ASK scheme, as shown in Figure 8.19. For this reason, PSK or QAM are typically preferred in practice over ASK.

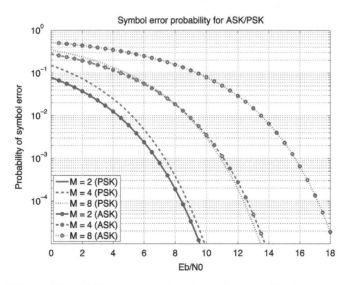

Figure 8.19 SER for ASK and PSK schemes. *M* denotes the number of levels or constellation points for a given modulation scheme.

8.6 Communication Through Dispersive Channels

In many modern communication systems, the channel is dispersive, i.e. it smears the transmitted signal in time so that the receiver cannot immediately recover the transmitted message without additional conditioning and/or processing. If dispersion is basically constant over a period of time, the system model is well approximated by the linear time-invariant model

$$r(t) = (h * s)(t) + n(t)$$

where * denotes the convolution operator.

Until now, we have only dealt with non-dispersive channels, which makes things simple. If the channel is dispersive, something must be done to ensure the receiver can collect and align the correct signals to perform demodulation, detection and decoding. The receiver could take into account the dispersive effects and form an ML estimate of the transmitted signal as was done in the non-dispersive case. However, the complexity of this detection operation is, in practice, often prohibitive. Alternatively, the transmitter and/or receiver can attempt to mitigate or remove the dispersive effects from the signal. This can be done through a process known as *equalisation*. Here, we provide a basic overview of equalisation techniques that can be used to combat channel dispersion. We begin with an overview of time-domain equalisation techniques. Following this exposition, we will turn our attention to frequency-domain techniques, which form the backbone of the physical layer in modern telecommunication systems.

8.6.1 Time-domain Equalisation and Detection

Dispersive channels cause *intersymbol interference* (ISI), which must be mitigated at the receiver. This interference manifests through the filter-like nature of the channel (i.e. the convolution in the system equation given above). If left unchecked, ISI caused by the channel will lead to poor performance. *Time-domain equalisation* (TDE) can be used to remove the ISI completely or otherwise reduce its deleterious effects. Many techniques exist for equalising a received message in the time domain. Time-domain equalisers range from simple linear finite impulse response (FIR) filters to non-linear trellis-based solutions.

Let us begin with the optimal approach. The optimal time-domain equaliser is a trellis-based technique derived from the BCJR algorithm,[10] which was originally presented as a method of decoding convolutional codes. The equaliser adaptation of the BCJR algorithm utilises the FIR properties of the dispersive channel and the finite alphabet of the transmitted symbol constellation – which may be, for example, BPSK – to find each symbol that maximises the a posteriori probability of transmission conditioned upon the received message. Consequently, this equaliser is usually referred to as a maximum a posteriori (MAP) equaliser.

Another trellis-based equaliser is the *maximum likelihood sequence estimator* (MLSE). In practice, the MLSE is implemented via the Viterbi algorithm, which can also be used to decode convolutional codes. This method differs from the symbol-by-symbol MAP equaliser in that it assumes all symbols are transmitted with equal probability. Furthermore, the MLSE determines the *sequence* of symbols that was most likely transmitted rather than operating on a symbol-by-symbol basis. The MLSE and the MAP equaliser perform similarly when they are used as conventional equalisers and no a priori information is known about the transmitted symbol distributions. This behaviour results from the fact that the MAP criterion reduces to the maximum likelihood (ML) criterion when all transmitted symbols are equally probable.

Although trellis-based equalisers such as the MLSE and the MAP equaliser achieve optimal performance, they are quite complex. Each of these techniques can be viewed as a state machine with M^L states where M is the size of the symbol constellation (e.g. $M = 4$ for QPSK) and L is the discrete memory order of the FIR channel, i.e. the amount of dispersion caused by the channel measured in symbol intervals. Consequently, these equalisers are extremely complex when the channel impulse response (CIR) is long or when large symbol constellations are used. For example, consider a system operating in a channel with a memory order of $L = 5$ that employs a QAM scheme with $M = 16$ symbols in the constellation. The implementation of an MLSE in this case would require the inspection of $16^5 \approx 1{,}000{,}000$ states, which would be impractical. Consequently, it is sometimes beneficial to employ suboptimal equalisers that require significantly fewer computations.

The suboptimal linear transversal equaliser (LTE) is the most basic of time-domain equalisers. These equalisers are typically FIR filters as depicted in Figure 8.20. The computational complexity of an LTE increases linearly with the channel memory order L. Thus, LTEs are generally much less complex than the trellis-based methods discussed above.

In Figure 8.20, the equaliser co-efficients $\{g_n\}$ can be spaced at the inverse of the symbol rate to give a *symbol-spaced* equaliser, or at shorter intervals, which results in a *fractionally-spaced* equaliser. Whatever the spacing, the filter co-efficients are typically chosen such that they satisfy a specific criterion. One popular criterion that is used to design an LTE is the *peak distortion criterion*. The term *peak distortion* refers to the peak value of ISI in the received signal that is caused by the channel. An equaliser that is designed such that it removes all ISI from the received message, thus leaving only the desired signal and noise, is known as a *zero forcing* (ZF) equaliser.

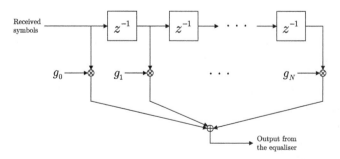

Figure 8.20 Linear transversal equaliser. The parameters $\{g_n\}$ are the equaliser co-efficients.

Mathematically, suppose the received signal can be represented as

$$y_k = \sum_{m=-\infty}^{\infty} a_m x_{k-m} + v_k$$

where a_k represents the transmitted constellation symbols, x_k denotes the response of the channel (including transmit/receive filtering), and v_k is noise. Passing this through a linear infinite impulse (IIR) equaliser with filter co-efficients g_k gives

$$\hat{a}_n = \sum_{k=-\infty}^{\infty} y_k g_{n-k}.$$

This is just the convolution of the composite channel co-efficients x_k, the equaliser co-efficients g_k, and the transmitted symbols a_k. Taking Z-transforms, we obtain

$$\hat{A}(z) = G(z)X(z)A(z) + G(z)V(z)$$

which uses the property that the Z-transform of the convolution of two sequences is the product of their individual transforms. Hence, if we wish to fully suppress the distortion caused by ISI, we must choose the equaliser co-efficients g_k such that $G(z) = 1/X(z)$.

There are two problems with this approach: 1) if the composite channel response $X(z)$ is poorly conditioned, then this equaliser can amplify noise considerably, and 2) IIR equaliser filters are impractical, and approximations to FIR filters could be poor. Frequency-domain equalisation solutions get around the second problem (see the next section). The first can be treated by considering a different objective for the equaliser design.

One such alternative approach considers the popular mean-square error (MSE) criterion for equaliser co-efficient selection. In this case, the equaliser co-efficients are designed such that the MSE between the desired signal and the signal at the output of the equaliser is minimised. An equaliser that satisfies this criterion is known as a *minimum mean-square error* (MMSE) equaliser. The filter co-efficients that satisfy the MMSE criterion can generally be computed directly with ease. Without going into details, the MMSE equaliser IIR transfer function is

$$G(z) = \frac{X^*\left(z^{-1}\right)}{X(z)X^*\left(z^{-1}\right) + N_0}$$

where N_0 is the variance of the complex noise samples at the output of the demodulator.

Suboptimal equalisers need not be linear. Indeed, the decision-feedback equaliser (DFE) is a popular suboptimal non-linear equaliser. DFEs make hard decisions on equalised symbols, then use these decisions to remove interference caused by the dispersive channel from subsequent received symbols. A block diagram of a basic DFE is depicted in Figure 8.21. From this figure, it is apparent that the DFE uses a feedforward transversal filter to perform an initial equalisation procedure on a received symbol. The output of this filter is then added to the output of a feedback transversal filter, which attempts to remove any residual ISI from the signal. Finally, the output of this feedback step is passed through a symbol detector where a hard decision is made on the equalised symbol. The process is

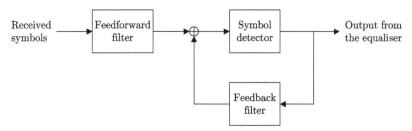

Figure 8.21 Decision-feedback equaliser structure.

then repeated for the next symbol. Note that the filter co-efficients for a DFE can be designed to satisfy either the ZF criterion or the MMSE criterion.

DFEs suffer from error propagation when incorrect decisions are made on the equalised symbols since the feedback process in a DFE relies on the quality of these decisions. Even with error propagation, however, DFEs typically outperform linear ZF and MMSE equalisers. It should be noted that although the DFE provides a significant performance improvement over these linear techniques, it does not perform better than the MLSE.

8.6.2 Frequency-domain Equalisation

Frequency-domain transforms, such as the efficient hardware implementation of the DFT known as the fast Fourier transform (FFT), can be used to perform channel equalisation in the frequency domain as an alternative to the TDE methods discussed above. This approach has many benefits, including an implementation complexity that is even lower than linear TDE as illustrated in Figure 8.22.

8.6.2.1 Multicarrier Modulation
In this section, we will see that it is often beneficial to treat channel dispersion in a holistic way through a combined modulation, coding and equalisation approach known as *coded orthogonal frequency-division multiplexing*, or C-OFDM.[11] OFDM is a *multicarrier* scheme, whereby a large chunk of bandwidth is partitioned into tens or hundreds of narrowband subchannels called *sub-carriers*. This modulation/equalisation scheme is perhaps the most important digital communication technique currently in use. Many current standards and products are based on OFDM. For example, products that draw on the current digital audio/video broadcasting (DAB/DVB) specifications utilise an OFDM air interface. Also, the IEEE 802.11a/g/n/ac standards for WLANs all specify OFDM as the modulation technique of choice. And, of course, fourth-generation cellular standards stipulate OFDM as the modulation on the downlink (base station to user equipment), and 5G New Radio also exploits OFDM.

OFDM is based on simple, fundamental principles, but is a holistic solution to the problems posed by dispersive channels. In addition, it is optimal in the information theoretic sense (i.e. it achieves capacity); it admits an elegant linear algebraic description; it is endlessly flexible; and it yields a straightforward, efficient implementation using the FFT, which is the architecturally and computationally beautiful method of implementing the discrete Fourier transform.

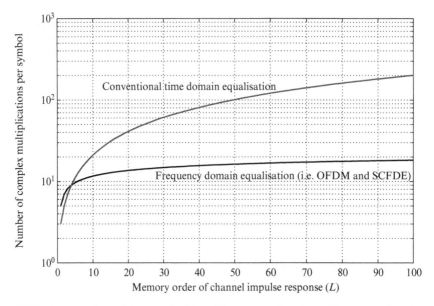

Figure 8.22 A comparison of the complexities of linear time-domain and frequency-domain equalisers where the size of the FFT used by the FDE techniques is assumed to be four times the memory order of the channel.

Conceptually, an OFDM waveform can be viewed as the superposition of N conventionally modulated signals (e.g. ASK, PSK, QAM) using a rectangular pulse of period T, which we again denote by $g(t)$ where each signal has its own carrier frequency. Working with the low-pass equivalent signal model, the constituent signals are offset by multiples of some nominal frequency difference Δf, which is known as the *sub-carrier spacing*. Thus, we can express the signals as

$$s_0(t) = \tilde{x}_0 e^{i2\pi t(0\cdot\Delta f)} g(t)$$

$$s_1(t) = \tilde{x}_1 e^{i2\pi t(1\cdot\Delta f)} g(t)$$

$$\vdots$$

$$s_{N-1}(t) = \tilde{x}_{N-1} e^{i2\pi t((N-1)\cdot\Delta f)} g(t)$$

where $\{\tilde{x}_n\}$ are the constellation symbols.[12] An *OFDM symbol* is then created by adding these signals together, as shown in Figure 8.23.

Consider the spectral characteristics of the signal $s_n(t)$. Assuming $g(t)$ is a rectangular pulse with period T centred at $t=0$, it is straightforward to show that the Fourier transform of $s_n(t)$ is

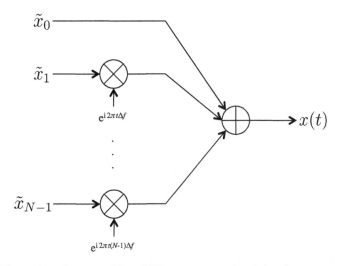

Figure 8.23 Illustration of superposition of *N* frequency translated signals.

$$\tilde{s}_n(f) = \tilde{x}_n T \frac{\sin \pi (f - n\Delta f) T}{\pi (f - n\Delta f) T}.$$

This function crosses the real line at $f = k/T + n\Delta f$ for $k = \pm 1, \pm 2, \dots$, and is equal to $\tilde{x}_n T$ at $f = n\Delta f$. If we set $\Delta f = 1/T$, then the peaks of each of the functions $\tilde{s}_0(f), \dots, \tilde{s}_{N-1}(f)$ will coincide with zero crossings for the $N - 1$ other functions. This is illustrated in Figure 8.24. This sub-carrier spacing creates an *orthogonality* condition whereby we can tune to a given peak to observe the corresponding signal without any corruption from neighbouring signals.

Returning to the expression for the superposition of the *N* signals (i.e. one OFDM symbol), we can write

$$x(t) = \tilde{x}_0 g(t) + \tilde{x}_1 e^{\frac{i2\pi t}{T}} g(t) + \cdots + \tilde{x}_{N-1} e^{\frac{i2\pi t(N-1)}{T}} g(t)$$
$$= g(t) \sum_{n=0}^{N-1} \tilde{x}_n e^{\frac{i2\pi t n}{T}}.$$

The summation in the equation above is just the continuous-time inverse DFT of the sequence of constellation symbols $\tilde{x}_0, \dots, \tilde{x}_{N-1}$. This points to a nice method of implementing the OFDM transmitter. In fact, by noting that the bandwidth of the signal $x(t)$ is about N/T, we can sample the signal at this rate to obtain

$$x_k := x\left(\frac{kT}{N}\right) = \sum_{n=0}^{N-1} \tilde{x}_n e^{i2\pi kn/N}.$$

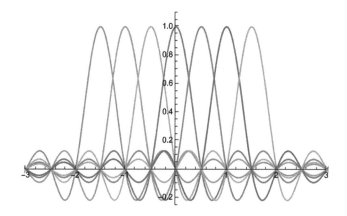

Figure 8.24 $\tilde{s}_0(f),\ldots,\tilde{s}_6(f)$ with $\Delta f = 1 / T$.

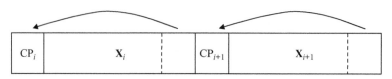

Figure 8.25 Example of a cyclic prefix guard interval extension.

Recall that the channel is dispersive. Thus, OFDM symbols transmitted contiguously will experience ISI, which is, perhaps, more appropriately named *inter-block interference* (IBI) in this case since we are speaking of interference between blocks of constellation symbols. Thus, a *guard interval* must be inserted between each OFDM symbol to mitigate IBI. But the guard interval does more than this. In fact, if designed correctly, it creates the illusion of periodicity in the transmitted signal and separates adjacent transmitted blocks in such a way as to allow independent equalisation of each block at the receiver. A standard approach to implementing the guard interval is to append the last few samples of an OFDM symbol to the front of the same symbol, thus extending its length. This is known as the *cyclic prefix* (CP). Figure 8.25 provides an illustration of this approach to implementing a guard interval.

Turning our attention to the receiver, the guard interval is first removed from each received block and a DFT is applied to each resulting block (Figure 8.26). These steps effectively convert the dispersive channel to multiple, orthogonal, non-dispersive subchannels. Therefore, each original constellation symbol is simply transmitted over its corresponding subchannel and is, ideally, not affected by ISI or IBI. The block structure combined with (inverse) DFT processing converts a dispersive channel into a *parallel channel*, and we can modulate each carrier independently of all others. Moreover, we can equalise the received message on each sub-carrier independently. Hence, OFDM provides a way of simply and elegantly mitigating the effects of ISI using a combination of preprocessing (IDFT), a guard interval (which adds modest overhead), and postprocessing (DFT).

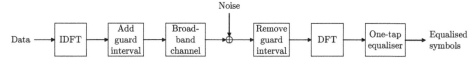

Figure 8.26 Block diagram of an OFDM system.

8.6.2.2 Single-carrier Modulation

A technique that is related to OFDM but has several fundamental differences is the *single-carrier with frequency-domain equalisation* (SC-FDE) approach. The only architectural difference between OFDM and SC-FDE is the order in which the IDFT operation is carried out. As shown in Figure 8.27, an SC-FDE system can be constructed by simply moving the IDFT from the transmitter of an OFDM system to the receiver. By transmitting data in the time domain and using the DFT/IDFT operations to transform received data into the frequency domain for equalisation, the benefits inherent in time-domain transmissions can be retained whilst the low complexity of OFDM is gained.

For example, a major drawback of OFDM transmissions is that they exhibit high peak-to-average power ratio (PAPR) properties. This disadvantage is inherent in all multi-carrier techniques since they are fundamentally constructed by adding many signals together. Since each constituent component of the resulting multi-carrier signal is, itself, random, a central limit effect takes hold and the amplitude of the resulting signal has a Gaussian distribution with a potentially large variance. In practice, a high PAPR must be mitigated through the use of complex signal processing schemes or by backing off the output power of the transmit power amplifier (PA) such that it operates well within the linear region of the transfer characteristic, thus preventing large signal peaks from saturating amplifier rails.

Single-carrier systems, however, do not suffer from PAPR problems. As observed in Figure 8.27, there is no linear preprocessing (e.g. an IDFT) at the transmitter. Hence, the signal conveyed to the (PA) is simply the sequence of conventionally modulated signals (e.g. ASK, PSK, QAM). Signals of this type are typically much better behaved, and the PA can be driven further towards the nonlinear operating region, thus resulting in more efficient amplifier operation.

OFDM and SC-FDE systems both equalise the received signal in the frequency domain (cf. Figure 8.27). As with TDE methods, ZF, MMSE, and DFE criteria can all be used to

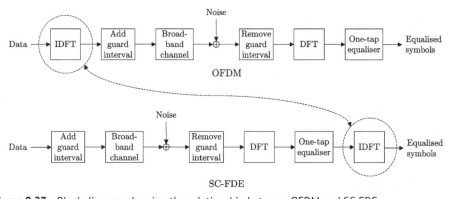

Figure 8.27 Block diagram showing the relationship between OFDM and SC-FDE.

design equalisers for these FDE approaches. In fact, for the linear equalisers (ZF and MMSE), the equaliser structures for OFDM and SC-FDE are exactly the same. However, the resulting performance differs markedly. For OFDM, ZF and MMSE equalisers perform identically. Because data is conveyed on a single sub-carrier, neither of these two equalisation techniques offers any advantage over the other. Yet, in single-carrier systems, data is localised in time, and hence it is spread across the frequency domain. As a result, it is inherently robust against *frequency selectivity* in the channel,[13] i.e. the common scenario whereby some of the bandwidth utilised for communication exhibits more favourable properties (e.g. less attenuation) than other parts of the spectrum. ZF equalisers invert the channel frequency response. If the attenuation is high, this inversion will lead to an amplification of noise in the ZF equaliser. On the other hand, MMSE equalisers take into account the average error in the entire block of transmitted signals by design. Thus, the effects of channel inversion are in some way reduced. Consequently, MMSE equalisers perform better than ZF equalisers in SC-FDE systems.

The coexistence of OFDM and SC-FDE in some point-to-multipoint systems has been proposed. For example, in 4G cellular networks, the uplink is approximately single-carrier, whilst the downlink is based on OFDM. This configuration theoretically provides several advantages:[14]

1) Most of the signal processing complexity is concentrated at the base station where one DFT/IDFT pair is required at the receiver and one IDFT is required at the transmitter. In contrast, the mobile terminal is only required to have one DFT at the receiver.
2) The mobile terminal utilises SC transmission; therefore, expensive, highly-efficient power amplifiers and/or PAPR reduction techniques are not a requirement, whereas they are a necessity in OFDM transmitters. This advantage primarily reduces the cost of the mobile devices.
3) By employing overlap-save or overlap-add techniques at the base station, the overhead due to the guard interval can be eliminated for the SC uplink. As a result, short-medium access control (MAC) messages can be transmitted efficiently from the mobile terminal; whereas if OFDM were used, the lengths of these messages would be constrained to multiples of the DFT size.

A block diagram of a hybrid OFDM/SC-FDE system is illustrated in Figure 8.28.

8.7 Multiple Access: A Second Look

Now that we have a firmer grasp of single-user digital communication theory and techniques, it is natural at this point to extend the discussion to multi-user systems. In earlier chapters, we provided a fairly qualitative overview of several multiple access techniques in the context of various generations of cellular systems. Here, we revisit some of these approaches from a slightly more quantitative perspective. Specifically, we will elaborate on the theoretical ideas that underpin multiple access in third-, fourth-, and fifth-generation cellular networks.

8.7.1 CDMA and 3G

During the first half of the twentieth century, the USSR and US militaries (and others) worked on a method of *spreading* a comparatively narrowband signal over a large bandwidth to make

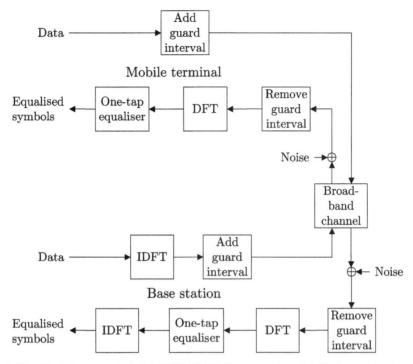

Figure 8.28 Block diagram of a hybrid OFDM/SC-FDE system where SC-FDE is used on the uplink and OFDM is used on the downlink.

it appear as noise, and thus be difficult to detect (Figure 8.29). This is known as *direct-sequence spread spectrum* (DSSS). The original approach that was taken was to modulate each data symbol with a high-bandwidth *spreading sequence*, which itself was a pseudorandom sequence (Figure 8.30). Of course, if one were to correlate the signal with the same sequence (i.e. a matched filter), then it would be possible to cover the message.

Let us take a closer look at the spreading and correlation (a.k.a. despreading) operations. Suppose we have a length-N pseudorandom sequence $\mathbf{c} = \left(c_0, \ldots, c_{N-1}\right)^T$. The elements $\{c_n\}$ of this sequence are called *chips*. We also have a length-M data sequence $\mathbf{x} = \left(x_0, \ldots, x_{M-1}\right)^T$. The spread spectrum sequence is given by

$$\mathbf{s} = \left(x_0 c_0, \ldots, x_0 c_{N-1}, x_1 c_0, \ldots, x_1 c_{N-1}, \ldots, x_{M-1} c_0, \ldots, x_{M-1} c_{N-1}\right)^T.$$

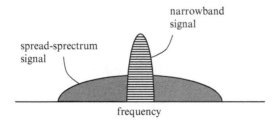

Figure 8.29 Illustration of direct-sequence spread spectrum.

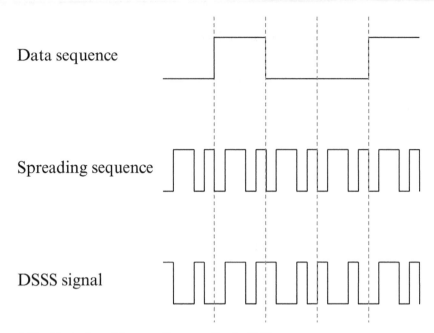

Figure 8.30 Illustration of the spreading operation for DSSS.

We see that through the spreading operation, each data symbol is replaced with a copy of the spreading sequence weighted by the data symbol, itself. Each symbol x_n has period T, but each chip c_n has period T/N. Typically, N is much larger than one. This leads to spectral spreading, since the data signal has bandwidth roughly $1/T$, but the spread signal has bandwidth N/T.

If the signal is transmitted through an AWGN channel along with signals from other users (interferers), the received signal is effectively the sum of all users' spread signals plus Gaussian noise. Note that different users apply different spreading sequences to their data signals. In a synchronous system, the received signal can be correlated with a given user's spreading sequence to recover that user's message. If the spreading sequences have the property of *orthogonality* – i.e. for any two different sequences represented by the vectors $\mathbf{a}$ and $\mathbf{b}$, $\mathbf{a} \cdot \mathbf{b} = 0$ – the receiver would simply take the dot product of the received signal vector with the appropriate spreading sequence. Due to the orthogonality property, the correlation process would eliminate interference from the signal and leave the desired signal corrupted by noise. At this point, detection can take place in the usual way.

The SNR at the output of the correlator at the receiver is simply the SNR that would be observed if only one single user had been transmitting in the channel. Note that, in much of the literature on DSSS and in practice, engineers sometimes refer to a *spreading gain* or *processing gain* when discussing spread-spectrum techniques. This refers to the ratio of the signal bandwidth after spreading to the original signal bandwidth (in this discussion, the spreading gain is N). But this is not, in fact, a gain, since the SNR is equivalent to an AWGN channel without interference.

It is the suppression of interference that led researchers and engineers to consider DSSS for multiple access. This gave way to *code-division multiple access* (CDMA), which dominated third-generation cellular standards. Typical spreading factors in 3G (UMTS) range from 4 to 512.[15] CDMA operates on exactly the same principle as DSSS. However, there has been some debate in the community about what the purpose of CDMA actually is. On the one hand, information theorists have conventionally taken the view that spreading should be performed with orthogonal spreading sequences, thus enabling the receiver to distinguish perfectly between users' signals (provided channel dispersion is properly dealt with). On the other hand, practising engineers have recognised that the despreading process, even if not based on orthogonal sequences, turns interference from other users into Gaussian-like noise. To understand this point of view, note that the despreading operation results in a summation of a large number of random interfering signals. As a result, by the central limit theorem, interference looks Gaussian at the output of the despreader. Hence, the *interference channel* becomes a *Gaussian channel*, and all of the results we have already discussed for Gaussian channels apply, including capacity expressions, channel coding, and error rate analysis.

A final point to note about CDMA is that it is an *interference limited* multiple access scheme, meaning that if codes are not completely orthogonal (or channel dispersion renders orthogonal transmissions non-orthogonal) then, as the number of users accessing the medium increases, the call quality decreases. The trick is to set the power of each user (in the uplink) such that the minimum SNR required to achieve a specified level of service is achieved, but no more. This is a complicated task that also requires consideration of transceiver dynamic range. For example, if the signal from user 1 is received at 10 times the power of user 2's signal, then the ADC will yield significant quantisation errors from user 2's perspective because the dynamic range has to be set high enough to cover user 1's signal. To overcome this problem, a few companies developed key *power control* techniques, whereby signals for different users would be transmitted at different powers to ensure that despreading could be performed efficiently and all signals could be recovered.

8.7.2 OFDMA/SC-FDMA and 4G

In Section 8.6.2, most of the discussion about FDE methods has been placed in the context of single-user systems. Now that we have a more formal understanding of multi-carrier modulation, we can explore how FDE methods play a part in multi-user systems. In fact, this is a simple leap, although it may, on the face of it, seem rather complicated. Recall that in Chapter 5, we introduced the idea of frequency-division multiple access (FDMA), whereby different transmissions are separated in frequency. At that point, we alluded to the use of so-called *orthogonal frequency-division multiple access* (OFDMA) and *single-carrier frequency-division multiple access* (SC-FDMA) in modern 4G cellular networks. Here, we explore these two schemes briefly by drawing on our newfound knowledge of how frequency-domain block modulation schemes work.

Let us begin with OFDMA. In our discussion of OFDM, we treated the scenario where each sub-carrier held a data symbol. However, this does not have to be the case. One can easily imagine a base station scheduling transmissions for two different users in a single

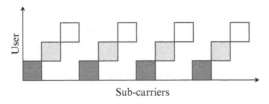

Figure 8.31 Illustration of distributed OFDMA.

OFDM symbol by assigning the even-indexed sub-carriers to the first user and the odd-indexed sub-carriers to the second user. This is the essence of OFDMA.

In the run-up to fourth-generation cellular systems, there was a debate in the community about whether OFDMA should be employed using this sort of *distributed* sub-carrier allocation (Figure 8.31), or whether a *localised* allocation – i.e. contiguous sub-carriers are assigned to a given user (Figure 8.32) – should be employed. Both techniques have advantages and disadvantages. The distributed sub-carrier allocation strategy can be used to mitigate the effects of poor channel conditions in much the same way as SC-FDE. Of course, a single data signal is localised to a given sub-carrier, and if that sub-carrier experiences severe attenuation, the signal could be lost. However, FEC can be employed across the entire OFDM symbol, thus creating dependence amongst the signals transmitted on different sub-carriers. The resulting signal is well conditioned for use in frequency selective channels. The main disadvantage of the distributed sub-carrier allocation scheme is that it potentially requires more overhead in the way of training signals in order to estimate the channel and correct for phase errors (see Section 8.8). This, again, is a direct result of the fact that most OFDMA systems are designed to operate in frequency selective channels.

In contrast to the distributed scheme, localised sub-carrier allocation is well suited to systems where low training overhead is desired since a given user's data is confined to a small portion of bandwidth, which may not experience frequency selectivity. The drawback of this approach can be seen by considering the case where this small portion of bandwidth is severely attenuated, which would potentially result in a loss of *all* data for that user.

In the context of 4G standardisation, localised sub-carrier allocation won. Localised OFDMA is now used in the downlink of modern 4G systems.

Sub-carrier allocation to support multiple access is not limited to the class of multi-carrier schemes. Inspired by SC-FDE, researchers and engineers recognised in about 2007 that many of the benefits inherent to single-carrier transmissions could be exploited in the uplink of 4G systems by adding another DFT to the usual OFDMA transmitter architecture (Figure 8.33). In this way, the time-domain input symbols could be mapped to sub-carriers in the

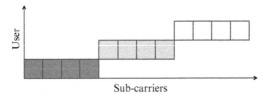

Figure 8.32 Illustration of localised OFDMA.

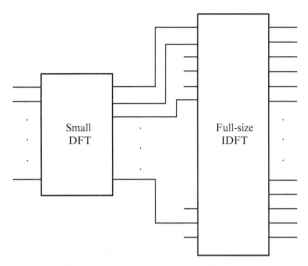

Figure 8.33 Illustration of DFT spreading operation in SC-FDMA (4G transmitter architecture). Input pins on the IDFT not connected to the output of the DFT are either hardwired or logically connected to ground; these sub-carriers are not loaded.

frequency domain, but then the final waveform would be transmitted as a single-carrier signal (in the time domain). This is the essence behind what is now known as *SC-FDMA* or *DFT-spread OFDMA*.

SC-FDMA does not retain *all* of the benefits of single-carrier waveforms. However, the most striking benefit that it does retain is that its PAPR is several dB lower than equivalent OFDMA signals. The drawback is that a DFT and an IDFT are needed at the transmitter. Also, the receiver (base station) requires a further IDFT to convert the equalised signal back into the time domain for FEC. Although far from perfect, SC-FDMA was adopted for the uplink of 4G systems.

8.7.3 NOMA and 5G

In the first four generations of mobile technology, multiple access was achieved through orthogonal signalling schemes. System resources were divided up into noninterfering segments. In GSM, the physical unit of each resource segment was time (TDMA). In UMTS and W-CDMA, resource slicing was performed in the code domain. In LTE, the frequency domain was partitioned to allow users to transmit concurrently in an interference-free manner. However, for decades, network information theory has shown that none of these approaches is optimal in the sense of maximising the collective rate of the users' transmissions. The optimal multiple access scheme is, surprisingly, *nonorthogonal*.

This may seem like a counterintuitive notion. Indeed, how can a user's rate be improved relative to the orthogonal case by *allowing interference to exist*? To conceptually understand this apparent paradox, consider the following simple example. Two users wish to transmit simultaneously to a single base station. User 1 experiences a good channel, whilst user 2 experiences a worse channel. One approach would be to orthogonalise the resources available by assigning half of them to user 1 and the other half to user 2. This, of course, limits the amount of information that both can transmit in a given period of time. Another approach would be to allow *both* to transmit using the *same* time/frequency/code resources. In this case, the base station can exploit the inherent differences in the qualities of both channels. It could first decode user 1's transmission (the higher quality transmission) whilst treating user 2's message as noise. In this way, user 1 can reliably communicate up to $\log_2(1 + \text{SINR}_1)$ bits of information, where SINR_1 is the signal-to-interference-plus-noise ratio of user 1's transmission, with the interference coming from user 2's transmission. If user 2's channel is sufficiently bad, user 1 should be able to achieve a rate that is close to the single-user rate *whilst utilising the full amount of resource available*. This rate is potentially much higher than that which can be achieved using only half the available resources (i.e. the orthogonal slicing approach). But what about user 2? Once the base station has user 1's signal, it can simply subtract it from the original received signal, leaving only user 2's signal corrupted by noise. Hence, user 2 can communicate at a rate that is close to its single-user limit $\log_2(1 + \text{SNR}_2)$, again using all available resources. This idea can be extended to any number of users to achieve a more favourable set of communication rates for the system.

Although this *successive interference cancellation* scheme was known for many years, it was not widely adopted due to its high complexity. More recently, however, advances in the semiconductor industry and system integration engineering have meant that multi-user detection schemes have become practical alternatives to orthogonal resource slicing. Many different versions of so-called *nonorthogonal multiple access* (NOMA) now exist, but all rely on the fundamental principles discussed above, i.e. the total available physical resource is *not* sliced into orthogonal chunks, with only one user allowed to access each chunk, but rather the physical resource is more fluidly assigned to all users with the receiver typically performing joint detection of signals in some manner (Figure 8.34).

NOMA schemes considered by 3GPP for 5G systems are complex and have not been finalised at the time of writing. However, the proposed schemes typically involve various symbol-level interleaving and spreading operations to randomise interference to other users. Aspects of these features have been chosen to facilitate multi-user detection at the receiver. In addition to (or irrespective of) these signal properties, power adjustments can also be specified to assist with user separation. Receiver algorithms matched to the NOMA transmitter schemes range from single-user detection to multi-user detectors with and without additional successive interference cancellation capability.

8.8 System Impairments

Until now, we have mostly considered systems where the channel state, carrier phase, symbol timing offset and carrier frequency are perfectly known at the receiver. In practice, these parameters are typically not known a priori. The lack of information about the channel state, carrier phase and other system parameters can impair system performance due to sampling misalignments, incorrect equaliser tap adjustments etc. It is important to understand the basics of how engineers overcome this issue in practice.

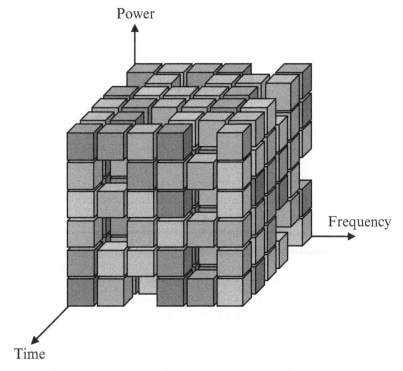

Figure 8.34 Illustration of resource allocation in basic NOMA transmission. Different shades represent transmissions from different users.

8.8.1 Carrier Phase Estimation

When coherent demodulation is employed (e.g. for QPSK), the carrier phase of the transmitted signal must be recovered accurately. To see why, let us consider the case where the transmitted band-pass signal is

$$s(t) = A(t)\cos(2\pi f_c t + \phi)$$

where ϕ is the actual phase of the signal. Neglecting noise for the time being, suppose we have a perfect replica of this signal at the receiver with zero delay (so ϕ does not change). We would typically convert the signal to a low-pass equivalent form by multiplying $s(t)$ by a carrier

$$c(t) = \cos\left(2\pi f_c t + \hat{\phi}\right)$$

where $\hat{\phi}$ is our estimate of ϕ. If the estimate and the actual phase are equal, the product $s(t)c(t)$ yields the baseband waveform $A(t)/2$ and a high-frequency signal, which can be filtered out. If, however, the estimate differs from the actual phase, the product becomes (after filtering)

$$s(t)c(t) = \frac{A(t)}{2}\cos\left(\phi - \hat{\phi}\right).$$

The cosine term introduces an attenuation that directly affects the SNR of the system. To put some numbers to this, when the estimate is $30°$ in error, the signal power loss is 1.25 dB. An error of $45°$ leads to a 3 dB power loss.

Theoretically, it would be easy to obtain an estimate of the phase by transmitting a *training signal*, which is known to both transmitter and receiver, and deducing the unknown phase from this signal. As discussed earlier in the chapter, the received signal is typically affected by random noise. This makes the estimation task a little more difficult. The optimal approach to obtaining an accurate estimate for the phase is to formulate the a posteriori distribution of the signal and determine the phase that maximises this expression, i.e. one would employ a MAP estimator. As with signal detection, if we have no prior knowledge of the distribution of the phase, then we must assume it is uniform. In this case, the MAP estimate is equivalent to the ML estimate.

It is often desirable to be able to track phase variations as transmission progresses. The ML estimator can be used for this purpose if it is efficiently implemented using a *phase-locked loop* (PLL). In its most basic form, the PLL is a simple feedback control system that takes the received signal as the input, forms an error based on the output of a voltage controlled oscillator (VCO), and uses this error signal to drive the VCO with the aim of locking onto the carrier phase (Figure 8.35). Numerous variations on the basic PLL structure exist, including circuits that can be used with modulated waveforms. Many of these techniques are derived from an ML estimator formulation; the PLL circuits are simply efficient ways of implementing the estimation process in real-time.

8.8.2 Timing Recovery

The demodulator of any digital communication system must sample the signal at the precise timing instants $t = kT + \tau$, where T is the symbol period, τ is the delay imposed by the channel, and k is the symbol index. Hence, a clock signal must be acquired or extracted at the receiver to aid demodulator sampling. The process of generating such a signal is known as *symbol synchronisation* or *timing recovery*.

As with carrier phase estimation, timing recovery can be achieved by employing a training phase or simultaneously with symbol demodulation. The general approach is similar for both in that the estimation method tends to be based on ML criteria. Moreover, it is possible (and often practical) to implement ML timing recovery using a PLL-like circuit. Further information on this topic can be found in Proakis (2000)[5].

8.8.3 Channel Estimation

In much of the discussion above, we concentrated on the case where the communication channel co-efficient is unity, i.e. $h = 1$. One of the biggest problems in estimation theory as it relates to communication systems is the acquisition of the channel state – a process known as *channel estimation* – so that coherent demodulation and decoding can take place.[16] As with carrier-phase estimation and timing recovery, the principle of ML

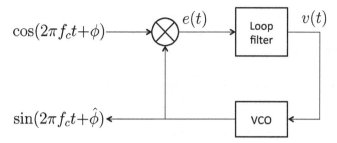

Figure 8.35 Block diagram of a PLL.

estimation can be invoked in the case of channel estimation. The main difference here is that estimation is performed on the sampled signal instead of the continuous-time signal. The primary reason for this is that the estimated channel is used for equalisation and detection, both of which are typically done at the sample level.

It will suffice to consider a simple example, which we will later relate to a cellular system. Without making reference to time (for the sake of simplicity), the canonical, discrete-time, complex-valued system model is given by

$$y = hx + w$$

where w is complex Gaussian noise, x is the transmitted symbol with power P, and h is the complex-valued channel state. Note that this is a sampled, low-pass equivalent model, the details of which can be found in numerous graduate level textbooks on communications (see Section 8.9). The received symbol y must be used to obtain an estimate of the channel. Typically, x is taken to be known at the receiver, i.e. it is a *training symbol*. This knowledge can be used to form the ML estimator. Note that systems such as Wi-Fi and 4G (Long-Term Evolution) reserve a portion of each transmission for training, where effectively a sequence of x symbols known to the receiver are transmitted in order for the receiver to perform channel estimation. The portion of each data packet that contains these training symbols is known as the *preamble* in Wi-Fi since it comes at the start of the packet. In 4G cellular systems, these are known as *reference signals*. Using the simple AWGN model given above, one can show that the ML estimate can be computed by simply dividing the received symbol by the known training symbol, i.e.

$$\hat{h} = \frac{y}{x}.$$

The ML channel estimator in this case is linear. In other words, the estimate is obtained by simply multiplying the received signal by some value. Another linear estimator is the MMSE estimator, which explicitly accounts for the amount of noise present in the system. The MMSE channel estimate for our model is given by

$$\hat{h} = \frac{E_h x^* y}{E_h |x|^2 + N_0}$$

where E_h is the average energy of the channel co-efficient h and N_0 is the variance of the Gaussian noise. In fact, practical systems usually use some form of linear channel estimator as a basis for implementation.

The simple model discussed here may seem to be rather limiting. However, the observant reader will have recognised that the same ML or MMSE estimators can be used in OFDM and SC-FDE systems since each sub-carrier can be modelled using this simple, complex-valued system equation. Hence, dispersive systems can be treated using these simple methods to estimate the channel frequency response coefficients. If the impulse response is required, one can simply transform the frequency response into the time domain with an inverse DFT. In fact, the same linear principles discussed here can be applied to estimate the channel in much more complicated systems, including those with multiple transmit and receive antennas. These systems will be covered in Chapter 9.

8.9 Further Reading

As discussed at the beginning of the chapter, the exposition of digital communication systems presented herein is by no means exhaustive. Interested readers are encouraged to consult other, more thorough texts on the subject. Many such texts exist. Arguably, the most widely used reference book for this field is the text by J.G. Proakis [5]. Two other references, which are a little more accessible to the novice, are the books by B. Sklar [6] and S. Haykin [7]. For those interested in the finer points of information theory, the book by T.M. Cover and J.A. Thomas [8] is a great place to start. Error correction is covered somewhat in the books mentioned above, but these should not be substituted for the works by S. Lin and D.J. Costello [9] or W.E. Ryan and S. Lin [10]. Although wireless communication was not specifically covered in this chapter, it is worth noting that many aspects of the subject follow naturally or are analogous to the systems and models discussed herein. Many excellent texts covering wireless communication systems exist, but the early chapters in the book by D. Tse and P. Viswanath [11] are a good place to start. More advanced topics are covered in the text by A. Paulraj, R. Nabar and D. Gore [12] as well as in the latter chapters of Tse and Viswanath's contribution [11]. We will turn our attention to some of these topics in the next chapter.

Notes

1 It is a *codeword* since we have used the repetition code to *encode* the original binary sequence.
2 Harry Nyquist was an engineer at Bell Laboratories in the middle of the last century.
3 The rules upon which language is based typically create memory and, thus, statistical dependence between characters. We will not discuss source coding for these cases here.
4 Claude Shannon was an American mathematician and engineer at Bell Labs. He is credited with many contributions to communications, and is recognised as the founder of information theory, which has led to the major technologies that are implemented in communication systems today.

5 Morse code uses sequences, or codewords, of dots and dashes, often represented audibly as short and long pulses. The letter 's' is frequently used in the English language, and is mapped to a codeword of three dots. The letter 'o' is less frequently used, and is mapped to a codeword of three dashes.

6 Note that the receiver can account for punctured bits through probabilistic decoding, specifically by treating each punctured bit as being a 0 with probability 1/2 and a 1 with probability 1/2.

7 Here, M is a power of two.

8 A Gaussian process is a stochastic process such that every finite set of random variables sampled from the process has a multivariate Gaussian distribution.

9 At room temperature, this decay occurs at terahertz frequencies.

10 The BCJR algorithm is named after L. Bahl, J. Cocke, F. Jelinek, and J. Raviv, the authors of the paper in which the technique was first presented; see Bahl et al [1]. Technically, the BCJR algorithm is not an equaliser, but a Bayesian estimator. For the purposes of our overview treatment of the subject, however, the classification of this method as an equaliser will suffice.

11 Typically, the 'C' is dropped, and the technique is referred to as OFDM.

12 Throughout this discussion, we use the tilde notation (e.g. $\tilde{a}$) to denote that we are referring to a quantity defined in the frequency domain. This will become clear as we go through the construction of OFDM signals.

13 OFDM transmissions can also be made to be robust against frequency selectivity by appropriately applying FEC to the data prior to modulation. Depending on the code that is used, this process spreads the energy of a single information bit across several sub-carriers.

14 In fact, some of these advantages are not fully exploited in LTE systems, as will be explained below.

15 Note that standards sometimes state spreading factors in dB (e.g. $N = 128$ corresponds to $N_{dB} = 21$ dB).

16 It is worth stressing that there are some cases where noncoherent demodulation is preferred. Usually, these are wireless applications with high-mobility requirements, in which case, the channel state changes so quickly relative to the symbol rate that estimation of the channel would require too much overhead or computation to be practical. Noncoherent schemes, such as differential PSK, typically exhibit a 3 dB (or thereabouts) performance loss relative to comparable coherent schemes. This is because the signal over two symbol periods must be received and processed for each bit of information. At present, although modern cellular systems are designed to work in mobile environments, noncoherent, differential modulation is not employed.

References

[1] Bahl, L., Cocke, J., Jelinek, F. et al. (1974). Optimal decoding of linear codes for minimizing symbol error rate (corresp.). *IEEE Trans Inf Theory* 20(2): 284–287.

[2] Shannon, C.E. (1948). A mathematical theory of communication. *Bell Syst Tech J* 27(3): 379–423.

[3] Gallager, R. (1962). Low-density parity-check codes. *IRE Trans Inf Theory* 8(1): 21–28.

[4] MacKay, D.J.C. and Neal, R.M. (1995). Good codes based on very sparse matrices. In: *IMA International Conference on Cryptography and Coding (ed. C. Boyd), 100–111.* Berlin, Heidelberg: Springer.

[5] Proakis, J.G. (2000). *Digital Communications*, 4e. New York, NY: McGraw-Hill.

[6] Sklar, B. (2001). *Digital Communications: Fundamentals and Applications*, 2e. Saddle River, NJ: Prentice Hall.

[7] Haykin, S. (2013). *Digital Communication Systems*. Hoboken, NJ: Wiley.

[8] Cover, T.M. and Thomas, J.A. (2012). *Elements of Information Theory*. Hoboken, NJ: John Wiley & Sons.

[9] Lin, S. and Costello, D.J. (2004). *Error Control Coding* 2e. Saddle River, NJ: Pearson.

[10] Ryan, W. and Lin, S. (2009). *Channel Codes: Classical and Modern*. Cambridge, UK: Cambridge University Press.

[11] Tse, D. and Viswanath, P. (2005). *Fundamentals of Wireless Communication*. Cambridge, UK: Cambridge University Press.

[12] Paulraj, A., Nabar, R., and Gore, D. (2003). *Introduction to Space-time Wireless Communications*. Cambridge, UK: Cambridge University Press.

9

Early Technical Challenges and Innovative Solutions

In Chapter 8 (Section 8.6), we saw that the output of a channel depends largely on the channel impulse response $h(t)$. In mobile telecommunication systems, the channel impulse response often exhibits interesting characteristics, such as randomness owing to changes in the environment (e.g. people moving in and out of the line-of-sight (LOS) path between the transmitter and receiver), which pose huge engineering challenges with regard to designing systems capable of recovering the transmitted message. In this chapter, we will explore some of the basic physics and modelling techniques that communications engineers must understand in order to develop working systems. We will then revisit multi-carrier modulation with a view to characterising the performance of such systems. This discussion will prepare us for a more sophisticated treatment of multi-antenna systems, which we will cover from a reliability perspective as well as a rate-enhancement point of view. Details of the former topic will manifest in this chapter as an explanation of *diversity methods*, whilst the latter will be covered in the section on so-called *multiple-input multiple-output* (MIMO) techniques.

9.1 Wireless Channels: The Challenge

A unique challenge inherent to mobile communication systems is the wireless channel. Conveying modulated signals wirelessly from a transmitting device to a receiver effectively exposes the message to interference from other devices, severe attenuation arising from the lossy characteristics of electromagnetic propagation, phase distortions that result from propagation delays, and even intersymbol interference (ISI) caused by the reception of multiple delayed copies of the same message. Here, we consider some of these effects and outline various models used in an engineering context to analyse them as well as to better design systems to cope with them.

9.1.1 Propagation

System performance ultimately depends upon the SNR measured at the receiver. Hence, it is important to understand how the power of the modulated signal decays as it propagates

The Technology and Business of Mobile Communications: An Introduction, First Edition.
Mythri Hunukumbure, Justin P. Coon, Ben Allen, and Tony Vernon.
© 2022 John Wiley & Sons Ltd. Published 2022 by John Wiley & Sons Ltd.

wirelessly to an intended receiver. This topic is covered in detail in Chapter 6; here, we review the basic notions of signal propagation before building a more sophisticated understanding of statistical wireless channel models.

Electromagnetic theory indicates that the amplitude of a wave propagating in free space decays with the reciprocal of distance. Hence, the power of a wireless signal decays like $1/r^2$, with r being the distance. More specifically, by accounting for the antenna characteristics, one can relate the signal power at the transmitter P_T to that at the receiver P_R by the following equation

$$P_R = P_T G_T G_R \left(\frac{\lambda}{4\pi r}\right)^2.$$

Here, G_T and G_R denote the transmit and receive antenna gains, respectively, and λ is the wavelength of the signal. The gain of an antenna is defined as the ratio of the maximum radiated power in the direction of the receiver to the radiated power of a hypothetical isotropic source. It is also possible to write the antenna gains as functions of the *effective apertures* $A_{e,T}$ and $A_{e,R}$ of the transmit and receive antennas in the following manner

$$G_T = \frac{4\pi}{\lambda^2} A_{e,T}$$

$$G_R = \frac{4\pi}{\lambda^2} A_{e,R}.$$

Note that $\lambda^2/4\pi$ is the aperture of a hypothetical lossless isotropically radiating element, since it is just the square of the wavelength normalised by the surface area of a unit sphere. Using these relations, we see that the relationship between the transmitted and received power can equivalently be written as

$$\frac{P_R}{P_T} = \frac{A_{e,T} A_{e,R}}{\lambda^2 r^2}.$$

A free-space model is not typically representative of propagation in terrestrial applications. The next simplest model we can consider is one where there is a direct LOS path as well as a single reflection. Figure 9.1 illustrates this scenario for a cellular system where the base station is located at a height y_T and the mobile terminal is positioned at a height y_R. In addition to the LOS component, there is a ground reflection.

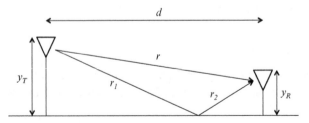

Figure 9.1 Illustration of a two-ray propagation model.

It is possible to construct a mathematical model that relates the transmitted power to the received power in this two-ray model, just as was done for the simple single-ray free-space model discussed above. Such an analysis leads to a striking observation: when the transmitter-receiver separation is large (i.e. $d \simeq r \simeq r_1 + r_2$), the received power decays like $1/r^4$. This additional signal attenuation effectively results from the nature in which signals that propagate over the two paths combine at the receiver. Due to the phase difference between the two signals, they somewhat cancel each other when they are summed at the receiving antenna. This result has practical implications for cellular network design: network planners must ideally ensure that the cell radius is small enough that the $1/r^4$ decay is not observed. There are also other practical advantages to having small cell (SC) radii, such as increased capacity per unit area, easier resource management to mitigate interference, and lower required up-link transmit power.

We can make the propagation model even more sophisticated by considering *multiple reflections* (Figure 9.2). Note that in this model, which is representative of reality, many copies of a single transmitted message travel along different paths. This is known as a *multipath* channel. Attenuated copies of the transmitted message arrive at the receiver with different delays relative to the first copy. Roughly speaking, if two copies arrive with a relative delay that is less than a symbol period, these signals cannot be distinguished, and the paths are said to be *nonresolvable*. On the other hand, if the relative delay between two paths is large compared to the symbol period, the paths are said to be *resolvable*. Nonresolvable paths contribute to a phenomenon known as *channel fading*, which will be discussed below; resolvable paths lead to ISI.

9.1.2 Fading and Multipath

The term *fading* refers to seemingly random fluctuations of wireless signal power as viewed at the receiver. It is easy to appreciate that the propagation environment is typically not well understood for every system. Macroscopic features such as path loss (see Chapter 6) can be fairly easily predicted or discerned, but mesoscopic and microscopic details will affect scattering, diffraction, angles of reflection, path delays etc.

Consider the following example. A transmitter conveys a message to the receiver wirelessly at a centre frequency of 2.3 GHz, and a received signal strength of –70 dBm[1] is

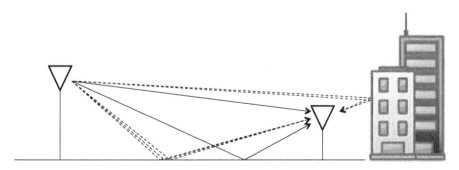

Figure 9.2 Multipath channel formed of two resolvable components (solid lines) and several non-resolvable components (dashed lines).

recorded. What happens when a transmitter moves to the left by 15 cm? At this centre frequency, the wavelength is 13 cm, so a translation of this magnitude would lead to changes in the scattering, angles of arrival of rays at the receiver, and path delays. This could all yield a severely attenuated signal, or even an increase in the received signal power.

Because of the somewhat nondeterministic nature of fading, engineers tend to model these phenomena statistically. Also, mobile systems are designed to operate in environments where the prevailing channel conditions change over time, space and frequency. It is customary to study performance metrics such as the *average probability of bit error* or the *average throughput* of the system. Data rate distributions are also of interest to vendors and operators alike. In these cases, the statistical modelling of fading is also a natural step to take.

9.1.2.1 Large-Scale Fading

Fading comes in several guises. The first we will discuss is *large-scale fading*, or *shadowing*, which is typically due to macroscopic obstructions in the communication medium (e.g. buildings, people). This type of fading was briefly discussed in Chapter 6. Due to the focus on macroscopic changes in the environment, large-scale fading primarily relates to changes in average received signal power (equivalently, the path loss). Figure 9.3 presents an illustration of the effects of large-scale fading on the receive signal power as a function of distance (i.e. transmitter–receiver separation).

Large-scale fading is statistically represented as additive Gaussian noise in the path-loss model measured in dB. For example, a simplified propagation model may represent the path loss $P_{L,\mathrm{dB}}$ in dB as

$$P_{L,\mathrm{dB}} = 10\log_{10}\left(\frac{P_T}{P_R}\right) = 20\log_{10} r - 10\log_{10} \Psi - \text{constant}$$

where r is the distance between the transmitter and receiver and the constant accounts for the antenna gains and transmission wavelength. The term $10\log_{10} \Psi$ is a Gaussian distributed random variable. In this case, the shadowing co-efficient Ψ has a log-normal

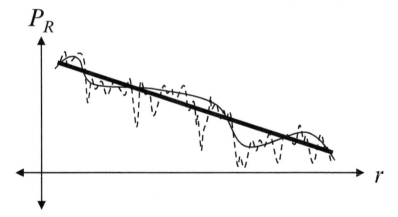

Figure 9.3 Fading effects: received power plotted as a function of distance indicating path loss (thick solid line), large-scale fading (thin solid line) and small-scale fading (dashed line).

distribution. Thus, engineers usually refer to *log-normal shadowing* in the context of modelling macroscopic fluctuations in signal power. The log-normal approximation to the distribution of large-scale fading has been verified empirically in a number of studies.

9.1.2.2 Small-Scale Fading
Small-scale fading refers to fluctuations in signal power due to constructive and destructive addition of nonresolvable multipath components. Referring to Figure 9.2, one can infer that rays in a nonresolvable multipath cluster will have similar amplitudes but random phases. Mathematically, we can model a cluster as a sum of complex random variables. The distributions of these random variables dictate the statistical nature of the signal fluctuations. Here, we discuss a few special cases.

Rayleigh Fading
When the random multipath components (equivalently, each complex variable in the sum noted above) are deemed independent and identically distributed, the system is said to experience *Rayleigh fading*. This is the canonical fading model that is considered in practice. Physically, it corresponds to the receiver observing a signal that has propagated through an environment that exhibits rich and diffuse scattering properties. Mathematically, the received signal envelope has a Rayleigh distribution. As a result, the SNR of the received signal has an exponential distribution. Figure 9.4 illustrates the probability density function for the SNR in this case. In this example, it is assumed the *average* SNR is normalised to one. In fact, the SNR measured in a Rayleigh-fading channel is below average for about 63% of transmissions. (The figure shows that the median is about 0.69). Hence, we can expect systems to perform worse in Rayleigh-fading channels than in channels that do not experience fading.

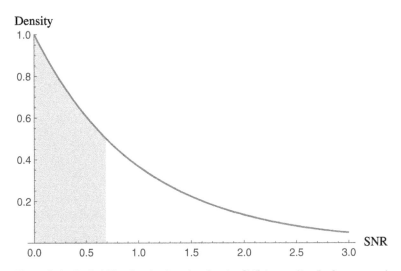

Figure 9.4 Probability density function for the SNR (normalised) of a system that experiences Rayleigh fading. The shaded area corresponds to the SNR values that fall below the median of **log2 ≈ 0.69.**

Rician Fading

In the previous example of Rayleigh fading, all nonresolvable rays are assumed to be identically distributed. This implies there is no dominant component present in the signal. We now consider the case where a single dominant component exists. In practice, this could arise through an LOS between the transmitter and receiver. Note that diffuse scattering may still exist. If it does, the channel is said to experience *Rician fading*. Mathematically, the received signal envelope adheres to a Rice distribution, and the received SNR follows a noncentral chi-square distribution. These distributions have more complicated mathematical expressions than were encountered in the case of Rayleigh fading owing to the fact that two parameters are required to define them.[2] The first parameter, which we denote σ^2, is the power of the channel co-efficient, i.e. the sum power of all multipath components. The second parameter, which is typically labelled K, is the so-called *Rician K factor*, and this succinctly characterises the significance of the dominant path. Physically, K is the power of the dominant path divided by the sum power of the diffuse paths. Usually, K is expressed in units of dB. Hence, a K factor of 10 dB indicates that the gain of the dominant path is ten times that of all other paths combined.

Nakagami Fading

A number of other small-scale fading models exist, many of which have been invented to describe empirical observations of channel measurements made in specific classes of environment (e.g. cluttered residential or office spaces and hilly terrain). One of the most important is the *Nakagami-m* model, named after M. Nakagami who conducted a measurement campaign that led to this model around 1960. The parameter m is known as a *shape* parameter for the distribution. This model subsumes the Rayleigh model, since by letting $m = 1$ in the former, we recover the latter. A larger value of m corresponds to skewing the distribution to higher SNR values. For example, consider Figure 9.5, which

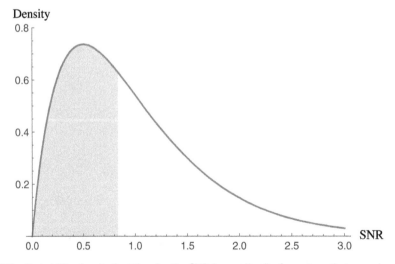

Figure 9.5 Probability density function for the SNR (normalised) of a system that experiences Nakagami-*m* fading, where $m = 2$. The shaded area corresponds to the SNR values that fall below the median of approximately **0.84**.

illustrates the received SNR for a signal that propagates through a Nakagami-m channel with $m = 2$. In this example, the probability of observing a signal with SNR less than the (normalised) average is about 59%. This is slightly better than the Rayleigh-fading scenario. For greater values of m, the situation improves monotonically (albeit slowly). Hence, we would expect a system operating in a Nakagami-m channel to perform better than one communicating via a Rayleigh channel as long as $m > 1$.

9.1.2.3 Frequency-Selective Fading

We now turn our attention to wideband channels. Consider a transmission with symbol period T_s where the channel delay spread is T_d. T_d is simply the time it takes for the channel impulse response $h(t)$ to die out (or at least reach a sufficiently small value). In previous discussions of small-scale fading, we have mostly been implicitly concerned with the case where T_d is much less than T_s, so that the received signal power is effectively confined to a particular symbol period, and different symbol periods can be treated independently. However, this does not have to be the case. When T_d is greater than T_s, the convolution property of the channel will lead to ISI at the receiver. There are several ways of dealing with this, including equalisation, multi-carrier modulation, and maximum likelihood detection.

The study of wideband channel characteristics is quite involved. For our purposes, we simply want to understand how the delay spread of a given channel relates to the definition of a *wideband* channel. In essence, one can understand this relationship by conducting a thought experiment in which the delay spread of the channel is fixed at T_d, and the system bandwidth is increased slowly, starting at DC ($T_s = \infty$). As bandwidth increases, T_s decreases. Eventually, we will have that T_d is greater than T_s. So, the notion of whether a channel is wideband or not is actually related to the transmission bandwidth. Note that when T_s is an order of magnitude smaller than T_d, ISI will start to significantly degrade performance for most systems unless it is properly dealt with.

Wideband systems experience dispersive channels. Thus, the system model involves a convolution, i.e. the signal is non-localised in the time domain. This gives rise to *frequency-selective fading*, which is, as the name implies, most easily understood by observing the spectrum of the channel response. An illustration of frequency-selective fading is shown in Figure 9.6. It is clear from this figure that the attenuation experienced by different spectral components of the transmitted signal varies dramatically across signal bandwidth. In practice, this condition may significantly affect performance if the system is designed such that information is located in a narrow bandwidth, since the attenuation may be large at that frequency. In this situation, the signal is said to experience a *deep fade*. To mitigate the deleterious effects of frequency selectivity in the channel, it makes sense to 'spread' the information content of the transmitted signal across a large portion of the available spectrum. In this way, the frequency selectivity is effectively averaged and system performance can be stabilised. In Chapter 8, we discussed a few ways in which spreading can be accomplished, namely through OFDM, SC-FDE, or CDMA.

9.1.2.4 Slow Fading

A slowly fading channel is one where the channel response does not change much over time. The fading models discussed above are implicitly slow-fading models, because it is assumed the channel does not change over several simple durations. This condition leads

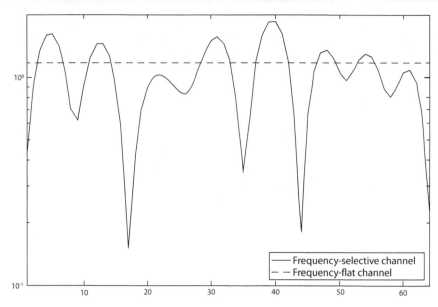

Figure 9.6 Illustration of frequency-selective channels and frequency-flat channels. The horizontal axis denotes the frequency index; the vertical axis indicates the magnitude of the channel response (i.e. attenuation of the signal).

to the simple linear time-invariant model that was described in Section 8.6. Many practical systems, such as Wi-Fi or fixed wireless access networks (for cellular backhaul links or 'last mile' connectivity), conform to the slow-fading model. Again, whether a channel can be labelled as 'slow fading' or not is relative to the symbol rate. In fact, it is often an engineering problem to design the system (e.g. modulation, coding scheme, packet structure, multiple access technique) to yield a virtual slow-fading environment. For example, system designers would typically want to ensure that the channel does not change significantly during a single OFDM symbol, otherwise, the cyclic nature of the channel would be destroyed and ISI would be introduced.

One final very important point that should be made about the notion of a slow-fading channel is that this *does not* mean that small/large-scale fading does not exist in these scenarios. This is sometimes an issue of semantics. Slow/fast fading refers to *temporal* changes in the channel state, whereas small/large-scale fading refers to fluctuations in the channel state that arise on a *spatial* scale. Hence, a channel might be considered to be static or slowly fading if the transmitter and receiver do not move. But, if a device is moved to the left a few inches and the channel is measured again, a completely different impulse response may be observed. This response may also remain static for a long period of time, but it is the spatial change (on a small scale) that has led to fading in this case. Hence, the channel can be said to be slowly fading, but affected by small-scale fading (scattering) only.

9.1.2.5 Fast Fading

If a channel changes rapidly (relative to the symbol period), then it is said to be a fast-fading channel. In this case, we have a linear *time-varying* model for the system, i.e. the channel impulse response changes with time. To characterise the statistics of the channel, we must consider the temporal randomness in the impulse response, which is quantified by the autocorrelation of the channel transfer function.

We will not go into the intricacies associated with such an analysis, but we will instead take a macroscopic view of some important parameters. The first is the *Doppler spread*. The phase difference associated with the ith multipath component of a signal is denoted by $\Delta\varphi_i = 2\pi f \tau_i$, where τ_i is the delay of the ith multipath component relative to the first and f is the transmission frequency in units of Hz. The quantity $D_i = f\tau_i$ is known as the *Doppler shift* for the ith path. Defining a 'significant' change in phase as $\pi/2$ or greater, it follows that such changes in phase occur over intervals of $1/4D_i$. When all (or most) paths experience significant changes in phase, the amplitude of the corresponding channel transfer function changes significantly due to the constructive/destructive addition of multipath components. The timescale over which this occurs is known as the *Doppler spread*, and can be quantified as

$$D_s = f\left(\tau_{\max} - \tau_{\min}\right)$$

where the delay difference is the maximum over all delays. Recall that the delay spread of a channel is given by this maximum delay difference, hence we have $D_s = fT_d$. Note that Doppler shift is somewhat arbitrary, since we could have defined a 'significant' phase change as $\pi/4$ or otherwise. The *coherence time* of a channel is generally defined as the time it takes for the impulse response to change significantly. Since this timescale is defined by the Doppler spread, it can be written as $T_c = 1/4D_s$.

Returning to the discussion of phase, we see that phase changes with frequency like $2\pi f\left(\tau_i - \tau_j\right)$ for some paths i and j. Hence, it follows that for frequency changes on the order of $\pm 1/2T_d$, the channel impulse response will change significantly. We can capture this notion in an analogous way to the coherence time, namely the *coherence bandwidth* of the channel, which is written as $W_c = 1/2T_d$.

We defined the notion of a *frequency-selective* channel earlier. Qualitatively, a frequency-selective channel is one where the channel bandwidth is much greater than the coherence bandwidth. Conversely, a frequency-flat channel is one where the channel bandwidth is of the order of, or significantly less than, the coherence bandwidth.

9.1.3 Signal-to-Noise Ratio in Fading Channels

Let us revisit the notion of signal-to-noise ratio (SNR) in the context of wireless channels that are subject to fading. The SNR, measured at the receiver, is a measure of the ratio of signal energy (power) to noise energy (power)

$$\text{SNR} = \frac{|\text{signal}|^2}{|\text{noise}|^2}.$$

Let us begin by considering the denominator. In practical communication systems, the noise is typically modelled as a Gaussian random process that has a white spectrum. Another way to think about this is that the existence of (infinitely) high-frequency components means that different noise samples measured in time are uncorrelated. Since they are

Gaussian, they are also statistically independent. We might also assume the noise process is *stationary* and *ergodic*. Roughly, this means that the statistics of the noise do not change over time, and an infinite time-average of the process would be equal to the ensemble average over the underlying distribution. Hence, by modelling noise as the random variable n, we would expect the expectation $\mathbb{E}\left[|n|^2\right]$ to well-approximate a time average of the noise power in practice.

Now, consider the numerator in the expression given above. In a similar manner to our treatment of noise, we can often assume that the continuous-time random process $r(t)$ representing the received information portion[3] of the signal is stationary. In modern digital communication systems, engineers are mostly interested in developing discrete-time models. Hence, we may shift our focus to the discrete-time equivalent process $r[k]$, which again can often be assumed to be stationary and ergodic. It follows that the numerator of the SNR expression is, in general, just the expectation $\mathbb{E}\left[|r[k]|^2\right]$.

Now, consider the fact the SNR is a measurement made at the *receiver*. Thus, $r[k]$ here denotes a signal measured *after* it is affected by the channel. But we learned above that the channel itself can be modelled as a random variable. By writing $r[k] = h[k]s[k]$, where $h[k]$ is a random variable representing the channel transfer function and $s[k]$ is the random transmitted constellation symbol, we are faced with the quandary of how to perform the average of $|r[k]|^2$. The key to knowing how to proceed is to think physically. By this, we mean that we should attempt to understand *what it is that we actually would wish to measure in practice*, and then model this mathematically.

For example, suppose we know that the channel transfer coefficient changes slowly with time, i.e. we have a slow-fading channel. Then, by definition, we would expect the signal $s[k]$ to vary much more quickly than $h[k]$. We could, in this case, take $h[k] = h$ to be a constant for all time. Then, performing the average would give the following mathematical representation of the SNR

$$\text{SNR}(h) = \frac{\mathbb{E}\left[|s[k]|^2\right]}{\mathbb{E}\left[|n[k]|^2\right]}|h|^2.$$

Hence, the SNR is actually a function of the channel (or, rather, the channel power to be precise). This measure of SNR is sometimes called the *instantaneous SNR* or *channel-dependent SNR*. If we then wanted to measure the *average* SNR over a long period of time, we would just take the average of the expression on the right-hand side of the equation given above with respect to the channel coefficient's distribution, i.e.

$$\overline{\text{SNR}} = \frac{\mathbb{E}\left[|h|^2\right]\mathbb{E}\left[|s[k]|^2\right]}{\mathbb{E}\left[|n[k]|^2\right]}$$

where the over-line notation signifies that this is the *long-term average SNR*.

9.2 Multicarrier Modulation: A Second Look

In the previous chapter, we explored multi-carrier modulation without saying much about how it combats frequency-selectivity. We now revisit this popular method with our new-found knowledge of fading channels.

9.2.1 Coded OFDM

Referring to Figure 9.6, one can imagine the case where a single OFDM symbol occupies the entire frequency-selective bandwidth. The basis of OFDM is the conversion of a disper-sive, wideband, frequency-selective channel into multiple narrowband subchannels (sub-carriers). Conventionally, a separate data symbol is transmitted on each sub-carrier. But, in a frequency-selective channel, some of these symbols will experience high attenuation, and the information they carry will be lost. An early approach to overcoming this problem was to first encode the binary data with an FEC code prior to modulation. The encoded bits would then be interleaved (permuted) according to a rule that is known to both the trans-mitter and receiver, and hence the permutation can be undone prior to decoding at the receiver. A diagram illustrating this set of operations is shown in Figure 9.7.

It is the combination of the FEC and interleaving processes that generates resilience to the frequency-selective fading channel in OFDM transmissions. First, by encoding the data, redundancy is added to improve performance in low-SNR conditions. Second, by interleaving at the bit level, coded bits that are highly correlated (because they belong to the same codeword, for example) are distributed throughout the available bandwidth. Thus, when one bit experiences a deep fade, related bits may be communicated in more favourable channel conditions. At the receiver, probabilistic, or *soft*, decoding can be per-formed to recover the transmitted message, whereby the quality of the channel that each bit is transmitted through is taken into account. On the whole, this procedure dramatically improves the performance of OFDM systems in frequency-selective channels relative to the case where no interleaving (or no coding) is employed.

9.2.2 Capacity and Adaptive Modulation

Although FEC coding and interleaving improves the reliability of OFDM transmissions, more can be done to maximise the rate of communication in frequency-selective channels. To explore this idea further, we must take an information theoretic view of an OFDM sys-tem. The way in which OFDM transforms a dispersive channel into a set of noninterfering subchannels suggests that such a system can be modelled as a *parallel channel*. Here, the input-output model of the channel can be written as

Figure 9.7 Block diagram illustrating the order of the FEC, interleaving, and symbol mapping processes in the transmitter of a coded OFDM system.

$$y_k = h_k x_k + n_k, k = 0,\ldots,N-1$$

where x_k represents the symbol transmitted on the kth sub-carrier, h_k is the corresponding channel transfer coefficient (in the frequency domain), n_k is the Gaussian noise sample, and y_k is the received signal. There are N sub-carriers in this model. In practice, we must constrain the total power transmitted over this channel so that the sum of transmit powers for all sub-carriers is at most P watts.

To study the maximum achievable rate, we simply add the capacities of the individual subchannels together, which gives

$$C = \max_{P_0,\ldots,P_{N-1}} \sum_{k=0}^{N-1} \log_2\left(1 + \frac{P_k|h_k|^2}{N_0}\right).$$

In this equation, P_k denotes the power transmitted on the kth sub-carrier and N_0 is the variance of the noise. The maximisation is taken over all possible sets of transmit powers subject to the constraint that their sum is at most P, as noted above. This optimisation problem is convex, and thus the theory of convex optimisation can be invoked to find the set of powers that yields the maximum. It turns out, there is a simple equation that defines these power levels, specifically

$$P_k = \max\left(0, \mu - \frac{N_0}{|h_k|^2}\right), k = 0,\ldots,N-1.$$

The parameter μ is chosen to satisfy the sum power constraint. The max function effectively ensures that negative power levels cannot be selected, since this would be a non-physical solution to the problem.

Thus, we have found that, as long as we have knowledge of the frequency-selective channel co-efficients, we can allocate power on each subchannel to achieve capacity. It is worth taking a closer look at the mathematical expression for the optimal power levels given above. Indeed, we can infer from the equation that stronger channels should be allocated more power (since the inverse channel-to-noise ratio $N_0/|h_k|^2$ is small), and very weak channels should be ignored completely. This strategy has come to be known as *water filling*, a term derived from the visual representation of the power allocation technique, as illustrated in Figure 9.8.

Of course, capacity is only achieved in a practical sense if we can increase the actual rate on each subchannel. This is done through the process of *adaptive modulation*, which is also referred to as *bit loading*. Specifically, for 'good' channels, we increase the modulation order – for example, we use a high-order QAM constellation – since the capacity of that subchannel is high, meaning it can support a higher rate transmission. For weak channels, we decrease the modulation order – for example, we might use BPSK – since we will require a large minimum distance between constellation points to decode the message correctly at the receiver. For very poor channels, the achievable rate is so low that we refrain from using them altogether, instead choosing to allocate more power to better sub-carriers. In this way, we can discretely alter the rate on a given subchannel to get as close to the capacity as possible. Of course, this assumes we use a capacity-approaching FEC code at the same time.

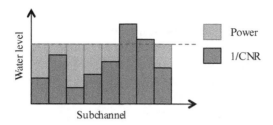

Figure 9.8 Illustration of the water-filling principle. The channel-to-noise ratio (CNR) is defined as $|h_k|^2 / N_0$. The dashed line represents the parameter μ, which is colloquially referred to as the water level.

One thing to note here is that we have assumed coding is performed independently for each subchannel. A natural question to ask is the following: can we increase capacity by coding across subchannels? It turns out that as N grows large, coding across subchannels does not help. However, for finite N, coding across subchannels can yield a lower error probability, which can be desirable in practice.

9.3 Diversity

One feature that wireless channels support, which is not common to other communication paradigms, is that of *diversity*. In its simplest form, we can think of diversity as the concept of receiving multiple copies of a message and collectively processing these copies to arrive at an estimate of the actual transmission. Here, we will discuss how a performance advantage is derived through *independently fading wireless channels*. It is the fading that yields a diversity advantage in wireless systems. This idea is related to the reasoning behind the use of interleaving in coded systems, which was discussed above. In this section, we will gently and gradually uncover a quantitative notion of diversity. This will be done by considering several important wireless techniques used in practice.

9.3.1 Macro Diversity

We begin with a brief discussion of macro diversity. This is the easiest to understand and does not require a sophisticated level of detail at a superficial level. Macro diversity deals with network-level diversity, i.e. multiple copies of a packet or signal are available at the network layer, and these can be processed to enhance reliability. This enhancement typically comes from a network layer decision that chooses the 'best' packet out of the copies. When 'best' is defined in the binary sense – i.e. a packet is determined to be the 'best' if it has been correctly decoded – then the choice is easy (and nonunique!). Macro diversity differs from micro diversity in that the latter is derived at the lower layers of the protocol stack (the physical layer in particular) through channel fading.

Macro diversity is experienced in mobile telecommunication systems every time a user moves across a cell edge from one cell to another. In this situation, the base stations in the two cells communicate with the mobile device to exchange important control and device ID information. Hand shaking occurs as the data/voice transmission is offloaded from one base station to the other. Of course, the two base stations belong to the same operator, and both are connected to the core network (CN) via high-capacity fibre backhaul links. Consequently, they can exchange information very quickly whilst the handover occurs.[4] As a user progresses from one cell to the next, the call is not instantaneously disconnected from one base station and reconnected to another. During this transition, both base stations receive the data/voice packets and the network layer can choose which packet to route onwards to the CN based on the perceived quality of the data. This is known as a soft handover.[5]

9.3.2 Time Diversity

For the remainder of this section, we will focus on *micro diversity*. The discussion begins with a theoretical exploration of how one can exploit diversity in time-varying channels. Suppose we construct a signal by repeating a message symbol N times to form a *repetition codeword*. In a simplified model, we can consider the case where these symbols are transmitted to the receiver over a fast-fading channel, such that each symbol in the codeword is conveyed through a different channel (Figure 9.9).[6]

At the receiver, a *maximum ratio combining* (MRC) operation can be performed to recover the transmitted message. This approach entails correlating each received codeword with the set of channel conditions that each symbol in the codeword experienced. Mathematically, this amounts to summing a weighted combination of codeword symbols, where the weights are the complex conjugates of the channel coefficients. The MRC receiver accomplishes two tasks: 1) it coherently combines all signal content related to useful transmitted information, and 2) it suppresses the noise. The first task is easy to see, but the second deserves further comment. Since the noise samples over different received codeword symbols are uncorrelated and have zero mean, summing these will reduce their impact in the detection process. Indeed, one can imagine the repetition operation being

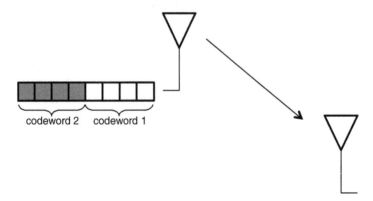

codeword 2   codeword 1

Figure 9.9   Illustration of time diversity using a repetition code.

performed over an infinite number of time periods (i.e. one symbol is repeated an infinite number of times), in which case, the MRC receiver would effectively sum an infinite number of noise samples. By the law of large numbers, and the fact that the noise has zero mean, this sum would tend to zero.

Perhaps more important to the current discussion is the fact that the MRC receiver combines signals that travelled through *different* channels, or at least were affected by different channel conditions. Ideally, these conditions would be statistically independent. In this case, it is possible to show that the error probability of this transmission at high SNR satisfies

$$P_e \propto \frac{1}{\mathrm{SNR}^N}.$$

Recall here that N is the number of symbols in the repetition code, i.e. the rate of the code is $1/N$. If no repetition operation is performed, we would simply have that the error rate decreases like the reciprocal of the SNR. But with each additional symbol repetition, we can logarithmically reduce the error rate by $\log \mathrm{SNR}$. On a log-log graph, the power N manifests as the negative slope of the curve that relates the error-rate to the SNR (Figure 9.10). Conventionally, N is referred to as the *diversity order* or *diversity gain* of the system. It is important to note that the full diversity gain can only be attained through the exploitation (through MRC) of *independently faded* channels in each codeword. If channels are correlated, the diversity order will be reduced.

This general concept is not only useful for understanding repetition coding in the time domain. It is applicable to many different system models, as we will see below. It is worth noting that most models are a variation on this fundamental theme, but some have practical advantages that have made them suitable for modern telecommunication systems.

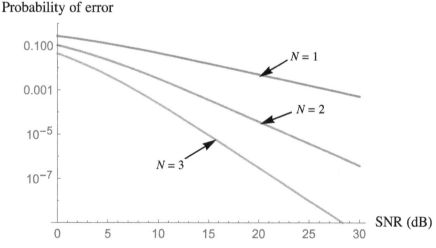

Figure 9.10 Probability of error plotted against SNR for a repetition coded system. The asymptotic (high SNR) slopes of the three examples increase as the code rate decreases.

9.3.3 Frequency Diversity

So far, we have discussed the case where the channel changes statistically independently over time (fast fading), and we used repetition coding to extract diversity from the channel, which manifests quantitatively as the negative slope of the log-log graph depicting the relation between the error probability and the SNR (Figure 9.10). It is also possible to do the same sort of repetition coding in the frequency domain. If the channel is frequency-selective, so much so that the individual channel coefficients are independent, then the system model will essentially be the same as the temporal repetition coding model discussed above, but where the channel coefficients will be defined in the frequency domain instead of the time domain. This equivalence can be understood by again recognising that OFDM can be used to transform the dispersive channel into parallel subchannels, in which case, repetition coding can be applied in the frequency domain. The diversity order in this case would be equal to the number of sub-carriers in the OFDM symbol.

Temporal dispersion of the order of only a few symbols also yields a diversity gain when appropriate signal schemes are employed. More complicated arguments have to be used to quantify the diversity order in this case. But the situation is easy to understand from a qualitative perspective. Suppose that the channel impulse response has L independent degrees of freedom. For example, the transmitted signal could propagate to the receiver over L resolvable rays, with each being affected by independent features in the medium. Even if multi-carrier modulation with N much larger than L is employed, the maximum diversity order of the system is dictated by the number of independent degrees of freedom L.

This example extends to many of the systems that have been covered thus far. Single-carrier transmissions over a large bandwidth inherently (in general) achieve a diversity order of L, since each data symbol is spread over the entire bandwidth. A similar result holds for single-carrier CDMA techniques. Even for systems that do not spread the content of each symbol over the bandwidth, coding and interleaving can be used to artificially achieve this result. Hence, it is possible to extract Lth-order diversity with coded OFDM.

9.3.4 Spatial Diversity

One of the most notable advances in wireless communications in the last 20 years has been the development of sophisticated techniques to exploit *spatial diversity* in the channel. This is done by transmitting and/or receiving information using multiple antennas. Antenna array technology is very well understood, and arrays have been used to design long-range and even deep-space communication systems for decades. The significant advances made in recent years have focused more on clever baseband signal processing and modulation techniques to yield diversity.

9.3.4.1 Diversity Reception
Receive diversity has been exploited in practical systems for many years. The basic concept is to collect different replicas of a transmitted signal at different spatial locations (Figure 9.11). If the separation of receive antennas is more than half of one wavelength, then the small-scale fading effects in the channel give rise to statistical independence amongst the channel gains. The receiver can select the best signal or combine the contributions from all paths to obtain a single signal that can be passed to the detector and decoder.

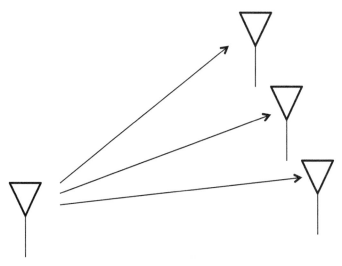

Figure 9.11 Illustration of diversity reception.

Maximum Ratio Receiver Combining

A number of different diversity reception techniques are employed in practice. A popular choice is *maximum ratio receiver combining* (MRRC). This is similar to the MRC techniques discussed in the context of temporal repetition coding above. Suppose a system employs a single transmitter and N receivers. The transmitter conveys a message symbol through the narrowband (frequency-flat) channel to the receivers. If the receive antennas are capable of coherently combining the received signals, the optimal approach (in terms of SNR maximisation) is to apply a matched filter, which, in this case, is equivalent to multiplying each received signal by a weight matched to the channel gain for that spatial branch and summing the outputs from each branch (Figure 9.12). The resulting signal is similar to the repetition coding system discussed earlier. The difference here is that transmission over multiple spatial channels constitutes the repetition operation. Thus, the rate of the transmission does not suffer, but Nth order diversity can be achieved.

There are drawbacks to MRRC. First, N antennas are required, which obviously increases the cost of transceiver hardware. In modern cellular systems, transmission frequencies are typically around 800 MHz, 1800 MHz, or 2300 MHz. At these frequencies, the wavelength of the transmission is in the order of a few centimetres, and hence antenna geometries are of the same order. Thus, manufacturers can pack several antennas onto a device without much trouble. Even if multiple antennas can be accommodated in a receiver, multiple receiver chains, along with all the required mixers and intermediate frequency (IF) circuitry, are also needed. Another disadvantage of diversity reception arises in the need to perform coherent demodulation (matched filter combining). This task becomes harder to accomplish as more channel state information (CSI) is required.

Selection Combining

To overcome some of the drawbacks of MRRC, one can design the multi-antenna receiver to have a single receiver chain and then select the 'best' antenna to use for reception. This

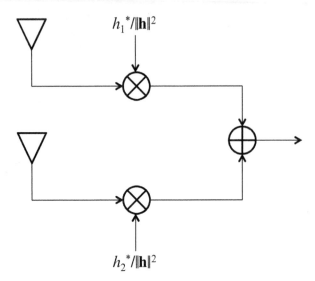

Figure 9.12 MRRC for a two-branch receiver.

approach is known as *selection combining*. It is natural to select the antenna that maximises the SNR at the receiver. Clearly, this approach has some advantages over MRRC: less CSI is needed to make the selection (only an indication of the best channel is needed), and considerably less receiver circuitry is needed. The question as to whether or not to use this approach becomes one of performance. Interestingly, it turns out that selection combining can achieve Nth-order diversity when N antennas are used. Thus, the high-SNR behaviour of this approach is similar to MRRC. However, the absolute performance is slightly worse. The diversity result can be understood by recognising that N independent channels are, in effect, used for transmission, i.e. the number of degrees of freedom for selection combining is the same as for MRRC, even though not all channels are being used simultaneously. The worse performance follows exactly from this difference; indeed, MRRC collects N copies of the signal, whereas selection combining only collects the 'best' signal out of N.

9.3.4.2 Diversity Transmission

The idea of transmit diversity (Figure 9.13) has been around for some time, mostly embodied as transmit analogues to receive diversity schemes. S. M. Alamouti proposed a simple transmit diversity scheme in 1998 [1]. Simultaneously, ideas from error correction were adapted for use in the spatial domain by V. Tarokh and colleagues [2, 3]. These initial proposals were called *space-time codes*. They are now used in a number of wireless systems. Here, we take a look at several transmit diversity schemes, including space-time-codes.

Maximum Ratio Transmission

The first technique that we will cover is *maximum ratio transmission* (MRT), which is essentially the same as MRRC, but where the weights are applied at the transmitter and the multiplexing nature of the channel is used to achieve the combining effect (Figure 9.14).

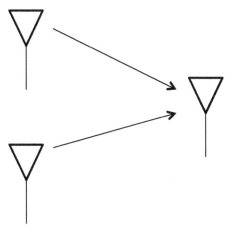

Figure 9.13 Illustration of diversity transmission.

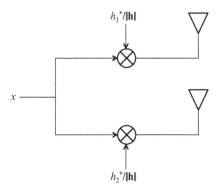

Figure 9.14 Illustration of MRT.

Specifically, suppose there are N transmit antennas, and each knows a priori the channel coefficients $\{h_n\}$. Then, each transmitter branch can apply a weight tailored to that branch such that a diversity gain is achieved.

There are some practical considerations to note. The main advantage of MRT (apart from Nth-order diversity) is that it works well in an asymmetric network, such as a cellular system, where one side of the link may be more able to accommodate several antennas and the associated complexity, but the other cannot. The main disadvantages centre around the fact that the transmitter must have information about the current channel state, which can be obtained via a feedback channel of some sort. However, it is important to recognise that this is an overhead that will reduce the data rate of the system. Also, unequal weighting of different branches will potentially lead to power fluctuations at the transmit side, which is undesirable due to power amplifier constraints. Indeed, if one channel is particularly strong, the PA may be driven into saturation, resulting in nonlinear effects that will ultimately result in a degradation in performance.

Transmit Antenna Selection

To overcome some of the drawbacks of MRT, we can consider a transmitter selection strategy in a similar manner as was considered above for diversity reception. If a limited amount of information is available at the transmitter, it may be possible to determine which antenna corresponds to the channel with the highest gain. With such information available, we could design a transmitter with several antennas but a single radio-frequency (RF) chain and select the antenna that would yield the best performance (highest SNR). Figure 9.15 illustrates the basic principle of *transmit antenna selection*. Clearly, this approach is analogous to selection combining at the receiver. The system model is exactly the same as for selection combining, and hence performance will be identical. Of course, there are practical advantages over selection combining, such as the ability to locate the complexity at the transmitter side and have simple receiver processing.

Delay Diversity

The concept of *delay diversity* involves the transmission of delayed copies of the same signal. In a non-frequency-selective channel, adding a delay in this manner effectively induces dispersion, and thus transforms the flat channel into a frequency-selective channel. The method is practically related to transmit diversity, particularly through its implementation in LTE. Specifically, LTE uses a particular form of delay diversity known as *cyclic delay diversity* (CDD), first proposed by A. Dammann and S. Kaiser in 2001 [4]. With this scheme, an OFDM symbol transmitted from one antenna is replicated and *cyclically shifted*; this version of the symbol is conveyed simultaneously from the second antenna (Figure 9.16). The cyclic shift is applied in the time domain prior to adding a cyclic prefix (CP) to the signal. Basic Fourier theory suggests that this shift induces a linear phase ramp on the signal in the frequency domain. Recall that the modulated signals are conveyed in the frequency domain in an OFDM transmission. Hence, the constellation symbols transmitted in the shifted OFDM symbol are related to the symbols in the non-shifted version by a simple phase shift. Physically, the effect is to increase the frequency selectivity in the effective channel observed by the receiver.

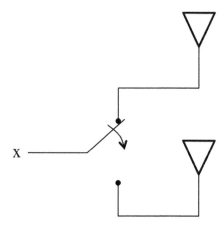

Figure 9.15 Simple illustration of transmit antenna selection.

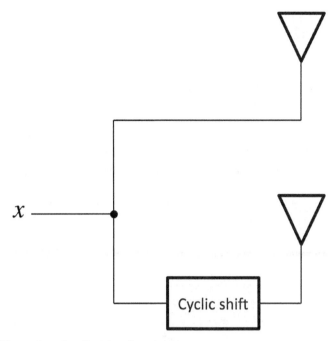

Figure 9.16 Illustration of cyclic delay diversity.

The advantage of CDD is that no knowledge of the scheme is required at a single-antenna receiver in order to decode the transmitted message. The receiver would simply estimate the channel in the usual way. The observed channel would be a composition of the two physical, spatial channels, but it would appear as a single frequency-selective channel at the receiver. Since a single OFDM symbol was effectively transmitted from the two antennas, the receiver would simply perform frequency-domain equalisation in the usual way to recover the transmitted message. The system experiences a diversity advantage by converting spatial diversity into frequency diversity.

Space-Time/Frequency Coding

We saw with CDD that no knowledge of the channel was required at the transmitter to extract diversity from the channel. Another method that achieves a diversity gain under the same restriction is *space-time coding*. This broad class of techniques was invented around 1998 in a couple of different forms: trellis coding and block coding. Here, we focus on the latter, since this approach found its way naturally into several standards, including LTE.

The basic motivation behind space-time block coding was to construct a method of exploiting spatial (transmit) diversity without having to resort to costly feedback channels to obtain channel state information, which is needed in MRT and selection schemes. In the simplest scenario, this can be accomplished in the following manner. Consider the case where we wish to transmit two symbols, x_1 and x_2, over two time periods, and thus yield a rate of one symbol per transmission. We can map these two symbols to the 2×2 matrix

$$\mathbf{X} = \begin{pmatrix} x_1 & -x_2^* \\ x_2 & x_1^* \end{pmatrix}$$

the row and column dimensions of which can be viewed as the space and time dimensions, respectively. Hence, the first column of this code is transmitted from two antennas during the first symbol period, and the second column is transmitted during the second symbol period (Figure 9.17). At the single-antenna receiver, a simple linear combining operation can be employed to decouple the two signals x_1 and x_2. Channel knowledge is required at the receiver for this step in the process, but this is not usually a significant issue to acquire in practice.

The 2×2 scheme discussed here can achieve second-order diversity. It is insightful to compare this approach with another second-order diversity scheme, such as a system that exploits time diversity using 1/2-rate repetition coding. Two time slots are employed with both schemes. To make a comparison, let us consider an example where the space-time coding system transmits BPSK, giving one bit per transmission on average. The repetition coding system must transmit 4-PAM to achieve the same rate. But the minimum distance of 4-PAM is less than BPSK. Thus, to achieve the same minimum distance, the repetition coding system must increase its power by a factor of five. To compare with the space-time coding system, which yields an SNR that is half that of repetition coding, this factor can be reduced to 2.5. In other words, the repetition coded system would need to operate at an SNR that is 4 dB more than the space-time coding scheme to obtain the same rate and minimum distance.

Extensions and generalisations to the 2×2 scheme outlined above exist. Codes that can be used with more transmit antennas have been designed, although the only rate-one code is the one presented here. Perhaps more importantly, from a practical point of view, has been the adaptation of space-*time* block coding to space-*frequency* block coding, which is implemented in LTE. This approach is exactly the same as space-time block coding, but the

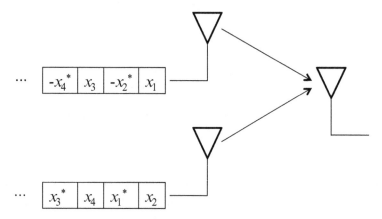

Figure 9.17 Illustration of the 2×2 code over four time slots.

roles of time and frequency are reversed. The exact encoding scheme used in LTE is similar to that shown in Figure 9.17, but where each symbol is assigned to a sub-carrier in an OFDM transmission instead of a single symbol period in the time domain. Hence, each group of two adjacent sub-carriers conveys a space-frequency codeword. It is important that the coherence bandwidth is large enough to ensure that adjacent sub-carriers (approximately) observe the same channel. Without this condition, the linear combining operation that is carried out at the receiver would not completely decouple the signals.

9.4 Multiple Input Multiple Output (MIMO)

In the late 1990s, a new information theoretic view of multiple-antenna systems was developed by G. Foschini and M. Gans [5], and separately by E. Telatar [6]. Instead of focusing on diversity techniques in systems with multiple transmit antennas and multiple receive antennas, a simple question was asked: how much information would such a *multiple-input multiple-output* (MIMO) channel support? Through an information theoretical argument, it was shown that the capacity of a MIMO channel increases linearly with the minimum number of antennas used on the transmitter or receiver side of the link (Figure 9.18). Until then, it was thought that more antennas meant more co-channel interference, which suggested great difficulties would be encountered in decoupling signals at the receiver, and the performance would potentially be quite poor even after this decoupling is carried out. Not only did the new result debunk this myth, but the discovery was actually made in a constructive manner, i.e. the argument demonstrated *how* to implement MIMO effectively in practice. MIMO is now used in most high-rate communication systems. The first widespread integration of MIMO techniques in cellular systems came with the adoption of its use in LTE. In this section, we will summarise the beautifully simple result that led to the MIMO revolution, and we will investigate some practical MIMO solutions that are implemented in practice.

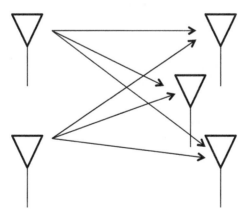

Figure 9.18   Illustration of a MIMO channel.

9.4.1 Capacity

We begin with a fundamental study of the capacity of MIMO channels. We are concerned with a MIMO system with M transmit antennas and N receive antennas. It will suffice to mostly deal with the simple flat-fading (narrowband) MIMO system. Later, we will consider how MIMO can be extended to frequency-selective channels by using multi-carrier techniques. The basic input-output system model for this system can be written as

$$\mathbf{y} = \mathbf{Hx} + \mathbf{n}$$

where $\mathbf{y}$ is the length-N vector of received symbols, $\mathbf{n}$ is the length-N vector of noise samples at the receiver, $\mathbf{x}$ is the length-M vector of transmitted symbols, and $\mathbf{H}$ is an $N \times M$ matrix of complex channel transfer coefficients, which account for the attenuation and phase adjustment induced in the transmitted signal as it propagates over each of the spatial channels.

To obtain an expression for the capacity of this system, we will require a fundamental result from linear algebra. Specifically, every linear transformation can be decomposed into a rotation, a scaling, and finally another rotation. With regard to the channel-transfer matrix $\mathbf{H}$, this is characterised as the *singular value decomposition* (SVD)

$$\mathbf{H} = \mathbf{U\Lambda V}^{H}$$

where $\mathbf{U}$ and $\mathbf{V}$ are matrices comprising the *left- and right-singular vectors* and $\mathbf{\Lambda}$ is the $N \times M$ diagonal matrix of *singular values*. Geometrically, the matrix $\mathbf{V}$ effectively rotates the transmitted signal constellation to align the transmission to a parallel set of *singular modes*, which form a parallel channel. Conceptually, this is similar to the way OFDM uses the Fourier transform operation to convert a dispersive channel to a set of parallel subchannels. The diagonal elements of the matrix $\mathbf{\Lambda}$ represent the attenuation experienced on each of the parallel singular modes. Finally, the matrix $\mathbf{U}$ represents another rotation of the signal at the receiver side of the channel.

If one were to redefine the signal vectors according to the model shown in Figure 9.19, we could express the input-output MIMO channel model as

$$\bar{\mathbf{y}} = \mathbf{\Lambda}\bar{\mathbf{x}} + \bar{\mathbf{n}}$$

where $\bar{\mathbf{n}}$ is the noise rotated by the matrix $\mathbf{U}^{H}$, which has the same Gaussian properties as the original noise vector $\mathbf{n}$. Hence, in this new model, we have a set of parallel channels with input-output characteristics described by the equation

$$\bar{y}_k = \lambda_k \bar{x}_k + \bar{n}_k, k = 0,\ldots,\min\{M,N\}-1$$

where λ_k denotes the singular value (attenuation) corresponding to the kth subchannel.

$$\bar{\mathbf{x}} := \mathbf{V}^{H}\mathbf{x} \longrightarrow \boxed{\text{``Channel''}} \longrightarrow \bar{\mathbf{y}} := \mathbf{U}^{H}\mathbf{y}$$

Figure 9.19 Equivalent representation of a MIMO system model.

This formulation is remarkably similar to what we encountered in our discussion of OFDM. Hence, it is a natural step to consider that the capacity of a MIMO channel will have a similar mathematical form to the capacity of an OFDM channel, namely

$$
C = \max_{P_0, \ldots, P_{\min\{M, N\}-1}} \sum_{k=0}^{N-1} \log_2 \left(1 + \frac{P_k \lambda_k^2}{N_0} \right).
$$

The units of this capacity are bits per second per Hz.[7] The optimal power set satisfies the water-filling solution

$$
P_k = \max \left(0, \mu - \frac{N_0}{\lambda_k^2} \right), k = 0, \ldots, \min\{M, N\} - 1
$$

where the water-level parameter μ is chosen to satisfy a constraint on the total power transmitted from all antennas.

Owing to the interpretation of the water-filling principle, one can see that MIMO capacity is maximised by assigning more power to the better singular modes (parallel subchannels) than to the highly attenuated ones. At low SNR, it can be shown that this approach manifests as only assigning power to *one* singular mode, with all others being deemed to offer no enhancement to the achievable rate of the system. At high SNR, the situation is reversed. Not only is power allocated to *all* subchannels, it is fairly *evenly distributed* amongst the subchannels.

The capacity expression for a MIMO channel is a sum of capacities of the subchannels. The maximum number of observed subchannels relates to the numbers of transmit and receive antennas employed. Specifically, the number of subchannels cannot exceed the *minimum* number of antennas used at either the transmitter or receiver side of the link, and hence the capacity of the channel scales linearly with this number. One way to think about this is to consider the following thought experiment. Suppose we constructed a MIMO transmitter with M transmit antennas and a MIMO receiver with N receive antennas, with $M < N$. We then positioned the transmitter on one side of a set of M waveguides, and we place the receiver at the other side. This is done such that each antenna is 'assigned' to its own waveguide. At the receiver, M out of N antennas can be 'assigned' uniquely to a waveguide, but the remaining antennas must either share a waveguide with another or remain 'unassigned'. During transmission, no interference will occur between the M spatial transmitted signals. Each signal will be received and decoded by its corresponding receive antenna(s). Thus, we have designed a MIMO system with M parallel subchannels, where we can increase the sum capacity of the subchannels by increasing the SNR on each, but not by adding more antenna elements at the receiver. A similar argument applies if we let $M > N$.

9.4.2 MIMO Transmission Techniques

Having sufficiently motivated the use of MIMO through a study of capacity, we now turn our attention to practical transmission schemes. Typically, these can be divided into *beamforming* and *multiplexing* techniques.

We have already alluded to beamforming in the discussion of capacity. In the context of MIMO communication, *beamforming* typically refers to unitary rotation by the right-singular vectors of the channel matrix. This is sometimes known as *SVD beamforming* or *digital beamforming* owing to the fact that it is performed at baseband through a linear precoding operation. Note that this is different from *analogue beamforming*, a term that usually refers to microwave design techniques using phased arrays of antenna elements to direct a physical antenna beam pattern in a prescribed way. Modern research has also begun to combine the ideas of digital and analogue beamforming into so-called *hybrid beamforming* in an effort to manage complexity and facilitate the scalability of MIMO systems for 5G applications. In contrast to purely analogue beamforming, SVD beamforming aims to direct information down parallel singular channels. It should be noted that the physical radiation pattern is altered by such beamforming, but the functional relationship is not straightforward to characterise.

The derivation of MIMO capacity implicitly gives a constructive manner that can be used to *achieve* capacity. If the transmitter and the receiver both know the singular vectors of the channel, they can pre- and post-rotate the signal as shown in Figure 9.20. Specifically, binary data is first encoded (FEC) and subsequently mapped to, say, QAM symbols. A vector of symbols $\mathbf{x}$ is then rotated through the linear operation $\mathbf{Vx}$. This signal is then upconverted and transmitted over the channel. At the receiver, the symbol vector $\mathbf{y}$ is rotated through the linear operation $\mathbf{U}^H\mathbf{y}$. The result is that the QAM symbols are conveyed through a set of decoupled subchannels. Each can be detected independently of the others upon reception.

There are practical issues with SVD beamforming that must be considered in real-world implementations. The most obvious is that capacity can only be achieved if the channel is estimated perfectly and rotations are applied correctly. Without perfect channel estimation, the singular vectors will not match the channel, and the conversion of the system to a parallel channel will not be possible. Cross-talk will exist amongst the subchannels, which will degrade performance. A related issue is that channel information must be available at the transmitter and the receiver. If the receiver estimates the channel on the downlink, a feedback channel is required in the uplink to convey the relevant singular vectors so that the transmitter can apply an appropriate rotation. As LTE was being standardised, a significant amount of effort was expended researching efficient methods of providing this feedback. For example, at the time, it was not clear how accurate the singular vectors needed to be, since all implementations used fixed-point arithmetic and thus the

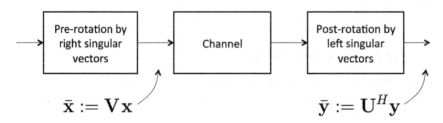

Figure 9.20 SVD beamforming in a MIMO system.

rotations would be imperfect anyway. Vector quantisation schemes were developed for limited feedback channels, and numerous ways to optimise these schemes were proposed and integrated into the standard.

It is possible to exploit the high throughput nature of MIMO channels without resorting to SVD beamforming. Techniques range from alternative precoding/beamforming approaches that do not rely on full knowledge of the channel state being available at the transmitter to the simple method of transmitting data simultaneously from all antennas without implementing any sort of pre-coding operation. This MIMO transmission strategy is known as *spatial multiplexing*. It is very simple to implement at the transmitter; the receiver, on the other hand, must decouple the signals. This can be computationally cumbersome. Several methods of carrying out this task will be detailed in the next section.

9.4.3 MIMO Reception Techniques

In the previous section, we discussed SVD beamforming. In doing so, we touched upon the receiver processing needed to decode the transmitted message, namely the rotation of the received signal according to the left-singular vectors of the channel. Here, we will focus on other techniques that can be used with general spatial multiplexing schemes.

Let us begin with the optimal approach: maximum likelihood (ML) detection. Consider the MIMO system model

$$\mathbf{y} = \mathbf{H}x + \mathbf{n}.$$

Let us assume the channel can be estimated accurately. With knowledge of the channel available at the receiver, and accounting for the fact that the noise is Gaussian distributed, it is possible to mathematically show that the optimal (ML) detection rule selects the signal vector $\hat{\mathbf{x}} = \mathbf{x}$ that minimises the squared Euclidean distance between $\mathbf{H}x$ and $\mathbf{y}$ in M dimensional complex space, where $\mathbf{x}$ must be a valid vector of constellation symbols. This last point is important from a practical design perspective. To carry out the comparison, one would need to consider all possible symbol vectors, compute the squared distance, and choose the vector that corresponds to the lowest. The number of these vectors is the Mth power of the size of the constellation. Thus, the comparison is simple for small systems and low-order modulation schemes. For example, for a system with two antennas that uses BPSK, the ML receiver only needs to consider four possible transmitted symbol vectors in each detection operation. However, for more complicated systems, the complexity of the ML received is prohibitive. Indeed, the ML receiver for a four-antenna system that employs 64-QAM would need to search through more than 16 million possible transmitted vectors to perform detection.

Several approaches to mitigating this complexity issue have been developed. In the additive Gaussian channels that we have been dealing with, noise tends to perturb the signal to within a radius (related to the standard deviation of the noise) relative to the received constellation points. It makes sense that receivers should focus most of their effort in comparing candidate signals that are *nearest neighbours* to the received signal. If a detector can draw a sphere of radius, say, d (in M-dimensional space) and capture these neighbours, it can perform its search over that space only. Of course, if the lattice for the transmitted signal is regular (e.g. QAM), the received signal without noise will be a skewed version of this lattice owing to the rotation and scaling induced by the channel (Figure 9.21). Hence, the

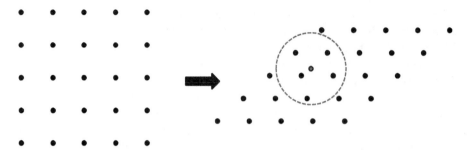

Figure 9.21 Illustration of the sphere decoding principle. Left: original lattice; right: lattice distorted by the channel matrix. The received signal is indicated on the right as the shaded dot with the 'sphere' surrounding it. The sphere indicates the reduced detection space.

choice of this sphere is not a trivial problem. This technique is known as *sphere decoding*. A number of *reduced lattice decoders* have been implemented based on this principle to yield excellent performance in practice relative to ML detection.

Another way to simplify detection is to relax the constraint that the output of the detector should be a valid transmitted symbol vector. Instead, we can design the detector to give the most likely vector of symbols belonging to the complex field. With this goal in mind, it is possible to derive a *linear* detector that gives the *least squares estimate* of the transmitted vector. Conceptually, detection in this case amounts to reversing the effects of the channel (for example, by multiplying the received vector by the inverse of the channel matrix) and selecting the constellation point closest to the least squares estimate. In practice, this is known as a *zero-forcing (ZF) receiver*. Although simple to implement, this receiver can actually amplify the receiver noise. Noise amplification occurs when the channel matrix is rank deficient or ill-conditioned (in the sense that the smallest singular value of the channel matrix is close to zero). In this case, multiplying the received vector by the inverted channel matrix will still decouple the transmitted signals, but the SNR will be relatively low.

To overcome the noise enhancement issue that arises in ZF detection, one can instead implement the so-called *minimum mean-square error* (MMSE) detector. The MMSE detector is a linear estimator that minimises the mean-square error between the output of the receiver and the actual transmitted message. Since the *mean*-square error forms the objective function, the noise statistics are accounted for. Hence, the MMSE detector naturally accounts for the fact that noise could be enhanced when the channel matrix is ill-conditioned.

In practice, the construction of ZF and MMSE detectors can pose problems since their complexity scales roughly with the cube of the matrix size. Battery powered devices cannot afford to use too much energy carrying out such calculations. Additionally, short, simple computation routines are desirable from a latency perspective so that the maximum admissible data rate can be attained. Efficient implementations of these linear detectors have been developed, which are typically based on QR decomposition and sequential updating as the channel changes over time.

9.4.4 MIMO vs Multicarrier

So far, the discussion of MIMO technology has focused on narrowband systems (flat-fading channels). This gives rise to the simple system model discussed in Section 9.4.1. Of course, it is possible to transmit wideband signals in MIMO systems. The ISI that results from the dispersive channel will need to be accounted for in the detector along with the inter-channel interference that arises naturally in MIMO systems. In single-carrier wideband systems (e.g. CDMA) employing ML detection, the receiver would need to consider $M^{L \times Q}$ possible transmitted symbol vectors, where M is the number of transmit antennas, L is the memory order of the channel measured in symbol periods, and Q is the size of the symbol constellation. The complexity of ML detection is, hence, prohibitive in most systems. Linear equalisation techniques can be employed to simplify the receiver architecture. A more popular approach, however, is to invoke OFDM on each transmitted stream. By synchronising the OFDM symbols transmitted from all antennas, the dispersive MIMO channel can be converted to a parallel MIMO channel, with the system model on the kth sub-carrier (out of a total of K sub-carriers) being defined as

$$\mathbf{y}_k = \mathbf{H}_k \mathbf{x}_k + \mathbf{n}_k, k = 0, \ldots, K-1.$$

Notice that the subchannels here relate to different frequencies; this is not the same as the parallel singular modes that were discussed above for SVD beamforming. Indeed, it is possible to perform SVD beamforming on *each* subchannel, thus giving a simple, scalar input-output model for $K \times \min\{M, N\}$ noninteracting subchannels. By implementing OFDM in this manner, MIMO detection can be performed separately on each sub-carrier (Figure 9.22). Thus, all techniques discussed above can be employed on a sub-carrier-by-sub-carrier basis.

9.4.5 Multi-User and Massive MIMO

At this point, it is important to consider how MIMO will be implemented in 5G networks. So far, we have covered a few of the fundamental properties of and approaches used in MIMO communication. All discussion has been focused on single-user transmission. As 4G matured, MIMO techniques were developed for use in *multi-user* links. Typically, multi-user MIMO

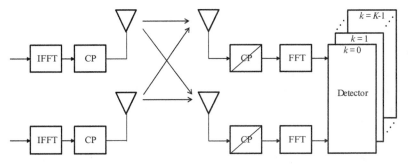

Figure 9.22 Basic block diagram of a 2×2 MIMO-OFDM system. MIMO detection can be performed independently on each sub-carrier.

(MU-MIMO) communication is accomplished in the downlink by 1) obtaining estimates of the MIMO channels to all served users in the cell, then 2) applying a linear pre-coding operation to the transmitted message vector to direct each user's signal in the appropriate way. The pre-coder operation is the key element of MU-MIMO. Many different approaches can be taken. For example, a ZF or regularised ZF (similar to MMSE) method can be invoked, whereby procedures related to the linear detectors mentioned above can be applied at the transmitter to condition the signal before transmission. In these cases, each receiver ideally observes only the message intended for it, i.e. its message arrives having already been decoupled from the other messages. More sophisticated nonlinear techniques based on the information theoretical notion of *dirty-paper coding* have been considered for future cellular systems, but it is unclear whether these will become viable approaches in practice.

A particularly interesting pre-coder is based on the *matched filter* or *maximum ratio combining* strategy, as pointed out by T. L. Marzetta in 2010 [7]. Here, the symbol vector at the base station is multiplied by the conjugate of a channel matrix that encompasses all users' MIMO channels. Hence, the set of receivers (i.e. the users' received signals) observe a signal that appears to have been transmitted through a quadratic channel matrix $\mathbf{HH}^{H}$. An interesting effect occurs when the number of antennas at the base station becomes large. Since the channel co-efficients between the base station and different users are random and uncorrelated (owing to scattering processes in the environment), the quadratic channel matrix approximates a diagonal matrix. In other words, the matched filter operation naturally facilitates an averaging process in the channel that suppresses all interfering users' transmissions when observed by a single user. This is a first-order averaging process, though. So, the number of base station antennas has to be very large for it to work well. This general principle has become known as *massive MIMO*. Many variations have been considered, ranging from the simple scheme discussed here to millimetre-wave (mmWave) techniques that beamform to pinpoint accuracy. All of these methods, in some form or another, are likely to be adopted for use in 5G systems.

Notes

1 Recall that a decibel is a relative unit of measurement. If we take the reference to be 1 mW, then we can talk in dB *milliwatts*, or dBm. Similar units exist when the reference is a watt (dBW) or, in the case of antenna radiation patterns, when the reference is the theoretical isotropic signal (dBi).

2 Rayleigh-fading channels only require one parameter: the mean signal power.

3 Here, $r(t)$ is not a function of noise.

4 In fact, 4G systems have special high-capacity links called X2 connections that join neighbouring base stations directly, which enable the base stations to execute sophisticated signal processing schemes and transmit simultaneously to mobile terminals, which increases downlink capacity considerably.

5 Soft handover came to prevalence in third-generation systems using CDMA, such as IS-95 and CDMA2000. It can be used in the uplink of those systems as well.

6 In practice, this would not be a good transmission strategy owing to the incredibly small coherence time of the channel; however, one can envisage the case where a *packet*, complete with preamble and/or pilot signals for channel estimation and synchronisation, is repeated in time, with each packet experiencing different channel conditions.

7 Technically, when the rate is normalised by the bandwidth, we are considering *spectral efficiency*. In practice, this distinction is assumed to be understood. Note that modern cellular systems are designed with a spectral efficiency goal in mind.

References

[1] Alamouti, S.M. (1998). A simple transmit diversity technique for wireless communications. *IEEE J Sel Areas Commun* 16(8): 1451–1458.

[2] Tarokh, V., Seshadri, N., and Robert Calderbank, A. (1998). Space-time codes for high data rate wireless communication: Performance criterion and code construction. *IEEE Trans Inf Theory* 44(2): 744–765.

[3] Tarokh, V., Jafarkhani, H., and Robert Calderbank, A. (1999). Space-time block codes from orthogonal designs. *IEEE Trans Inf Theory* 45(5): 1456–1467.

[4] Dammann, A. and Kaiser, S. (2001). Standard conformable antenna diversity techniques for OFDM and its application to the DVB-T system. *Globecom* 5: 3100–3105.

[5] Foschini, G.J. and Gans, M.J. (1998). On limits of wireless communications in a fading environment when using multiple antennas. *Wirel Pers Commun* 6(3): 311–335.

[6] Telatar, E. (1999). Capacity of multi-antenna Gaussian channels. *Eur Trans Telecommun* 10(6): 585–595.

[7] Marzetta, T.L. (2010). Noncooperative cellular wireless with unlimited numbers of base station antennas. *IEEE Trans Wirel Commun* 9(11): 3590.

10

Small Cells – an Evolution or a Revolution?

10.1   Introduction

With the proliferation of wireless connected devices (e.g. smartphones and pads) and the popularity of media-based services such as iTunes, YouTube and others, mobile operators have experienced an exponential traffic growth in recent years. The academia and industry of mobile communications have been making efforts to meet an estimated 1000x increase requirement in traffic capacity for the forthcoming 5G networks. Although it is difficult to predict when the 1000x traffic growth will happen, tremendous growth in mobile data traffic has been observed in the past and is expected to increase steadily in the future. As observed in [1], global mobile traffic has grown steadily at a compound annual growth rate (CAGR) of 50 to 66 percent from 2012 to 2017, which means an 8- to 13-fold traffic increase over 2012. According to a recently published report [2], global mobile data traffic will increase sevenfold between 2016 and 2021 at a CAGR of 46 percent, reaching 48.3 exabyte (EB) per month by 2021. In addition, by the year 2021, there will be 4.6 billion internet users, 27.1 billion connected network devices globally, and the total network traffic will reach 3.3 zettabyte (ZB) annually, 82% of which will be video. Evolving the radio network to meet this traffic demand trend brings the greatest challenge for the mobile communication ecosystem.

To overcome this 1000x traffic-growth challenge, some key enabling 5G technologies have been proposed, including massive multiple input multiple output (MIMO), millimetre-wave (mmWave) transmission, dense small-cell (SC) deployment, filtered orthogonal frequency-division multiplexing (OFDM), full duplex communications, polar codes, spatial division multiple access (SDMA), native support of machine-to-machine (M2M) communications etc. These technologies can be classified into three main categories [3]: air interface improvements, more spectrum allocation and network architecture improvements. Firstly, air interface improvements are expected to provide 3x–5x capacity gain for mobile networks, which not only focuses on improving the traditional link-level spectrum efficiency by fundamental techniques (e.g. massive MIMO, new modulation schemes, or new coding schemes), but also the capability of allowing cooperation amongst different network nodes for better interference management to exploit opportunistic gains from the

The Technology and Business of Mobile Communications: An Introduction, First Edition.
Mythri Hunukumbure, Justin P. Coon, Ben Allen, and Tony Vernon.
© 2022 John Wiley & Sons Ltd. Published 2022 by John Wiley & Sons Ltd.

perspective of system level, such as coordinated multipoint (CoMP) transmission and reception, non-orthogonal multiple access (NOMA), and three-dimensional (3D) MIMO. Secondly, it is expected to provide 5x–10x capacity gain by granting cellular networks access to the unlicensed spectrum or higher frequency mmWave bands. Last, but not least, the network architecture improvements, e.g. the heterogeneous networks (HetNets) with SCs and support of device-to-device (D2D) and M2M communications, are expected to contribute mostly in capacity enhancement, which can provide a 90- to 160-fold improvement of network capacity.

Amongst the network architecture improvements, densification of SC deployment is the most promising contribution to the capacity enhancement for the 5G networks, which is envisioned to increase linearly with respect to SC density. Historically, densification of traditional macro base stations (BSs) is only possible up to a certain extent because of the limited deployment space in cities and the high capital expenditure (CAPEX) of associated infrastructures (e.g. monitoring systems and cooling systems). Consequently, it is difficult to ultra-densely deploy traditional macro BSs for capacity improvement. In addition, the SC BSs can be treated as the simplified traditional macro BSs with lower transmitting power, which need almost no modification in the air interface design. Furthermore, since 50% of phone calls and 70% of data services occur indoors [4], the SC BSs, especially the femto BSs that are designed for indoor usage, can provide better indoor coverage as compared with the macro BSs. Therefore, performance analysis and optimisation for the SC deployment becomes essential. Note that the densification of SCs is also greener than the macrocells as it can shorten the serving distance between SC BSs and users, which can also extend the battery life of user equipment (UE) by saving energy consumption for the uplink transmission.

Due to the deployment flexibility, lower maintenance costs and the capability to boost the network capacity, mobile operators are experiencing a huge growth of SC deployments [5]. From the point view of the mobile operator, SC deployment is able to reduce the peak load by offloading users from high-traffic burden cells to low-traffic burden cells. Additionally, by shortening the distance between the user and its serving cell, better downlink quality of service (QoS) can be provided to retain customers. According to the SC forum, around 67% of worldwide operators have already deployed indoor SCs (i.e. femtocells), and the number of deployed femtocells has considerably increased from 4.3 million to 36.8 million in recent years. In addition, more than 40,000 outdoor SCs (i.e. picocells or metrocells) had been deployed in the USA by the end of 2015 [6]. SCs have already been deployed in New York, Chicago, Atlanta and San Francisco since 2016. In UK, over 50 outdoor trial sites of SCs have already been deployed [7]. However, the deployment of SCs is still facing three main challenges: interference management, mobility management and backhaul design.

The rest of this chapter is organised as follows: Initially, a brief history of SC development and its concepts are introduced. Then, the network architecture of HetNet is illustrated. After this, the above-mentioned three main challenges in SC deployment are discussed. Case studies for SC deployment in outdoor and indoor scenarios are provided using a network planning tool [8] afterwards in Section 7. Finally, before concluding, some future SC evolutions are represented.

10.2   Small Cells Concept Formation

The concept of SCs was originally developed from home BSs in 1999, which aimed to deploy Global System for Mobile communications (GSM) home BSs to connect existing standard GSM phones to a mobile operator's network [9]. The first product of complete 3G-based home BS was claimed at Swindon in 2002. Until around 2005, the term femtocell had been widely accepted for a standalone, self-configuring home BS, and the corresponding products were demonstrated at the 3GSM conference in February 2007. Five months later, the femto forum was formed to market these products worldwide in 2008. Home NodeB (HNB) and Home eNodeB (HeNB) were first introduced in Release 8 of 3rd Generation Partnership Project (3GPP) standard to describe a 3G femtocell and an LTE femtocell, respectively. The scope of femtocells was then extended into outdoors, to include outdoor urban areas [10]. This increased scope led to an evolution into the term 'small cell (SC)' in January 2012.

The SCs are with smaller coverage areas served by low-transmit-power BSs as compared with those served by macro BSs. Based on the range of transmit power, the SCs are typically classified into three categories: microcells, picocells and femtocells [5]. Different types of SCs may vary in terms of some specific usages, such as outdoor or indoor deployment, open or restricted access. The detailed description of the SCs thereof are summarised as follows:

- Microcells are the coverage areas provided by a relatively low-transmit-power operator-installed micro BSs as compared with that provided by macro BSs. With the same backhaul and access features as macrocells, each micro BS uses power control to limit the radius of its coverage area. Typically, the maximum coverage radius of a microcell is two kilometres with a transmit power of 2 W (33 dBm), and a maximum of several hundred users can be supported by each microcell. Microcells are usually utilised to increase the outdoor network capacity for the hot-spot scenarios, such as a shopping centre, a stadium or a transportation hub. Considering these aspects, microcells can be considered as an evolution from traditional macrocells.
- Picocells are difficult to precisely distinguish from microcells, but picocells are usually smaller than microcells, serving a few tens of users within a radio range of 300 metres or less. The transmit power of the pico BSs ranges from 200 mW (23 dBm) to 2000 mW (33 dBm). Picocells can be deployed for capacity improvement in traffic-intense outdoor scenarios, and are also frequently deployed indoors to improve poor wireless coverage within a building, such as a shopping mall, an office floor or a retail space.
- Femtocells are the serving areas provided by low-transmit-power access points with low CAPEX, which are also known as home BSs. They are typically user-installed to improve the coverage for a small vicinity, such as a home office or a dead zone within a building, and offloading data traffic using consumers' broadband connection (e.g. digital subscriber line (DSL), cable, or fibre). Note that, in this case, the operator does not have to bear the cost of backhaul and operational expenditures (OPEX) like the cost of power consumption. The transmit power of femto BSs is typically 20 mW (13 dBm) and the radius of a femtocell is less than 50 m. Unlike picocells and microcells, femtocells are designed to support only a dozen active users and are only capable of handling a few simultaneous calls. With the above differences, it can be argued that femtocells represent a revolutionary shift from traditional macrocell deployments.

Microcells and picocells are typically operated in open access mode, and femtocells can be operated in either open- or restricted-access modes. The restricted-access mode is also namely closed subscriber group (CSG) access mode. Additionally, microcells and picocells are planned-deployed by mobile operators, whilst femtocells are unplanned-deployed by users. Therefore, locations of microcells and picocells should be carefully designed, and self-organisation becomes a critical component in femtocells to control them automatically. Strictly, it is worth mentioning that operator-deployed relays do not belong to the SC family, which route data from macro BSs to end users and vice versa. This is because relays are independent from network architecture, acting like a coverage extender, which cannot improve the spectrum reuse spatially. However, they still cause interference to SCs, which cannot be ignored in SC deployment.

Despite the aforementioned types of SCs, some distributed BS systems (DBSSs) operate resembling the principles of SCs. The DBSS separates the radio function unit, also known as remote radio head (RRH), from the digital function unit, i.e. baseband unit (BBU), by optical-fibre communication. These deployments fall under centralised and cloud RAN (C-RAN) and we cover the technical aspects in detail in Section 5.9. Digital baseband signals are transmitted over the optical-fibre cables between RRHs and BBUs using open base station architecture initiative (OBSAI) [11] or common public radio interface (CPRI) standards [12]. The RRH is advocated to be installed close to the antenna, eliminating power losses in the cable from the RRH to the antenna. Furthermore, DBSS architecture enhances the flexibility of network deployments for operators that face site acquisition challenges and/or physical limitations. There are some commercial products adopting the DBSS scheme. The Lampsite BS [13] consists of a BBU, multiple pico RRUs (pRRUs) and RRU hubs (RRU HUBs), extending the pico BS that has only single RRU into multiple RRUs based on the principles of DBSSs. The Radio Dot System architecture [14] resembles an active DBSS since optical-fibre communication is used to connect the BBU to indoor RRUs, and uses local area network (LAN) cables to link the RRU to the antennas. Both solutions can support indoor multimode deployment in large- to medium-sized sites such as office buildings, venues, transportation hubs and semi-enclosed sports stadiums. Nevertheless, dedicated cabling infrastructure is required in the DBS which increases installation cost and the deployment period as compared with traditional SCs.

10.3 Multi-tier Cellular Networks/HetNets Architecture

The multi-tier cellular networks, also namely HetNets, are networks with various kinds of SCs or relays underlying traditional macrocells. The architecture of multi-tier cellular networks are illustrated in Figure 10.1. It shows that SCs provide the last mile wireless accesses to the mobile Internet users by core networks (CN) via backhauls. Different SCs may have different coverage areas, resulting from different transmit powers. However, the deployment of SCs still faces the following challenges:

10.3.1 Interference Management

Interference management is a critical issue for SC deployment. In Release 10–12 of 3GPP standards, the licensed spectrum band, which is already used in macrocells, is advocated to

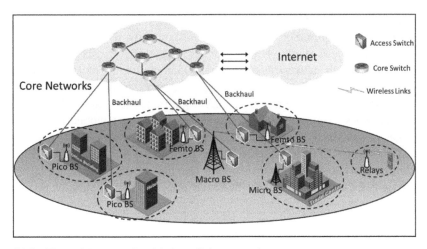

Figure 10.1 The architecture of multi-tier cellular networks.

be utilised in SCs. As a consequence, for the downlink transmission, users who are associated with macrocells will suffer additional interference from SC BSs and vice versa for the users attached to SCs. For the uplink transmission, macro BSs will suffer interference from SC users and vice versa for SC BSs. These two kinds of interference belong to the downlink and uplink *cross-tier* interferences, respectively, where cross-tier interference is defined as the interference generating from other-tier devices. In Release 13 of 3GPP standard, the 5 GHz unlicensed spectrum band was expected to be used in SCs for increasing downlink capacity, and may be extended to uplink transmission in Release 14 of 3GPP standard with the aid of carrier aggregation. This can inhibit cross-tier interference between macrocells and SCs, but cross-tier interference between SCs and the incumbent Wi-Fi network will become another unavoidable issue. Therefore, cross-tier interference is inevitable for SC deployment. Additionally, SCs of the same tier use the same licensed or unlicensed spectrum band, thus *co-tier* interference, which is defined as the interference received from devices of the same tier, should also be mitigated, especially in an ultra-densely deployed scenario. It is worth mentioning that the mmWave spectrum band has also been proposed for use in SCs to avoid cross-tier interference with the incumbent networks (i.e. macro cellular networks and Wi-Fi networks). Moreover, due to the characteristics of highly attenuated path loss and low penetration for the mmWave spectrum, the co-tier interference can be effectively retained. Nevertheless, the mmWave transmission can only provide decent performance in specific scenarios with line-of-sight (LOS) links. Furthermore, the additional hardware equipment (e.g. radio frequency (RF) unit, MIMO etc.) to use the mmWave spectrum band will increase the CAPEX of SC BSs.

10.3.2 Mobility Management

Mobility management, i.e. the handover process for mobile UEs, becomes much more complicated in multi-tier cellular networks, because mobile UEs may trigger frequent handovers when they move across the smaller coverage areas of SCs. The handover process aims to transfer the active links of a UE from its serving cell to a target cell connectively whilst satisfying the requirement of QoS [6]. Conventionally, the same set of handover parameters, e.g. hysteresis margin (HM) and time-to-trigger (TTT), can be used throughout the whole homogeneous network as the coverage radii of different macrocells are nearly equal. Unfortunately, using the same set of handover parameters may severely degrade the performance of UEs as the coverage areas differ significantly for macrocells and SCs (i.e. microcells, picocells, femtocells and relays). For instance, fast mobile macrocell UEs (MUEs) may already have located in the centre coverage areas of SCs before the TTT expires, causing handover failure due to the degraded signal to interference plus noise ratio (SINR). Furthermore, it is unnecessary to perform handover if the fast mobile MUE quickly passes through SCs. Therefore, cell-specific handover parameter optimisation is needed in multi-tier cellular networks.

10.3.3 Backhaul

As shown in Figure 10.1, backhaul is defined as the links between the radio access networks and the CNs. It is claimed in [15] that approximately 56% of operators treat backhaul design as one of the major bottlenecks for SC deployment, because wired backhaul connections may not always be available for SC BSs, and deployment of new wired or wireless backhaul infrastructures for each of the SC BSs in an ultra-dense scenario will significantly increase the CAPEX of SC deployment.

10.4 Interference Management and Modelling in Small cell/HetNets

10.4.1 Interference Management

In the literature, interference management can generally be divided into two categories: interference cancellation and interference coordination. The basic mechanism of interference cancellation is to regenerate the interfering signals and subsequently subtract them from the received signal [16]. This can improve the SINR level of users by cancelling the extracted interference from the received aggregate interference. Specifically, successive interference cancellation (SIC), a well-known physical-layer interference cancellation technique, enables the receiver to successively extract the strongest signal from the residue signal after subtracting the extracted one from the combined signals. As a result, the desired signal and each interfering signal can be obtained iteratively. However, there is a trade-off between the extracting number of interfering signals and the time consumption in SIC, thus the extracting number of interfering signals is limited. Other interference cancellation techniques can be referred to in [16] for details.

Although interference cancellation techniques can cancel the interfering signals, the computing consumption scales exponentially with the number of users served at the BSs. Additionally, real-time information exchange is required between BSs to achieve interference cancellation. Consequently, interference coordination techniques can provide supplementary efforts to improve the SINR performance, the aim of which is to reduce or mitigate the suffered interference of UEs or BSs by intelligently allocating radio resource (i.e. frequency, time, power and space). In traditional homogeneous networks, fractional frequency reuse (FFR) is the most widely adopted mechanism for the downlink inter-cell interference coordination (ICIC). The main idea of the FFR is to divide the whole spectrum band into several subbands, which tries to avoid the edge UEs in neighbouring macrocells using the same subband. For example, strict FFR divides the whole spectrum band into $N + 1$ subbands. One of the subbands is commonly allocated to the centre UEs in each macrocell as centre UEs are insensitive to the interference, and one of the other N subbands is distributed to the edge UEs in a typical cell with the rule that the edge region in neighbouring cells must use different subbands, as shown in Figure 10.2. The number of subbands N is typically configured as *3* for a hexagonal cellular network. This is because an increase of the subband number will decrease the spectrum efficiency of the whole network and *3* is the minimum value to satisfy the rule that the edge region in neighbouring cells cannot use the same subband. To improve spectrum efficiency, soft FFR is proposed, in which the whole spectrum band is divided into N subbands ($N = 3$). The edge UEs in a typical cell can only access one of the subbands with full transmit power, and the centre UEs in each cell use the left *N-1* subbands other than the one used in the edge regions with a reduced transmit power. In order to further improve spectrum efficiency, an adaptive FFR is advocated to adjust the FFR configuration (e.g. soft or strict, the value of N) based on interference levels and scheduling users with a specific subband based on channel quality measurements [16]. Additionally, cognitive radio is also an effective tool to enable BSs to access the unused spectrum band to improve network capacity [17]. Power control, which is initially employed in cellular networks to minimise near-far dynamic range effects for uplink transmission, has also been deployed for the downlink transmission to improve the coverage performance for edge users. In addition, CoMP transmission can effectively turn inter-cell interference from neighbouring cells into useful signals by exploiting the spatial diversity gains. Moreover, sectorial antennas are also used to mitigate the interference as compared with the omni-directional antenna from the point of view of space. From the above discussion, we can conclude that the ICIC in homogeneous networks are exploited mainly in forms of frequency, power and spatial domains.

In multi-tier cellular networks, the above-mentioned ICIC techniques can also be utilised, but some adaptations may be required, which may lead to greater complexity as compared to those applied in homogeneous networks. Besides, there are also many existing

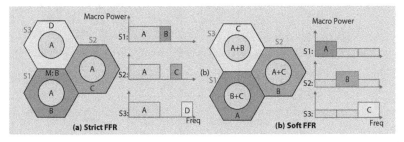

Figure 10.2   The subband allocation in (a) strict FFR and (b) Soft FFR.

works studying the ICIC in multi-tier cellular networks. For femtocell networks, dynamic frequency planning (DFP) for WiMAX femtocell networks was proposed in [18], which optimises the subchannel allocation to each cell based on the user requirement and interference level in a CSG femtocell network to decrease interference and improve network capacity. In [19], the power tuning, which adjusts the transmit power on the data channel, minimised the coverage area of each SC cell whilst ensuring an acceptable received signal strength for its farthest attached user. A detailed account of interference management techniques for femtocell networks can be referred in [20]. For picocells with cell-range expansion (CRE), which extends coverage area of each picocell without increasing the transmit power to attract more UEs to be served by picocells for load balancing. Enhanced inter-cell interference coordination (eICIC) was proposed in Release 10 of 3GPP standard. This scheme insisted that almost blank subframes (ABSs), with no power transmitting on data and control channels at some coordinated time slots, should be leveraged in macrocells to mitigate the interference to CRE users. A brief summary of aforementioned ICIC techniques is illustrated in Table 10.1.

In this remaining paragraph, we will mainly introduce how eICIC works in a two-tier picocell network. As shown in Figure 10.3, the coverage area of a picocell is much smaller

Table 10.1 A brief summary of mentioned ICIC techniques

ICIC techniques	
Homogeneous Networks	**Heterogeneous Networks**
Strict FFR, Soft FFR, Adaptive FFR, Cognitive Radio, Power Control, CoMP	
Sectorial Antennas	DFP, Power Tuning, eICIC

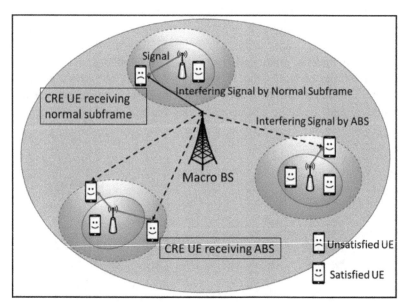

Figure 10.3 An illustration of interference in the eICIC scheme.

than the macrocell, leading to fewer associated UEs, which is undesirable when macrocells are overloaded. To overcome this load imbalance, the CRE is adopted in picocells to extend their coverage areas without increasing the transmit power of pico BSs. Unfortunately, in this CRE region, the UEs become more vulnerable to interference from macro BSs. Consequently, as shown in centre regions, the picocell CRE UEs can receive full power subframes (FPSs) in the same time slots as ABSs without suffering from significant interference from macro BSs [5]. Nevertheless, there are still two major problems for the eICIC scheme. One is that ABSs cause significant capacity losses in macrocells and the other is that the interference in reference signals still exists. Therefore, in Release 11 of 3GPP standard, the further enhanced inter-cell interference coordination (FeICIC) scheme was proposed. On one hand, as shown in Figure 10.4(b) to reduce the capacity loss, reduced power subframes (RPSs) were proposed. Whilst serving macrocell-centre UEs with a relatively low transmit power as compared with FPSs, the RPS managed to mitigate the interference to picocell CRE UEs [21]. On the other hand, interference cancellation techniques, such as SIC, are expected to solve the interference in reference signals. Intuitively in reality, the effects of the duty cycles of ABSs or RPSs, the reduced transmit power, and the range expansion bias for CRE on the network performance should be investigated before the deployment of ABSs or RPSs.

10.4.2 Interference Modelling

In recent years, stochastic geometry has become a popular and effective mathematical tool to analyse large-scale network performance [22] for several reasons. Firstly, the actual network deployment varies significantly from different cities, the conventional spatial models for BS locations, such as the hexagonal model, become over-simplified for performance analysis. Secondly, analytical results of statistical performance metrics can be obtained, which are more effective in time and resource consumption than the Monte Carlo simulation. Thirdly, the influence of specific parameters may be observed in analytical results, and some system design insights can be provided. Consequently, stochastic geometry

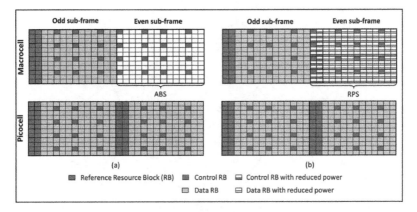

Figure 10.4 Illustration of (a) ABSs and (b) RPSs that are used for time-domain (F)eICIC in 3GPP heterogeneous networks.

becomes powerful in investigating interference management schemes and analysing the influence of the corresponding parameters.

To apply the stochastic geometry into the analysis of RPSs for the FeICIC scheme, the entire process can generally be divided into the following four steps:

a) Determine what point process (e.g. Poisson point process, Poisson cluster process, and Poisson hole process) should the spatial locations of macro and pico BSs follow. A point process can be defined as a countable random collection of points that reside in some measure space, usually the Euclidean space [23]. Specifically, Poisson point process (PPP) [23] is the most widely used point process, and a realisation of macro and pico BSs following two independent PPPs is illustrated in Figure 10.5. In the realisation of PPP, number of BSs follows Poisson distribution and locations of BSs uniformly distribute on the two-dimensional space.

b) Since open-access mode is adopted in picocells, user association strategy between macrocells and picocells, and resource allocation strategy in each cell needs to be established. The circle around each SC BS, i.e. blue triangles in Figure 10.5, denotes its coverage area. There are two circles around pico BSs because CRE is adopted. Then, the user association probability should be derived based on the proposed strategy. The black circle around each macro BS (i.e. red dots in Figure 10.5) denotes the centre region of the corresponding macrocell.

c) Derive the analytical coverage probability and validate this theoretical result by the Monte Carlo simulation, which is used to model the probability of different outcomes in a process that cannot easily be predicted due to the intervention of random variables.

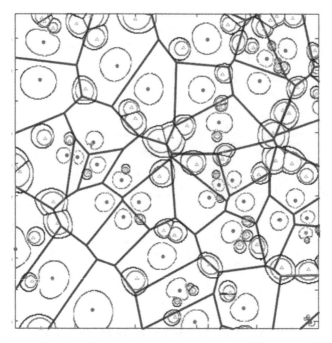

Figure 10.5 A realisation of spatial locations and coverage area of BSs in the FeICIC scheme.

Because of the randomness of BS locations, the probability density function (PDF) of the serving BS and the Laplace transform [22] of the received aggregate interference power, which captures the property of the aggregate-interference-power distribution, should be derived respectively to obtain the coverage probability.

d) Based on the coverage probability, other performance matrices, such as rate coverage probability, which is defined as the average fraction of users achieving a target rate, can also be derived. Additionally, the influence of parameters on the corresponding performance matrices (e.g. spectrum efficiency and energy efficiency) can be analysed numerically.

In Figure 10.5, the dots and triangles denote macro and pico BSs respectively. The inner and outer circles around each pico BS represent the original and the extended coverage area by CRE. The circle around each macro BS is the centre area of the corresponding macrocell.

Based on the system model as shown in Figure 10.5 and steps illustrated above, we investigate network performance using RPSs. The values of simulation parameters for Figures 10.6 and 10.7 are listed as follows unless otherwise specified:

- Densities of marco BSs and pico BSss: $1.27/km^2$, $3.81/km^2$;
- UE density following another independent PPP: $38.1/km^2$;
- Transmit powers of macro BSs and pico BSs: 43 dBm, 30 dBm;
- Path-loss exponents of the macro tier and the pico tier: 3, 3;
- Duty cycle: 0.5;
- Rayleigh-fading channel with exponential distribution factor on power: 1;
- Centre region factor: optimised values in [19].

In Figures 10.6 and 10.7, the effect of the RPS duty cycle β and the power reduction factor ρ of RPSs, respectively determining the probability of a subframe being the RPS and the transmit power of RPS, are illustrated in terms of the rate coverage probability. The rate coverage probability is defined as the probability of a typical user having a larger throughput than a target rate. Details of the numerical result can be referred in [21].

In Figure 10.6, a comparison of rate coverage probabilities obtained by RPSs and ABSs with range expansion bias B_p being 3 dB and 7 dB is illustrated versus the RPS duty cycle. The target rate ω is set as 100 kbps. Results show that the rate coverage probability with RPSs outperforms that with ABSs for typical duty cycles with static typical range expansion biases.

In Figure 10.7, the rate coverage probabilities are analysed with a variety of power reduction factors with typical range expansion biases and rate thresholds. Results show that a sharp increase occurs when ρ varies from 0, indicating that RPSs provide better rate coverage as compared with ABSs if appropriate transmit power of RPSs is configured. According to simulation results, the optimal power reduction factor is in the range of $[0, 0.1]$.

Note that these two results are obtained by assuming time synchronisation is achieved throughout the whole two-tier HetNet. However, strict-time synchronisation may not always be achieved because the backhaul, providing the control signal exchange between macrocells and picocells, may be congested in a high-density scenario. Moreover, due to random propagation delays, subframes transmitted from neighbouring cells may be misaligned. Accordingly, the effect of subframe misalignment was taken into consideration in [24]. In this work, the macro BSs form tier-1 BSs and pico BSs form tier-2 BSs. There are

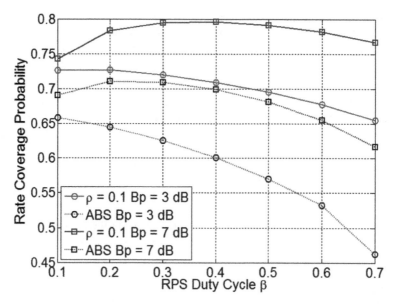

Figure 10.6 Rate coverage probability against duty cycle of RPSs with range expansion bias being 3 dB and 7 dB, and the target rate being 100 kbps.

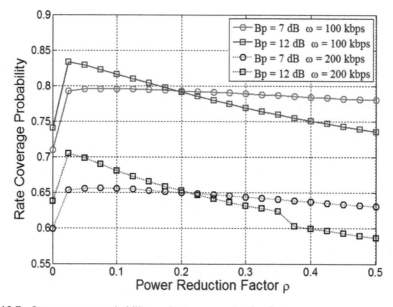

Figure 10.7 Rate coverage probability against power reduction factor.

some value changes in the simulation parameters, i.e. the transmit power and density of macro BSs is respectively 100 and 0.1 times of pico BSs, the cell range bias is 6 dB, and the path loss exponent of the pico tier is 4.

In Figure 10.8, the coverage probabilities of victim users (VUs) of both tiers are illustrated versus the maximum subframe misalignment factor, which determines the maximum value

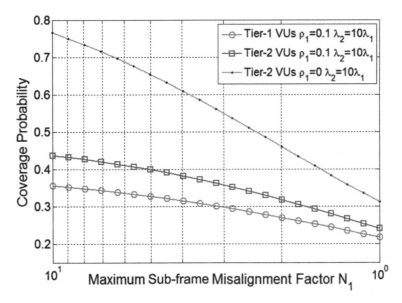

Figure 10.8 The theoretical coverage probabilities of VUs versus the maximum subframe-misalignment offsets.

of subframe-misaligned restricted range [24]. A typical macrocell centre user and picocell range expansion users are respectively referred to as a tier-1 victim user (VU) and a tier-2 VU, as they will suffer from increased interference due to misaligned subframes. The results show that misaligned subframes decrease coverage probabilities of VUs remarkably, especially in the case of ABSs, and using RPSs instead of ABSs can alleviate the subframe-misaligned effect on decreasing the coverage probabilities of VUs.

Due to the increasing complexity of multi-tier cellular networks, some of the existing interference management techniques are still under investigation before their deployment in reality, and new techniques or schemes, such as 3D MIMO, are heated research topics for forthcoming 5G networks.

10.5 Mobility Management

Mobility management, i.e. handover, aims to ensure the service continuation and quality of UEs when the current associated BS for a mobile UE cannot provide acceptable performance (e.g. signal strength). As mentioned in Section 10.2, the handover process has become more complicated in multi-tier cellular networks than that in traditional homogeneous networks mainly because of the smaller coverage area of SCs. Using the same set of handover parameters is no longer suitable for multi-tier cellular networks. In the handover process, defined by 3GPP LTE, the UE must disconnect the radio link to the source cell before establishing a new radio link to the target cell. Thus, the entire handover procedure can typically be divided into four main phases: measurement phase, preparation phase, execution phase and completion phase. This procedure is illustrated in Figure 10.9 [25], and although this procedure is defined for the conventional homogeneous networks, the signalling flow of this procedure still fits for the handover procedure of the multi-tier cellular networks, which may be affected by different handover decisions and/or different cell-selection algorithms.

Note that in the HetNet with SCs, we can focus on the case of intra-RAT intra-carrier handovers [6]. Typically, there are two matrices to evaluate the network handover quality: the average number of handover failures (HOFs) and ping-pongs. To define the HOF, it is a perquisite to understand the radio link failure (RLF). As shown in Figure 10.10, a radio problem, such as being out of synchronisation, is detected if the channel quality indicator (CQI) of the UE becomes lower than a threshold Q_{out}. The CQI of the UE is evaluated with a sliding window of 200 ms for a comparison with Q_{out}. After detecting this problem, the timer T310 starts counting down. Meanwhile, the CQI of the UE is evaluated with a sliding window of 100 ms to compare with another threshold Q_{in}. The T310 timer is terminated once the CQI of the UE is larger than Q_{in}, and no RLF is detected. If the T310 timer expires, an RLF will be declared. Based on this, the three entering conditions for counting an HOF are described as follows [26]:

- When the UE is connected to the source BS, the T310 timer has been triggered and is still running when a handover command is received by the UE.
- When the UE is connected to the source BS, RLF is declared in the time duration between the event A3 entering condition [6] and the handover command is successfully received by the UE.
- When the UE is connected to the target BS, at the end of the handover execution time, the downlink filtered average wideband CQI received from the target BS is smaller than Q_{out}, which is typically -8 dB.

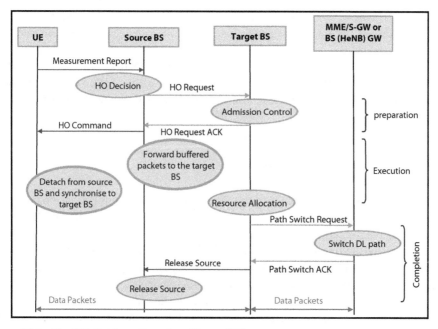

Figure 10.9 The LTE Handover Procedure. (*Source* [25]).

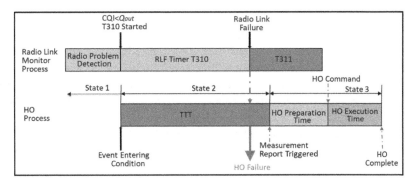

Figure 10.10 Timers in radio link monitoring and handover processes. (*Source* [26]).

Ping-Pong refers to unnecessary, repeated handovers between two base stations. The concept of time-of-stay is used to describe the Ping-Pong handover. The time-of-stay is defined as the time duration that a UE maintains connection to the new serving BS instantly after a handover, which starts when the UE sends a handover complete message to a BS, and ends when the UE sends another handover complete message to another BS. If the time-of-stay is less than a threshold (e.g. 1 s), and the new target BS is the same BS as the source BS in the previous handover process, then these two handovers are treated as a Ping-Pong handover, which is meaningless and wastes signalling resources. The value of TTT has great influence on the performance of handover. Small TTT values may lead to too-early handovers, increasing the average number of ping-pongs, meanwhile large TTT values may cause too-late handovers, increasing the average number of HOFs. Therefore, the optimisation of the TTT, according to UEs' velocities, carries capital importance in mobility management.

Potential solutions for the mobility management in HetNets are still under investigation. Based on the forthcoming software defined network (SDN) architecture, existing works have discussed the benefit of SDN-based mobility management. The key concept of SDN is the decoupling of data and control planes, and the centralisation of control plane [27]. Therefore, in the handover procedure, the SDN controller handles the necessary handover signalling and manages the handover decisions in the HetNet. The main advantage of this solution is the centralisation of handover decision, admission control and data forwarding. As a consequence, optimisation of the HO decision algorithms becomes critical.

In recent years, not only the average number of HOFs and ping-pongs, but also the energy, offloading and fairness are considered as optimisation goals for the HO decision algorithm in the multi-tier cellular networks [28]. Based on the primary decision criteria used in the HO, the algorithm can typically be classified into the five following groups:

a) **Speed based algorithms:** These algorithms compare the UE speed with a threshold, which may be related to received signal strengths (RSSs) of serving and target cells, the UE traffic type, the available bandwidth on target cells and the UE membership status, to obtain the HO decision.
b) **Cost-function based algorithms:** This group of algorithms obtains the HO decision by comparing outcomes of the cost-function for the serving and candidate target cells.

The cost-function can incorporate a wide range of parameters, such as battery lifetime, traffic type, cell load, RSS and speed.

c) **Interference-aware based algorithms:** These algorithms capitalise the measured results, e.g. received signal quality (RSQ), received interference power (RIP) and received signal power, and the interference constraints in cells or UEs to make the HO decision to improve the SINR performance. However, signalling procedures become more complicated for this group of algorithms.

d) **Energy-efficiency based algorithms:** The algorithms in this group achieve the HO decision by fulfilling the energy saving potential offered by SCs and saving energy consumption for UEs.

Details of these algorithms can be found in [28]. Nevertheless, existing HO algorithms target over-simplified single-macrocell single-SC HO decision scenarios, the HO algorithms are urgently needed for the multiple-macrocell multiple-SC scenarios, especially for the ultra-densely deployed SC scenarios.

To evaluate the performance of different HO algorithms, it is essential to model and analyse the handover performance in the multi-tier HetNet. One of the most critical research areas is user-mobility modelling. In the literature, user-mobility models include four types: the random walk model, the random waypoint (RWP) model, the fluid flow model and Gauss-Markov model [29]. However, only simplified user-mobility models, such as the random walk model or the RWP model, are utilised to analyse network performance. Besides, the impact of human tendency and clustering behaviours on mobility is still under investigation. Consequently, it is important to investigate how to tractably incorporate user mobility models into the modelling and performance analysis in the complicated multi-tier HetNet.

10.6 Backhaul

The backhaul for SCs is different from that deployed in macrocells mainly in terms of the CAPEX. Conventionally, optical-fibre cables are usually deployed in macrocells to support a large volume of traffic aggregated from hundreds of UEs. Nevertheless, the optical-fibre cables used in macrocells cannot meet the cost requirement of SCs. Consequently, wired or wireless based backhauls should be considered in terms of both cost and feasibility. In addition, different SCs may vary in terms of the number of associated UEs. For example, the SC deployed in a traffic hotspot (e.g. stadium or office building) serves more UEs than that deployed for home UEs, thus the backhaul of SCs should be designed based on the customised capacity requirement for different SCs. Furthermore, the backhaul traffic throughput for the same SC can vary significantly during high and low demand hours in a day. In peak hours, there occurs a higher probability that SC users (all having good SNR conditions) exploit high data rate services. As a consequence, the SC backhaul should have the capability to dimension the backhaul allocation for high- and low-load conditions, which may be achieved by traffic aggregation with point of presence (PoP). An abstract view of the SC backhaul architecture is illustrated in Figure 10.11. PoP refers to a central access point where the traffic from different cells is aggregated. Macro BS sites can act as

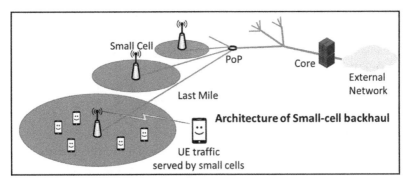

Small Cell

PoP

Core

External
Network

Last Mile

Architecture of Small-cell backhaul

UE traffic
served by small cells

Figure 10.11 Backhaul network architecture. (*Source* [30]).

PoPs to serve SC BSs, with a PoP density of around 9 sites per km^2 assuming an inter-site distance (ISD) of 500 m.

The main technical challenges and possible solutions for backhaul design of SCs can be summarised as follows:

a) **Cost:** Cost is the most important issue for SC backhaul design, which includes CAPEX and OPEX. CAPEX for backhaul includes hardware expenses, comprising of cables, antennas, waveguides, the Ethernet switch, Routers etc. OPEX includes site rental, maintenance, upgrades, and monitoring of SC backhauls. In [31], the SC backhaul cost is analysed based on a three-layer model, which separates the whole network into BS, UE and backhaul layers, and divides the total backhaul cost into three categories: infrastructure, capacity and physical connection costs. The infrastructure cost denotes the deployment cost of an infrastructure, such as cell BS and backhaul node. The capacity cost refers to the cost involved in connecting two adjacent layers subject to meeting required capacity. The physical connection cost represents the cost of physically connecting two points in the whole network. Results show that there exists a range of backhaul-node density to minimise backhaul-deployment costs whilst satisfying the requirements of UE throughput. However, the variance of traffic volume in terms of time and space is not discussed as well as the OPEX. Consequently, how to reduce backhaul total cost in terms of the CAPEX and OPEX is still under investigation.

b) **Backhaul Capacity Dimension:** Because personalised traffic volume for SCs are in different locations or different periods, the adjusting capability of the SC-backhaul capacity increases CAPEX efficiency as compared with the worst-case oriented backhaul capacity dimensioning. Traffic aggregation may be a promising technique to improve cost efficiency of SC backhaul, which aggregates traffic from several SCs and sends this aggregated traffic through a common backhaul. For example, neighbouring SCs with inverse temporal traffic patterns can be aggregated to use a common backhaul to reduce the CAPEX. Aggregating backhaul traffic through a linear backhaul topology may be the easiest to realise, but single points of failure may become an issue. As a result, other robust but more expensive backhaul topologies, e.g. star and mesh, are advocated to be adopted in SC backhaul deployment.

c) **PoP Deployment:** Although traffic aggregation can be an effective way to reduce CAPEX in terms of traffic variance and provide control and user plane functionalities to lower the signalling load on the CN components [31], large planning efforts are required for the PoP deployment because various existing backhaul solutions (i.e. wired and wireless) have their own pros and cons. The PoP deployment with wired solutions focuses on the cost of its own deployment and the connecting sockets, whilst those with wireless solutions focus on the coverage area of the PoP, which is defined as the area where SC BSs can connect with the PoP. In terms of wired solutions, due to the high costs of installation of new wired connections, existing infrastructures, such as macro BSs and edge clouds, are the most desirable choices to be the PoPs. Nevertheless, this may lead to undesired PoP placement from the perspective of traffic load balancing. In terms of wireless solutions, since the signal is expected to transmit at an ultra-high frequency, such as mmWave communication, the desirable transmitting distance of such signals should firstly be investigated. Since free space loss has its most significant impact on links of over 1 km, this may not be a bottleneck to SCs if located within distances of 100 metres from the PoP. Secondly, the availability of LOS or non-LOS (NLOS) links has great influence on the maximum backhaul capacity for UEs. A LOS link is defined as a direct non-blocked link and a NLOS link refers to a situation where the direct radio transmission path is blocked by a physical object. On one hand, LOS links for the SC backhaul may not be available as SCs are always deployed at the edge of macrocells. On the other hand, NLOS links may have to be used in dense urban areas, as they enable more SC deployment locations. Thirdly, the choice between point-to-point (P2P) and point-to-multipoint (P2MP) transmission should be made for different application scenarios. P2P means one-to-one communication between the PoP and SC BS, and P2MP represents a one-to-many communication between the PoP and multiple SC BSs, which resembles NLOS conditions and low-frequency bands (e.g. sub-6 GHz) and is able to overcome signal obstructions with a lower capacity. In the P2MP transmission, the PoP acts as a unique data sink and can be equipped with either an omni-directional antenna or a number of directional antennas pointing in different directions, e.g. antenna arrays with static beams, large scale antenna systems (LSAS) [32]. Therefore, the P2P is suitable for high-loaded SCs and P2MP is suitable for multiple low-loaded SCs. Based on the above discussion, perhaps a hybrid solution for the PoP deployment may be effective for scenarios where wired solutions will become too expensive, such as streets. In the hybrid solution, an access point should be deployed near the targeted scenario. There is a wired backhaul, such as an optical fibre, connecting the access point and CN. Wireless equipment should also be deployed at this access point, which provides the last-mile wireless backhaul access for UEs in the targeted scenario to reduce installation cost. The Vectastar solution [33], uses such a hybrid solution to provide backhaul services with more flexibility and lower cost than other forms of microwave backhaul.

In order to exploit the diversity of wireless solutions, LSAS and optical wireless broadband (OWB) may also be incorporated in future SC backhauls. LSAS scales the conventional MIMO by utilising antenna arrays consisting of tens to hundreds of antennas, which can simultaneously serve hundreds of SCs. It can generate numerous static or semi-static directional beams transmitting at different directions, which is ideal for achieving P2MP communications at PoPs. The perfect channel state information (CSI) is required which can be extracted from the uplink pilots in a time division duplexing (TDD) system. The LSAS is suitable for SC backhaul because of its scalability without the requirement of LOS links. However, optimal configuration of LSAS is still under study. OWB combines the high-capacity optical fibre and low-cost RF to provide communication for distant end points. Optical fibre extends the communication distance and RF technology provides the last-mile access. This OWB solution can improve the link throughput whilst reducing cost and expediting the deployment process as no licensed RF frequency is required [15]. Additionally, visible light communication (VLC) can also be incorporated in the backhaul structure, which transmits data via visible light between 400 and 800 THz. It exploits the light emitting diode (LED) to modulate information at visible light frequencies taking advantage of existing lighting infrastructure. However, this technique is not suitable for outdoor environments because of the interference caused by sunlight, weather conditions (e.g. rain) and undesirable street lights. Therefore, with all the possible solutions, the optimisation of backhaul deployment becomes more complicated. To comprehensively analyse the performance and cost of backhaul deployment, the cost, performance and constraints of each solution should be investigated and compared. Moreover, the spatial and temporal traffic patterns of the network for a given area can be predicted due to the development of machine learning (ML) in recent years. Based on predicted traffic patterns, the total backhaul cost may be minimised with guarantee of UE requirement by optimising the number, locations and types of backhaul nodes.

In Release 15 of 3GPP standard, integrated access and backhaul (IAB) has been proposed [34]. Compared with fixed-bandwidth schemes for backhaul design, IAB incorporates flexibility into the bandwidth allocation between backhaul and access links in each SC to meet the dynamic traffic demand. All backhaul and access links are expected to transmit in the mmWave spectrum band. As the mmWave spectrum band has a large bandwidth in each channel, there may exit redundant bandwidth under a fixed-bandwidth scheme if the traffic load is low. IAB can provide access to the redundant bandwidth to improve spectrum efficiency. Furthermore, by using directional beamforming, all backhaul and access links can share the same wireless channel without interference. Note there are two major problems under investigation for the IAB: One is to determine the optimal fraction of bandwidth split between access and backhaul links in each SC, e.g. to maximise the number of UEs that receive required downlink throughput. Another is the partition of total backhaul bandwidth in the anchored BS to its wireless-backhaul-connected SCs.

10.7 Small-Cell Deployment

The main purpose of SC is to boost the capacity for populated areas, such as stadiums, stations, offices, hotels and shopping centres, which can also offload traffic from the macrocell to SCs for alleviating the backhaul burden of the macrocell. Since the self-organising network (SON) is still at its early stage, which aims to enable the network to solely organise itself and optimise network performance with resource management, the locations of the public access SC BSs need be carefully planned. In the following, we will firstly illustrate a

Figure 10.12   A dense urban scenario – Paris centre.

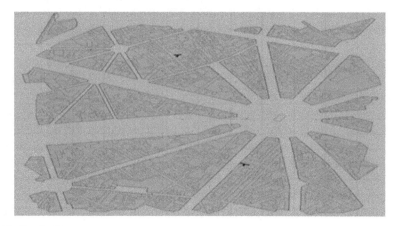

Figure 10.13   Traffic maps.

case study of outdoor SC deployment in a dense urban area of Paris's city centre and secondly we will investigate a case study of indoor SC deployment in a building. The network planning and optimisation tool that was introduced in [8] is used in both case studies.

The size of the outdoor scenario, i.e. a dense urban area of Paris's city centre, is approximately 1.56 (1.3*1.2) km^2, which is shown in Figure 10.12. The average height of buildings is about 21 metres.

The traffic map is illustrated in Figure 10.13, in which there are two traffic models defined by 3GPP. The green areas refer to low-traffic-intensity outdoor scenarios with 50 web browsing and 20 video users. The brown areas represent high-traffic-intensity indoor scenarios and

courtyards between buildings with 500 web browsing and 200 video users. Thus, a total of 770 users is considered in the case study scenario.

Only the macrocell and the multi-tier solutions are both investigated for the case study scenario, and the network and UE configurations for the simulation are listed as follows [27]:

- Transmission: FDD-LTE 2.6 GHz with 20 MHz;
- UE: antenna height: 1.5 m above ground or floors, gain: 0dBi, sensitivity: -106 dBm;

- Path-loss model: ray tracing tool RRPS (Ranplan Radio Propagation Simulator);
- MIMO: 2 × 2 MIMO with CLSM (Close-Loop Spatial Multiplexing);
- Scheduler: Proportional Fair (PF);
- Scheduling granularity: Physical Resource Block (PRB);
- Traffic models: web browsing and video defined by 3GPP;
- Shadowing margin: 3 dB;
- Simulation: Monte Carlo, 20 snapshots, TTI: 100;
- Configuration for macrocells:
    - two base-station sites, three sectors each BS with total bandwidth of 20 MHz, macrocells use channel 1 (centre frequency 2630 MHz);
    - BS Antennas: Height: 45 m, Tx power: 43 dBm, gain: 7dBi.
- Configuration for small cells:
    - 23 SCs use channel 2 (with centre frequency of 2670 MHz);
    - BS Antennas: Height: 6 m, Tx power: 33dBm, gain: 0dBi.

We assume the SC BSs will be installed on the lampposts, thus the locations of 23 SCs are automatically chosen by the intelligent cell optimisation (ICO) module in the planning tool, based on the locations of lampposts.

In Figure 10.14, the SINR distributions for the macrocell and multi-tier solutions are both shown. For the macrocell solution, the SINR performance of some users in shadowing areas is so low (below -15 dB) that even basic modulations will be exacerbated. These low-SINR users can be significantly reduced by the multi-tier (HetNet) solution, as different channels are used in macrocells and SCs which avoid mutual interference. Figure 10.15 shows the cumulative distribution function (CDF) of the SINR distribution for both solutions, in which SC deployment improves 9% of the voice coverage (SINR: -7 dB) from 85% to 94%, and enhances the SINR of cell-edge-area UEs (5% of UEs) by 9 dB.

In Figure 10.16, the cell capacity for both macrocell only and multi-tier solutions are both illustrated. For the macrocell solution, total network throughput is 57.4 Mbps, with an

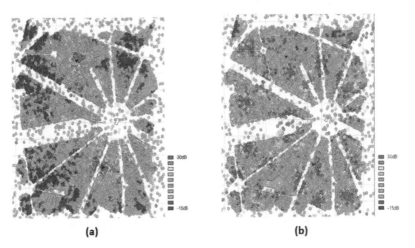

(a)   (b)

Figure 10.14 The SINR distribution for (a) the macrocell solution and (b) the HetNet solution.

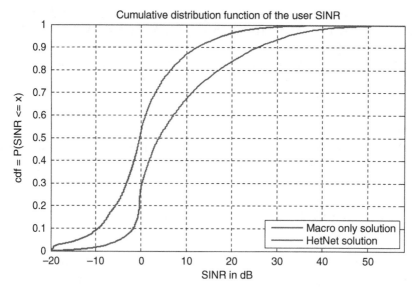

Figure 10.15 The CDF of the SINR distribution.

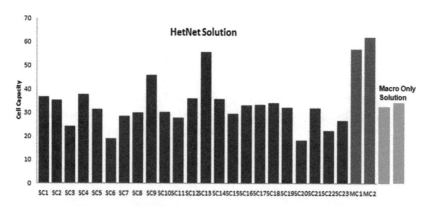

Figure 10.16 Cell capacity distribution.

average throughput of 28.7 Mbps in each macrocell. For the multi-tier solution, the total network throughput is 850 Mbps with the average throughputs of each SC and each macrocell being 31.8 Mbps and 59 Mbps, respectively. The total network throughput is nearly 14 times in the multi-tier solution as compared with that in the microcell solution, which has more than doubled with a second channel used in the SCs. This further shows the benefit of SC deployment. In addition, the average MC throughput has more than doubled in the multi-tier solution as the SINR of UEs increased substantially. In addition, the average UE throughputs are respectively 3.58Mbps and 155.3Kbps for video service and web service with the SC deployment, which is better than those (283.3Kbps for video service and 14.13Kbps for web service) with the macrocell only deployment. This indicates that the QoS can be substantially increased by SC deployment compared with the macrocell only solution.

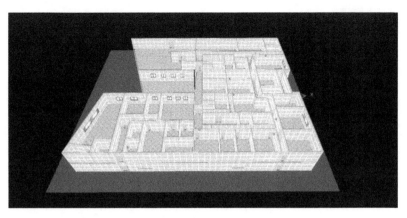

Figure 10.17 An indoor scenario – two-storey office building.

Next, the case study for indoor scenario will be introduced. We consider a two-storey office building with 2520 m^2, which is shown in Figure 10.17.

The network and UE configurations for the simulation are listed as follows:

- Transmission: FDD-LTE 2.6 GHz with 20 MHz;
- 13 small cells use channel 1 (with centre frequency of 2630 MHz);
- SCs: Height: 2.4 m, Tx power: 17dBm, gain: 0dBi;
- UE: antenna height: 1.5 m, gain: 0dBi, sensitivity: -106dBm;
- Path-loss model: RRPS;
- MIMO: 2 × 2 MIMO with CLSM;
- Scheduler: PF;
- Scheduling granularity: PRB;
- Traffic models: web browsing (50 UEs) and high-definition video (10 UEs) defined by 3GPP;
- Shadowing margin: 3dB;
- Simulation: Monte Carlo, 100 snapshots, TTI: 100.

The location of 13 SCs are automatically chosen by the ICO module in [8] based on the possible locations, such as the locations of Wi-Fi access points. As the second floor is similar in layout to the first floor, we focus on the SINR distribution at the first floor, in which six SCs are deployed as shown in Figure 10.18.

In Figure 10.19, the SINR distribution for the first floor and its corresponding CDF are illustrated. The result shows that the SC can provide decent communication quality for indoor scenarios, and locations of SCs should be carefully designed.

10.8 Future Evolution of Small Cells

To further improve deployment efficiency of SCs, the evolution of SCs is still undergoing. In the following, two promising techniques for future SCs: caching on the edge and drone SCs, will be introduced.

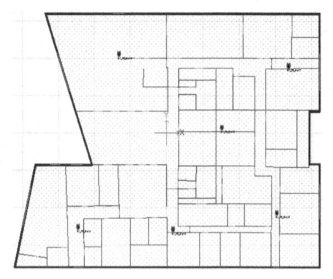

Figure 10.18 The layout and optimised SC BS locations of the first floor in a two-storey office building.

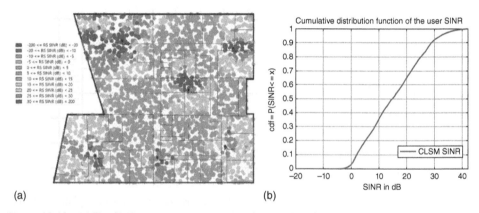

(a)   (b)

Figure 10.19 (a) The SINR distribution for the first floor (b) The CDF of SINR for the first floor.

- **Caching on the edge:** As mentioned in Sections 10.3 and 10.6, one of the main research topics of SC deployment is to reduce the cost of SC backhaul deployment. Consequently, there is growing interest in improving storage capacity (i.e. cache) of SCs (i.e. the edge of network). Firstly, it can alleviate the burden of the network backhaul by proactively storing the content in which serving users might be highly interested, in SCs at the off-peak time periods (e.g. midnight). Secondly, the price of storage devices, such as hard disk drive (HDD) and solid-state drive (SDD), has been decreasing significantly over the past few years and may decrease further in the near future. With the development of ML techniques in the artificial intelligence (AI) domain, users' profiles can be built through big data analysis on available data, which can be collected from social networks, browsing history etc., and predict future requests from corresponding users. Typically, there are three main influencing factors for the performance of caching on the edge [35]: number of requests, cache size, and popularity distribution. As a result, most existing works are studying and investigating the effect of these three parameters on the load of SC backhaul and the number of satisfaction requests. A comprehensive survey about caching has been investigated in [36], and future research directions include popularity prediction and joint caching for increasing storage efficiency, and mobility-aware caching to improve the quality of experience for mobile user.

- **Drone small cells:** Drone small cells (DSCs) are defined as low-attitude flying devices, such as unmanned aerial vehicles (UAVs), equipped with wireless BSs to provide wireless services, especially for unexpected events (e.g. earthquakes, floods, rural-area concerts, traffic congestions etc.) [37]. Unlike traditional SCs, DSCs are non-rigid which can be a complementary option to satisfy the diverse demands for 5G or beyond 5G networks. Therefore, the main advantage of DSCs can be concluded as its flexibility to provide wireless connections whenever, wherever and however needed. For example, DSCs are effective for seasonal rural traffic, e.g. ski resorts in winter and hiking routes in summer. Nevertheless, deployment of DSCs is still facing the following challenges [38]:

  a) **Efficient Design:** The usage of drones is a new concept in the communication field, therefore there is still a lack of optimised design of drones for carrying wireless BSs. DSCs would have unique requirements that may benefit from special-purpose designs, such as long-time hovering, high levels of endurance, robustness against turbulence, minimum wingspan allowing multiple-input multiple-output (MIMO), and provision of energy for transmission [31]. Additionally, DSCs have limited volume, weight, and energy, thus the pay-load of DSCs such as size, aerodynamics, and maximum take-off weight should be taken into consideration.

  b) **Backhaul/Fronthaul Connection:** The wireless backhaul/fronthaul, such as mmWave or free-space optical communication, are inevitable choices for DSCs since it is impractical and unaffordable to deploy optical-fibre connectivity. However, the quality of wireless connection is highly affected by weather and rapid channel variations.

  c) **Placement:** DSCs require quick and efficient placement thus it is critical to have a fundamental understanding of parameters affecting their performance

  d) **Cooperation and Management:** The dynamic nature of DSC requires cooperation of multiple DSCs to provide seamless service efficiently by interference management, and mobility management. In addition, to reduce the CAPEX and OPEX of DSC deployment, programmability in DSCs and controllability via wireless networks are two other fundamental components for the management, which may yield by network functions virtualisation (NFV) and software-defined network (SDN).

There is also some other research, such as the placement of network intelligence and applications in the network edge through NFV and edge-cloud computing, as well as development of a conceptually novel, adaptive and future-proof 5G mobile network architecture, all of which are under investigation as part of EU projects.

10.9 Conclusion

In this chapter, the concept of SCs and structure of multi-tier cellular networks were introduced first. The three main SC deployment challenges, i.e. interference management, mobility management and backhaul were discussed. Specifically, the influencing parameters of the FeICIC for the interference management have been investigated. Then, case studies of SC deployment in the outdoor Paris centre and, in the indoor office building were presented, indicating that SC deployment can significantly increase capacity of traditional homogeneous networks. Finally, two future research topics for the evolution of SCs (caching on the edge and drone SCs) were introduced. We could expect significant uptake in the deployment and capabilities of SCs in the 5G era.

References

[1] 4G Americas. (2013). *Meeting the 1000x Challenge: The Need for Spectrum, Technology and Policy Innovation*, Vol. 1. *4G Americas. Tech. Rep.*

[2] Cisco Visual Networking Index. (2019). Global mobile data traffic forecast update, 2017-2022. https://s3.amazonaws.com/media.mediapost.com/uploads/CiscoForecast.pdf (accessed 19 June 2021).

[3] Li, Q.C., Niu, H., Papathanassiou, A.T. et al. (2014). 5G Network Capacity: Key Elements and Technologies. *IEEE Veh Technol Mag* 9(1): 71–78.

[4] Lopez-Perez, D., Valcarce, A., De La Roche, G. et al. (2009). OFDMA femtocells: A roadmap on interference avoidance. *IEEE Commun Mag* 47(9): 41–48.

[5] Lopez-Perez, D., Guvenc, I., De La Roche, G. et al. (2011). Enhanced intercell interference coordination challenges in heterogeneous networks. *IEEE Wirel Commun* 18(3): 22–30.

[6] Lopez-Perez, D., Guvenc, I., and Chu, X. (2012). Mobility management challenges in 3GPP heterogeneous networks. *IEEE Commun Mag* 50(12): 70–78.

[7] BT whosale. Mobilising the outdoor small cells market. [ONLINE] Available at: https://www.btwholesale.com/assets/documents/products-and-services/mobile/mobilising-the-outdoor-small-cells-market-whitepaper.pdf. [Accessed 25 July 2018].

[8] Ranplan. RANPLAN PROFESSIONAL (iBuildNet). [ONLINE] Available at: https://www.ranplanwireless.com/gb. [Accessed 25 July 2018].

[9] Zhang, J. and De La Roche, G. (2011). *Femtocells: Technologies and Deployment.* Chichester, UK: John Wiley & Sons.

[10] Chambers, D. (2008). Femtocell History. https://www.thinksmallcell.com/FAQs/femtocell-history.html (acessed 19 June 2021).

[11] Beyene, Y.D., Jäntti, R., and Ruttik, K. (2014). Cloud-RAN Architecture for Indoor DAS. *IEEE Access* 2: 1205–1212.

[12] Saadani, A., El Tabach, M., Pizzinat, A. et al. (2013). Digital radio over fiber for LTE-advanced: Opportunities and challenges. *17th International Conference on Optical Networking Design and Modeling (ONDM)*, Brest, France (16–19 April 2013). IEEE.

[13] HUAWEI. Lampsite. [ONLINE] Available at: https://carrier.huawei.com/en/products/wireless-network-v3/Small-Cell/LampSite. [Accessed 25 July 2018].

[14] ERICSSON. (2018). Radio. https://www.ericsson.com/ourportfolio/radio-system/radio-dot-system?nav=fgb_101_0528%7Cfgb_101_0526 (accessed 19 June 2021).

[15] Jafari, A.H., López-Pérez, D., Song, H. et al. (2015). Small cell backhaul: Challenges and prospective solutions. *EURASIP J Wirel Commun Netw* 1: 206.

[16] Boudreau, G., Panicker, J., Guo, N. et al. (2009). Interference coordination and cancellation for 4G networks. *IEEE Commun Mag* 47(4): 74–81.

[17] Akyildiz, I.F., Lee, W.Y., Vuran, M.C. et al. (2008). A survey on spectrum management in cognitive radio networks. *IEEE Commun Mag* 46(4): 40–48.

[18] López-Pérez, D., De La Roche, G., Valcarce, A. et al. (2008). Interference avoidance and dynamic frequency planning for WiMAX femtocells networks. *11th IEEE Singapore International Conference on Communication Systems*, Guangzhou, China (19–21 November 2008). IEEE.

[19] De La Roche, G., Ladányi, A., López-Pérez, D. et al. (2010). Self-organization for LTE enterprise femtocells. *GLOBECOM Workshops*, Miami, USA (6–10 December 2010). IEEE.

[20] Zahir, T., Arshad, K., Nakata, A. et al. (2013). Interference management in femtocells. *IEEE Commun Surv Tutor* 15(1): 293–311.

[21] Hu, H., Weng, J., and Zhang, J. (2016). Coverage Performance Analysis of FeICIC Low-Power Subframes. *IEEE Trans Wirel Commun* 15(8): 5603–5614.

[22] Andrews, J.G., Baccelli, F., and Ganti, R.K. (2011). A Tractable Approach to Coverage and Rate in Cellular Networks. *IEEE Trans Commun* 59(11): 3122–3134.

[23] Haenggi, M. (2012). *Stochastic Geometry for Wireless Networks*. Cambridge, UK: Cambridge University Press.

[24] Hu, H., Zhang, B., Hong, Q. et al. (2018). Coverage Analysis of Reduced Power Subframes Applied in Heterogeneous Networks with Subframe Misalignment Interference. *IEEE Wirel Commun Lett* 7(5): 752–755.

[25] Li, Y., Cao, B., and Wang, C. (2016). Handover schemes in heterogeneous LTE networks: Challenges and opportunities. *IEEE Wirel Commun* 23(2): 112–117.

[26] 3GPP (2015). Mobility Enhancements in Heterogeneous Networks. https://portal.3gpp.org/desktopmodules/Specifications/SpecificationDetails.aspx?specificationId=2540 (accessed 19 June 2021).

[27] Assefa, T.D., Hoque, R., Tragos, E. et al. (2017). SDN-based local mobility management with X2-interface in femtocell networks. *22nd International Workshop on Computer Aided Modeling and Design of Communication Links and Networks (CAMAD)*,Lund, Sweden (19–21 June 2017). IEEE.

[28] Xenakis, D., Passas, N., Merakos, L. et al. (2014). Mobility Management for Femtocells in LTE-Advanced: Key Aspects and Survey of Handover Decision Algorithms. *IEEE Commun Surv Tutor* 16(1): 64–91.

[29] Ge, X., Ye, J., Yang, Y. et al. (2016). User Mobility Evaluation for 5G Small Cell Networks Based on Individual Mobility Model. *IEEE J Sel Areas Commun* 34(3): 528–541.

[30] NGMN (2012). Small cell backhaul requirements. https://www.ngmn.org/wp-content/uploads/NGMN_Whitepaper_Small_Cell_Backhaul_Requirements.pdf (accessed 19 June 2021).

[31] Suryaprakash, V. and Fettweis, G.P. (2014). An analysis of backhaul costs of radio access networks using stochastic geometry. *IEEE International Conference on communications*, Sydney, Australia (10–14 June 2014). IEEE.

[32] Yang, H. and Marzetta, T.L. (2013). Performance of conjugate zero-forcing beamforming in large-scale antenna systems. *IEEE J Sel Areas Commun* 31(2): 172–179.

[33] CBNG (2021). VectaStar platform introduction. http://cbnl.com/vectastar-platform-introduction (accessed 19 June 2021).

[34] 3GPP (2017). Study on integrated access and backhaul. *Tech. Rep. 38.874.*

[35] Bastug, E., Bennis, M., and Debbah, M. (2014). Living on the edge: The role of proactive caching in 5G wireless networks. *IEEE Commun Mag* 52(8): 82–89.

[36] Li, L., Zhao, G., and Blum, R.S. (2018). A Survey of Caching Techniques in Cellular Networks: Research Issues and Challenges in Content Placement and Delivery Strategies. *IEEE Commun Surv Tutor* (20(3): 1710–1732.

[37] Mozaffari, M., Saad, W., Bennis, M. et al. (2015). Drone Small Cells in the Clouds: Design, Deployment and Performance Analysis. *2015 IEEE Global Communications Conference (GLOBECOM)*, San Diego, USA (6–10 December 2015). IEEE.

[38] Bor-Yaliniz, I. and Yanikomeroglu, H. (2016). The New Frontier in RAN Heterogeneity: Multi-Tier Drone-Cells. *IEEE Commun Mag* 54(11): 48–55.

11

Today's and Tomorrow's Challenges

In this chapter, we will look at some of the main challenges mobile operators are currently facing. As we have illustrated in this book, mobile networks have evolved from their humble beginnings in 1G to the highly complex, multi-functional 4G/5G networks of today. Yet new challenges, largely borne out of its own success, now have to be overcome.

The success of mobile data in 4G has meant that there is now a significant increase in mobile data consumption year on year. We will look at how industry plans to tackle this 'capacity crunch', especially in the 5G era. Along with capacity demand and multiple supported services, the complexities of mobile networks have increased many fold. With 5G bringing in multiple new dimensions to these complex networks, we will analyse how network management tools have helped 4G and now are evolving to support 5G networks. With the Telecoms sector accounting for 2–3% of global energy demand and the greater capabilities of 5G likely to push this further, there is a renewed focus on promoting Green and lower EMF mobile networks. We will investigate this aspect as a current challenge in line with growing concern for our climate and natural environment. Finally, we will discuss the challenges and solutions to provide basic broadband connectivity to the billions of unconnected people. We will see how advances in technology and renewed investments in this area are trying to address this perennial challenge.

11.1 The Capacity Crunch

11.1.1 A Historical Perspective

Since the very early stages of wireless communications, there has been a continuous need to increase both the capacity and capabilities of these networks. This stems from two factors; firstly, as a service becomes popular, more and more subscribers join. The radio access part of the network extends geographically, expanding the radio coverage and densifying the number of radio stations (or 'cells' in mobile communications) in areas where there are already wide area cells. Whilst this popularisation in early wireless and mobile communication services (such as wireless telegraphy in the early 1900s and the GSM voice services in the 1990s) could largely be tackled by replicating the radio access nodes, this also promotes a need to increase the capacity of transport networks. In particular, the backhaul

The Technology and Business of Mobile Communications: An Introduction, First Edition.
Mythri Hunukumbure, Justin P. Coon, Ben Allen, and Tony Vernon.
© 2022 John Wiley & Sons Ltd. Published 2022 by John Wiley & Sons Ltd.

capacities at the aggregation points, which are usually handled by single-mode optical fibre in modern networks, need to be increased. Secondly, as newer, more advanced services are introduced, the capability limitations in the existing networks become exposed. This is the point where a new wireless generation (which has already been researched and standardised) really takes traction with mass deployments. So far, each of these wireless technology and generational transformations have witnessed significant increases in data capacities along with the capabilities. The early telegraphy services consumed only fractions of bits per second (bps), whilst the voice calls in GSM consumed only kilobits per second (kbps). Since the introduction of mobile data services in late 3G, the data demands from users have grown exponentially. Today, the immersive AR (Augmented Reality) and VR (Virtual Reality) 5G mobile applications can demand closer to gigabits per second (Gbps) data rates. This kind of demanded peak data rate growths will need a corresponding increase in data rates in the radio access network, together with a subsequent interconnection capacity leap in transport networks. Whilst densifying the network with more small cells (SCs) can help to ease capacity demand in a given area, real technological advancements in the radio access layers are needed to increase peak data rates. This can be in the form advanced radio-frequency (RF) technologies to handle wider bandwidths or advanced base-band processing to include more spectrally efficient air interface PHY and MAC implementations.

It can be argued that this relentless demand for new services and refined user experience is the factor that has driven the need to advance the technologies of mobile communications into new generations and provide more and more capabilities.

If we step back and look at the evolution of wired and wireless communications over the last 150 years, it is evident that the offered capacity in these networks has roughly doubled every two years up to the latter parts of last century, similar to Moore's Law [1] applied in computer science. A graphical illustration of this is provided in Figure 11.1, from [2].

Curve A shows the best reported capacities (including scientific publications) and curve B shows the switched capacities in the network core from commercial products. Curve C shows the data rates offered to consumers, which can be significantly lower than the best reported capacities under test conditions. The CAGR (Compound Annual Growth Rate) of these reported capacities has slowed in recent years, indicating what the communications systems can offer to consumers. If you compare this with mobile traffic growth as shown in Figure 1.3, which indicates consumer demand, the gravity of the 'capacity crunch' facing today's operators can be visualised.

11.1.2 Methods for Capacity Enhancement

Four basic methods (or four dimensions) are recognised as having the capability to increase the radio access capacity of a cellular network. Whilst each individual method may deliver only limited relief to the capacity crunch, significant capacity enhancements can be realised by combining them. These four dimensions, which can be utilised to optimise available radio access capacity, are described below.

11.1.2.1 Space

The idea is to limit the spatial footprint of a radio transmission only to intended users, so that it can be replicated more to provide higher capacity per area. This could be cell densification by reducing the distance separating cells, or the use of directional radio beams to

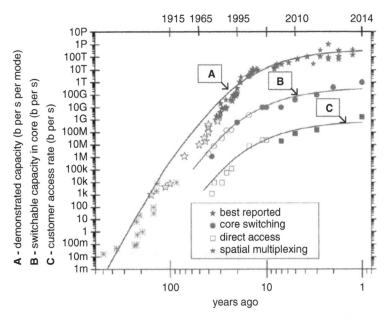

Figure 11.1 Capacity evolution in wired and wireless communication networks. (*Source* [2]).

separate transmissions. The use of SCs (covered in Chapter 10) in recent years has greatly increased the offered capacity in hotspots where there is higher demand. Using higher frequencies (mmWave) and beamforming in 5G will help push these capacities even further. The higher path loss in mmWave is beneficial in this respect, as the coverage footprint of each cell and a beam can be tightly controlled to reduce the interference effects.

11.1.2.2 Spectrum

Allocation of more of the radio spectrum for a service can increase the offered capacity. Whilst there is an enduring demand for spectrum from cellular services, there are many other services also vying for this limited resource, as discussed in Chapter 7. The general trend across newer mobile communication generations is to adapt ever higher frequencies for its use. This is driven by the better availability of these spectrum bands in higher frequencies. The mobile communications industry has shown remarkable technical innovation to overcome the various challenges posed in utilising higher frequencies over the years. Furthermore, spectrum regulators are opening up higher frequency spectrum in support of broadband connectivity. A recent example is the UK spectrum regulator, Ofcom, consultation on the 100–200 GHz band [3] potentially envisaging 6G operations. As operating frequencies increase, so does the available channel bandwidth, hence there is great interest in exploring these higher spectrum bands for supporting higher capacities.

11.1.2.3 Spectral Efficiency

Spectral efficiency basically translates as to how many (data) bits can be transmitted per every 1 Hz of spectrum used in the system physical layer. Incumbents face the challenge to ensure their services are as spectrally efficient as possible. There are several means of achieving this, such as data encoding, high-order modulation schemes, minimising protocol overhead etc. The MIMO (use of multiple transmit and receive antennas) techniques introduced at the turn of this century (discussed in Chapter 9) really opened up a new avenue for improving spectral efficiency. Recent advances in massive MIMO [4] represent the next step in this direction.

Traditionally, the uplink and downlink of a mobile channel should occupy two separate frequency channels or time slots, leading to Frequency Division Duplexing (FDD) or Time Division Duplexing (TDD) respectively. This is not spectrally efficient and spectrum efficiency can ideally be doubled if a means can be devised for both users to share the same frequency or time slot. This is referred to as division free duplex [DFD] or as Full Duplex (FD), which has attracted much research over recent years. Such schemes require a mixture of carefully designed antennas, RF hardware and software self-interference cancellation to achieve enough isolation between the uplink and downlink channels to be viable. Great strides have been made in this respect. More recently, research has been undertaken on utilising 'spatial modes' for multiplexing signals on the same frequency [5]. This remarkable technique has been shown to support transmission of 120 Gbit/s by multiplexing 11 data signals of 9.6 to 13.3 Gbit/s using 11 spatial Orbital Angular Momentum modes [6].

11.1.2.4 Power

This is the transmit power dimension that can be used to multiplex users in a non-orthogonal manner and increase the offered capacity. For example, two users can be scheduled to utilise the same time and frequency resources but can be separated with advanced signal processing at the receiver, if different transmit power levels are used. This is one tenet of the Non-Orthogonal Multiple Access (NOMA) schemes currently being developed, as discussed in Section 8.7.3. Generally, the power should be set to be just enough to sustain a link and where any more power would create unnecessary interference to other users and any less power would compromise link performance. Power is very often dynamically adjustable to follow changes in link propagation, or changes in the quality of service (QoS) level demanded by users. Higher QoS requires higher transmit powers (or lower path loss in the radio link) to support higher-order modulation schemes.

The overall goal is to manage Signal-to-Interference plus Noise Ratio (SINR) such that maximum data rate is achieved in an area (b/s/km^2). If the achieved SINR is below what is required in achieving the required QoS, then adjustments would have to be made to the system involving at least one of the dimensions. On the other hand, an SINR in excess of that required to achieve the target QoS would indicate a sub-optimum use of resources, and hence it would be possible to squeeze more from the network by making adjustments to at least one of the dimensions.

Innovating performance in terms of the above dimensions is essential for improvement of system efficiency and addressing capacity crunch. One highly innovative technique referred to as Linear Angular Momentum Multiplexing (LAMM) and published in [7] describes a means of supporting 9 Gbps in a bandwidth of only 15 MHz, i.e. in less bandwidth than that of a single Wi-Fi or 4G LTE transmission. The system makes use of a line of carefully spaced antennas that each have a carefully selected delay between them to create a number of independent transmissions. The longer the line, the more independent

Figure 11.2 Linear Angular Momentum Multiplexing concept.

transmissions can be supported. The transmissions are then received by a second line of similarly configured antennas. This second line may be mounted on a moving vehicle such as a train. The concept is shown in Figure 11.2 and can be thought of as a Line of Sight MIMO communications system, or alternatively derived from Orbital Angular Momentum Multiplexing as described in [5].

Signal compression has an important role to play in addressing the spectrum crunch. MPEG and other data compression techniques are widespread, including those in video and audio streaming services. They operate by only sending essential data, for example, by sending only video data that has changed from the previous image. A second area that also reduces the amount of data sent is carefully designing protocols that do not require excessive overhead for them to work. This could be in the design of addresses, acknowledgement messages etc. Perhaps a vision would be for having zero overhead transmission, where network routers do not require added data to correctly route packets.

11.1.3 Impact on Transport and Core Networks

The starting point for any examination of the impact of current-generation mobile networks on their internal transport networks is a reflection and consideration of what is actually being transported. In the now very-familiar world of 2G networks, where GSM and other air-interface standards were simply a means of accessing the ISDN (Integrated Services Digital Network; indeed, a mobile phone number is still referred to as 'MSISDN'), nearly all higher-functionality and value added services were provided by that ISDN. The mobile network was relegated to relatively simple call manipulation functionality such as divert-on-no-reply, divert-on-busy etc. So, the commodity being transported all the way from Uu air interface to ISDN handoff, were the raw voice bits (plus the then-tiny amount of data) which had a variable timeslot mapping from BTS to BSC, then fixed mapping from BSC to MSC (Please refer to Section 5.5 for details of these nodes).

With the advent of third-generation (3G) services, the global Internet had been firmly established for nearly a decade in the consumer domain, although widespread proliferation of the smartphone was still some way off. 3G anticipated the disparate transport requirements of voice and data with development of the 2G data handling model, where voice data flowed via the Mobile Switching Centre (MSC) and could be 'dumped' to Subscriber Trunk Dialling nodes at optimal locations depending on the two endpoints, and data flows which could be destined for/originating from any place in the Internet-connected world, flowed via the Serving GPRS Support Node (SGSN) onwards to one or more partner Packet Data Networks, as illustrated in Chapter 3. Network Function Virtualisation (NFV) was pre-figured in 3G, via many references to MSC-servers and SGSN-servers but, by and large, networks remained monolithic into the 4G era.

Traffic in the third generation still flowed via operator-provided links between the BSS comprising NodeB base stations and the 3G successor to the Base Station Controller, the Radio Network Controller (RNC) which had the added complexity of being instrumental in the inter-base site handover procedures. Differing QoS, latency and reliability for each flow could be defined and set up all the way from real-time connection-oriented voice/ video calls, to low-rate connectionless browsing, placing the least demand on latency and link resilience. However, as the physical links were still either owned or leased by the mobile operators themselves, granularity in all of latency, bandwidth, BLER and reliability were constrained within the bounds of the physical links themselves.

Only with fourth generation (4G) networks and the advent of true Multi-Operator Radio Access Networks (MORANs) did it become possible to implement a chosen service quality from end to end [8]. Dedicated spectrum resource is required for up to two operators in a MORAN, but S1-MME and S1-U links can be carried in common for up to six Evolved Packet Core Connections, i.e. a total of four Mobile Virtual Networks may share the MORAN at the core level with two parent operators, carried across common physical links either optically or via Gigabit radio (see Chapter 3 for a full discussion of MVNOs). Within the strictures of shared common transmission resource, each MNO/MVNO is free to designate specific QoS Class Indicators (QCI) to itself; one infrastructure provider, Nokia has extended the basic LTE QCI scheme, adding a further 21 non-guaranteed bit rate service designations that can be assigned to bearers set up over the shared S1 transmission resources, which will preferentially push the 'squeeze' on transmission resources to the non-GBR services such as browsing and downloading when the networks are all busy simultaneously.

With 5G, a solution to optimising shared transmission resources comes closer still via 'Network Slicing'. Much misunderstanding has crept into the mobile industry around this term. At its most basic, Network Slicing is an extension of the MVNO techniques introduced in Chapter 3 all the way to the 5G air interface [9]; whereas the virtualisation techniques previously discussed centred on shared or dedicated core resources and a shared BSS, with Network Slicing, the air interface itself is dedicated to an MVNO, with time and frequency resources completely compartmentalised between slices by a Network Slice Controller (NCS).

Network slicing in 5G offers the capability to create 'virtual networks' (from compartmentalised radio access, transport network to core network [CN] resources) that meet the very distinct KPIs demanded by emerging 5G verticals. Figure 11.3 shows an example for the implementation of four main network slices where, for very low latency applications such as automatic driving and gaming, the Centralised Unit (CU) i.e. the baseband aspect of the gNodeB, cannot tolerate any excess link delay and must be located in the local cloud at the base site, here designated as Central Office (CO). For voice interconnectivity, at least one vendor specifies that routed links shall be less than 50 Gλ between site and concentrator and less than 200 Gλ from concentrator to the User Plane (UP) of the Service Oriented Core (UP-SOC) instance running in the Local Data Centre (LDC). For free space and the typical 1400 nm wavelength (λ) of an infra-red laser diode used in the 0.7 velocity factor single-mode optical fibres of the transport network, this corresponds to less than 50 km and 200 km routed length (usually much shorter topologically, as the cable routes are not straight lines 'as-the-crow-flies'). However, bear in mind that typical DWDM (Dense Wavelength Division Multiplexing) links use light wavelengths down to 1260 nm so the

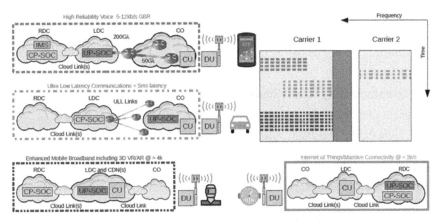

Figure 11.3   Type and Placement of Transmission Links for Network Slicing.

actual allowed routing lengths so as not to violate application latency maxima may be slightly shorter by a few km for site-to-concentrator and a few tens of km from concentrator-to-User Plane.

For voice communications, the User Plane of the Service Oriented Core can be implemented in instances running in the Local Data Centre thereby yielding economies of scale. However, for Ultra Reliable Low Latency Communications (URLLC) and Ultra Low Latency apps, the User Plane must terminate in the local site or cluster, with the Control Plane (CP) of the SOC located at the LDC, again both to drive down latency to below the permissible maxima.

In the case of Enhanced Mobile Broadband (eMBB) and Internet of Things (IoT) connectivity, the demands on transmission links are relaxed to the extent that the CU can, in both cases, be implemented as instances running in the LDC (for economies of scale) where, of course, the SOC User Plane must terminate. However, for bandwidth reduction reasons, Content Delivery Networks (BBC iPlayer, CBS Catchup, Prime Video etc.) will have a Point of Presence in the LDC. This would remove the need for popular content to be repeatedly drawn down all the way through the transport network. The reader might think that there should also be a buffer instance running at the Central Office further reducing bandwidth on the cloud transport links between LDC and CO, however, bear in mind that content flows are encrypted per-user, so buffering a flow to serve other users is futile.

IoT connectivity imposes the least demanding bandwidth and latency profile on the network, with inter-entity links being implemented over the cloud or optionally via the operator's dedicated (not critical) links. Both User and Control Planes of the Service Oriented Core can be implemented in the Regional Data Centre (RDC) albeit there will be separate and firewalled UP/CP instances in the RDC per IoT MVNO.

The above examples show how diverse and demanding the requirements have grown to be on the 5G Transport network and the 5G Core (5GC) based on application type. The Multi-access Edge Computing (MEC) or Edge processing is a topic receiving much research attention at the moment. This refers to placing computer processing capability at the edge of a network along with the content on the application servers which feed URLLC applications. We will likely see many implementations of MEC in coming years along with the 5G core for many low latency use cases.

11.1.4 Complementary Technologies

We mentioned in Section 11.1.2 that many other technologies, along with mobile communications, compete for spectrum. Some of these technologies can effectively be used in a complementary manner to offload some of the data demand from the mobile networks. The mobile devices having the capability to operate in multiple technologies, like Wi-Fi, Bluetooth and more recently UWB (Ultra Wide Band) make it easier to accommodate these technologies in the cellular domain.

Out of these data offloading technologies, Wi-Fi, technically known as the IEEE 802.11 series of standards [10], plays the pre-dominant role. In fact, Wi-Fi carries significantly more IP traffic than Mobile wireless. It is predicted that in 2020, Wi-Fi carried 59% of global internet traffic, when Mobile carried only 19% [11]. Wi-Fi was initially developed in the late 1990s, to enable wireless inter-operability between devices and operated in the 2.4 GHz band. A basic tenet in Wi-Fi is the use of un-licensed spectrum and, as many other users and technologies also has access to this spectrum, there is no guarantee of a QoS level. Wi-Fi employs a 'listen before talk (LBT)' mode for initial access, i.e. a user will transmit in a certain Wi-Fi band only after listening to it to ascertain it is free of other users. Due to the popularity of initial Wi-Fi standards 802.11a and 802.11b [10], a large ecosystem developed around the technology. As mobile data applications became popular in late 3G, the Wi-Fi connectivity in Smartphones became an essential feature, so users could access data services 'free of charge', whenever they are near a public Wi-Fi access point. Due to huge demand, Wi-Fi standard variants developed around more unlicensed bands in the 5 GHz spectrum and then a mmWave capable, very high capacity version in the 60 GHz spectrum. Wi-Fi became synonymous with WLAN (Wireless Local Area Networks), as no other technology could establish a presence in this WLAN market. Wi-Fi technology continued to develop addressing security vulnerabilities in the initial releases with encryption, for example. Wi-Fi alliance also developed the product Hotspot 2.0, which enabled such certified Wi-Fi networks to support user 'roaming' (without the need to repeatedly gain access rights to each Wi-Fi cell) when users move about. The latest developments are in the unlicensed 6 GHz bands and an 802.11 standard variant known as Wi-Fi 6E will be commercialised soon. It is estimated by the Wi-Fi Alliance that over 15 billion Wi-Fi capable products are used globally today [12]. This number will continue to rise as the number of IoT and MTC applications relying on Wi-Fi for their packet transmissions, expand.

Within this context, 3GPP started looking at how to integrate Wi-Fi as an intrinsic method of accessing IP data flows. In 3GPP Release 8, the basic Wi-Fi interworking solution made it possible to continue an IP data flow started with LTE, through Wi-Fi. Simultaneous IP connectivity through LTE and Wi-Fi was supported in Release 10. 3GPP Release 12 saw the inclusion of enhanced Wi-Fi selection policies in line with Hotspot 2.0 (noted above). More broadly, Release 12 heralded development of ANDSF (Access Network Discovery and Selection Function), which enables UEs to connect to non-3GPP access technologies including Wi-Fi. A practical implementation of this 3GPP Wi-Fi interworking is the O_2 Wi-Fi Extra service provided in the UK by the operator O_2 [13], where the mobile subscribers would automatically connect to one of the thousands of Wi-Fi access points they have installed, if this Wi-Fi signal is stronger than the cellular signal. In the 5G era, this non-3GPP selection and access has developed as a function, with a distinction between trusted and untrusted (where the 3GPP network applies IP security features IPsec for packet transmissions) non-3GPP networks. Relevant functions are named as TNGF (Trusted Non-3GPP Gateway Function) and N3IWF (Non-3GPP Interworking Function) respectively. A number of procedures, including user registration, have been made (optionally) possible in the 5G standards through these non-3GPP functions.

LTE unlicensed (LTE-U) is a more recent development where 3GPP attempts to develop a flavour of LTE which operates in unlicensed bands, primarily in the 5 GHz spectrum. This is developed as an LAA (Licensed Assisted Access) version, where the operator will have a traditional LTE cell with licensed spectrum and guaranteed QoS levels and will use the LTE-U with unlicensed spectrum as a secondary cell in the form of carrier aggregation. The secondary cells will be used in an opportunistic manner to boost capacity, but the QoS levels will not drop due to the primary cell. One of the main considerations in developing LTE-U is to ensure fair co-existence between LTE-U and Wi-Fi users, and also amongst LTE operators who would use LTE-U in the same geographic areas [14]. The transmission power levels and access attempts are controlled in a manner such that the addition of LTE-U secondary cells with LAA should not have more performance impact than adding a similar number of Wi-Fi cells in the same area. A conceptual diagram for LTE-U operation is shown below in Figure 11.4.

Another form of LTE and Wi-Fi link aggregation was developed as LTE-H (LTE-Hetnet) as a carrier aggregation solution, led by mobile operators. This is also known as LWA (LTE-WLAN Aggregation) in the 3GPP domain. The proof of concepts in the recent past have shown that LTE-H can more than triple peak data rates under favourable conditions [15].

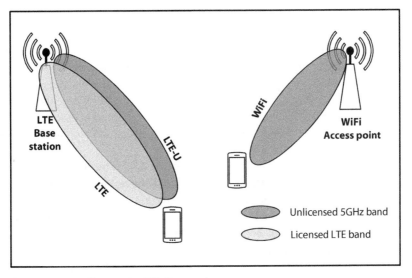

Figure 11.4 A concept diagram for LTE-U operation.

Ultra Wide Band (UWB) technology has been on and off the technology radar for a number of years. Technically, the use of very wide bands implies the ability to use very narrow pulses in the time domain. This can multiplex many low-power UWB signals in the time domain with very low cross interference. Recently, there has been a surge in interest in UWB as technology that can enable very high-resolution positioning and tracking of users. A number of new 5G Smartphones include UWB modules, which can provide positioning to within 10 cm of accuracy [16]. A growing ecosystem around this UWB positioning capability is building up with the Fira consortium [17]. The UWB signal can also support Mbps data rates, so it is likely that future applications of UWB within Smartphones will also make more use of this data capability.

An emerging technology that can provide extremely high data rates (Gbps) in indoor environments is LiFi. LiFi uses direct modulation of the light signal (this can also be 'invisible' light such as infra-red) to provide a data connection. An everyday indoor light source such as an LED bulb can be replaced with a LiFi-enabled bulb, so it can provide data connectivity to LiFi-enabled devices as well as being the light source. LiFi is claimed to be highly secure, as light can be physically contained within the premises without leakage and with use of secure encryption. Due to the direct-data modulation methods used, it is claimed that latencies can also be kept very low, making LiFi suitable for indoor applications like AR/VR and also industry 4.0 (covered in Chapter 12) applications in factories. LiFi could become very useful in environments where mobile RF signals are deemed to interfere with other vital equipment, such as inside airplanes or in hospitals. LiFi can be seen as highly complementary to 5G, where it can take up the capacity offloading in indoor environments, in keeping with the eMBB and URLLC requirements of 5G. A good introduction to LiFi technology can be found in [18].

11.2   Increasing Network Complexity

As mobile networks have evolved into providing greater capabilities and a wider variety of services, their complexity has also increased many fold. Up until the later stages of 3G, voice was the dominant application for mobile networks and the total capacity demand for a network would show a high correlation with the number of subscribers. Thus, the networks could be planned and deployed in a monolithic manner, with Macro cells (with reduced cell area for urban and dense urban environments) providing coverage and capacity for the entire network. Network maintenance and management would mostly be carried out with human interventions. Large teams of engineers and technicians would be deployed to carry out these manual tasks.

The real shift in network complexity and the need to deploy automated tools to manage the networks came about in the final stages of 3G, when mobile data became the dominant application. These trends accelerated with mobile data native 4G, where all-IP networks were built to support data applications exclusively. The exponential growth in mobile data demand, driven by Smartphone usage, made it necessary for mobile operators to develop new strategies to enhance their network capabilities. The deployment of SCs at capacity hotspots was a strategy adapted by many operators, leading to HetNets or Heterogeneous Networks (SCs and HetNets are covered in Chapter 10).

In terms of network management, HetNets provided a novel set of challenges, which needed complex automated tools. The manual network management processes cannot be scaled up to meet data demand growth and increasing complexity without an exponential increase in the costs involved. The mobile operators were facing multiple challenges during the early 4G stages (captured in Chapter 1.3) including a fall in ARPU (Average Revenue Per User), so the network operational costs needed to be cut back on all sides. The automated network management (or simply network automation) offered significant cost savings in the longer term for the operators and this was highly attractive in the above context.

The Self-Organising Network (SON) paradigm developed in-line with these requirements from 4G networks to the highly complex and effective tools we see in the networks today. The basic capabilities and development history of SON tools are covered in the section below.

11.2.1   The Self-Organising Networks

SON is a combination of automated features that supports and enhances the planning, deployment, operation and overall management of mobile networks. These automated features need to monitor specific network conditions and parameters continuously and make adjustments to fine-tune the network where necessary. The SON controller, with a number of these automated algorithms, can reside with the network's OSS (Operations Support System) as a single, highly complex entity. This is a centralised SON implementation. A distributed SON implementation is also possible, where individual network elements can carry different SON functionalities. The distributed SON version is quick to respond to changes in the network and carries no risk of a single point failure, but the implementation will have to be tightly coordinated with the vendors providing these network elements. A hybrid of the above two SON versions is another possible implementation.

The development of SON was initiated by the major mobile network operators, through their collaborative work in NGMN (Next Generation Mobile Network, an operator organisation for technological advancement). In a white paper published in 2008 [19], NGMN specified the basic requirements from a network management tool or SON. With LTE specification development, 3GPP also took up developing network management capabilities to adopt SON functionalities [20]. From LTE Release 8 up to Release 14, some of the major features and capabilities were developed to support the SON functionalities detailed below. Although SON functionalities are executed through proprietary algorithms of specific vendors, the LTE-specified SON supports multi-vendor implementations, mainly through standardised interfaces. We will discuss this aspect later in the section.

SON features are broadly divided into three categories – self-configuration, self-optimisation and self-healing. Whilst the commercial SON tools, developed by different vendors, can possess different combinations of these features under different names, the main generic SON features under these three categories are detailed below. Many of these SON features help with the complexities that come with SCs and HetNets, as explained therein. A broader overview of SON features can be found in [21].

11.2.1.1 Physical Cell Identity (PCI) Configuration

Automated PCI configuration is a self-configuration SON capability and one of the first SON features to be standardised by 3GPP for LTE [20]. The PCI is an identification number given to an eNodeB and, in LTE, it has to be selected from a set of 504 PCI numbers. As this is a limited set, the numbers are used repeatedly across the network, with a requirement that the PCI number should be unique within the local area. The PCI is broadcast by the eNodeBs and the UEs use this in reporting signal strengths of neighbour cells in handover and in cell re-selection processes. Hence, there is a critical need to maintain this unique number pattern within the local area.

This SON feature enables automated configuration of the PCI when new cells (eNodeBs) are added to an area where there are a number of existing cells. This is a common occurrence in HetNets, where SCs are added to urban areas to boost capacity, on top of an existing Macro cell network. This SON feature allows 'plug and play' self-installation of SCs and for them to become operational within minutes. On the other hand, a manual reconfiguration would be a very time-consuming process and could lead to significant delays in getting new cells operational.

11.2.1.2 Automatic Neighbour Relation (ANR) Configuration

ANR is a related function to PCI, where ANR develops a list of neighbour cell identities in each of the cells. The ANR functionality becomes necessary in handovers. The ANR configuration is a continuous process and tables need to be updated when new cells are added to the network. The automation is especially useful in HetNets where SCs can be added regularly to boost capacity demands in a particular area. As with PCI, ANR helps reduce the time needed for new cells to go live in the network, thus enabling the impact of newly added cells to be felt by consumers almost immediately. Even in the large scale roll-out of new cells, the ANR feature significantly shortens the time needed to make the eNodeBs go live from the point of deployment, when compared to manual configuration of neighbour lists.

Inter-RAT handovers are a useful feature in mobile networks, especially when a new cellular generation is in its nascent stages. Then, the new generation cells will provide services only in the urban capacity hotspots and, when users move out to wider areas, they have to be connected through the older generation cells. This was the case with LTE around the time of 3GPP Release 10 (2012), when 3G and 2G cells would provide the wider connectivity. Proper ANR lists with cells from all utilised RATs are very useful in this stage. The ANR SON function should be able to collect this neighbour data from all radio access technologies (RATs) involved and speed up the network configuration.

11.2.1.3 Mobility Robustness Optimisation (MRO)

This SON feature is a self-optimisation function, dealing with the handover of users from one cell to another (we discussed technical aspects of handovers in Section 5.2.7). If the handover failures occur for an active user, the voice or data connection will drop at the point of handover, impacting the overall QoS provided by the network. In a manual optimisation process, there needs to be a lot of drive tests across the cell borders to record and refine the parameters and ensure that the handover processes are properly executed. The SON feature does this automatically and wherever there are handover failures, the parameters such as the signal level thresholds are updated to remedy these. The MRO feature is vital in HetNets where this enables SCs to be added with very little planning effort. The handovers to and from SCs are optimised automatically as the network gathers data from the live network. As the network densification with more SCs increases with the maturity of a certain generation and increase of the capacity demand, the MRO feature becomes an essential component in the SON tool box.

11.2.1.4 Mobility Load Balancing

Mobility Load Balancing (MLB) is another self-optimisation feature, where the network load (in terms of capacity demand) in a certain area can be distributed to be optimally shared amongst the available nodes in that area. This involves changing the handover thresholds so some of the active users can be handed over to neighbour cells which are less 'crowded', from cells with high-capacity demands. The physical locations of cell borders thus change, making the less 'crowded' cells larger to capture more active users. This dynamic cell-size variation is also smetimes called cell breathing. The MLB procedures can also be utilised to off-load few active users in a cell to neighbour cells and switch-off cells for energy saving. This can be especially useful in HetNets, when few users of a Macro cell can be accommodated by the SCs within or by nearby Macro cells and it can be switched off, as discussed in this research paper [22].

11.2.1.5 Enhanced Inter-cell Interference Coordination

The Inter-cell Interference Coordination (ICIC) feature was developed in LTE standards to mitigate the interference cell-edge users will experience. For users to avoid causing and receiving interference whilst operating close to one another, the use of time and frequency resources at the cell edge should be controlled. ICIC provided such control mechanisms and this comes at the cost of not fully utilising all available radio resources at the cell edge. SON functionality could be utilised to optimise cell-edge performance under given operating conditions, using ICIC tools.

The enhanced ICIC (eICIC) was developed as part of LTE-Advanced standards (Release 10 onwards) and specifically deals with the issue of users connected at the cell edges of SCs receiving excessive interference from the Macro cells where the SCs are embedded. SCs in a planned network are particularly useful if they can support more Macro cell users with good signal quality and thus reduce Macro cell load. The eICIC supports this by dictating certain time-frequency radio resource parts (called sub-frames; please check Section 5.7) in that the Macro cell has to be almost blank (void of any data bits and containing only essential control bits). These are called Almost Blank Sub-frames (ABS). By allowing SCs to allocate these ABS resources to its cell-edge users, the eICIC allows the cell range of the SCs to be extended and serve more of the Macro cell users with better signal quality. Again, there is a cost-benefit trade-off, as the Macro cell has to sacrifice some of its radio resources in order to allow some users to be offloaded. The benefits are more when there are multiple, spatially dispersed SCs in the same Macro cell utilising the same ABS resources to support multiple off-loaded users. Yet, deciding the optimum levels of ABS allocations under different cell load and user distribution conditions become a highly complex problem. The SON optimisation tools become very handy in this situation. In a large urban network with many Macro cells each having multiple SCs, a distributed SON implementation to analyse and optimise the ABS allocation under local conditions would be computationally more efficient and would provide faster solutions.

11.2.1.6 Cell Outage Detection and Compensation

The cell outage (or non-operation) is one of the common problems mobile operators face. In the early days, mobile operators would have to rely on customer complaints to figure out that a particular cell was not functioning properly. Nowadays, however, technology has advanced to such an extent that by analysing a complex set of parameters over time through SON, the network is able to detect even minor malfunctions in its cell network. This SON self-healing feature will build-up a picture of the expected performance (the number of served users, cell throughput, number of handovers etc.) for each of the cells over time. If there is a deviation from this, the SON feature will conduct further tests to detect if a certain eNodeB is malfunctioning. In terms of short-term compensation for a fully non-functional cell, the SON feature can activate elements from load balancing to expand other neighbour cells to take up much of the cell load. In case of an SC outage, the Macro cell where the SC is embedded can be entrusted to stop offloading users to this particular SC. Obviously, there will be some performance degradation until the affected eNodeB is fully restored, but this self-healing SON feature allows mitigation of the full impact of the cell outage to a significant extent.

What we have detailed above are only some of the core features of LTE-SON capabilities, to give you an insight into how these features help in managing a complex network. Many of the features discussed are inter-related and a parameter change to fine-tune one feature can impact another. So, SON algorithms should also work in coordination, to yield the optimum network performance in a holistic manner. More detailed technical accounts of LTE-SON functionalities can be found in [23].

In implementing a multi-feature SON system, the SON system must be incorporated into the overall network management system. In its 4G LTE standardisation, 3GPP has defined Element Managers (EM) for each network element (NE-) such as eNodeB and an overall Network Manager (NM) to work with the Operations Support System (OSS). SON functions can be implemented in the EMs or alternatively at the NM (distributed or centralised implementations respectively, discussed above). Either way, there should be clear management interfaces between multiple NEs and EMs and with the NM. 3GPP has defined such specific interfaces as [24];

- Type 1 between the Network Elements (NEs) and the Element Manager (EM).
- Type 2 between the Element Manager (EM) and the Network Manager (NM).

Type 1 interfaces can be proprietary, as the NEs and the EMs are typically provided by the same vendor. In this sense, a distributed SON system will also have to be tightly connected with this proprietary interface, usually coming from the same vendor. 3GPP has spent a lot of effort to standardise Type 2 interfaces. As a result, a centralised, third-party SON function can reside in the NM, and can communicate with individual EMs. Thus, the centralised SON option is more readily accessible to multi-vendor implementations. The 3GPP management reference model, with the centralised and distributed SON options are shown below in Figure 11.5, from [25].

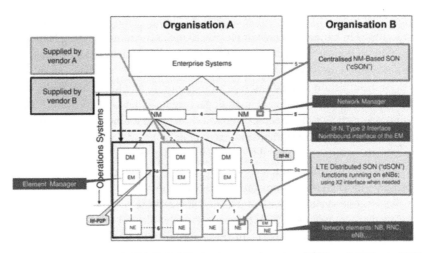

Figure 11.5 Centralised and Distributed SON implementations in a 3GPP model. (*Source* [25]).

11.2.2 Network Automation in 5G

As we move into the 5G era, the need for network automation is intensified by multiple factors. Technically, 5G will provide far more capabilities than an increased service quality for mobile broadband. It will offer a multitude of services fulfilling the requirements of enhanced Mobile Broadband (eMBB), Ultra-Reliable and Low Latency Communications (URLLC) and massive Machine-Type Communications (mMTC). Different vertical sectors will benefit from these services (we discuss the main vertical sectors in Section 12.2), where services will consist of different mixes of the above three categories with distinct quality and reliability metrics. The same 5G network can be split into different virtual network slices, to cater for these different verticals. Some of the slices will be needed only at certain times (such as emergency response) and thus these will have to be dynamically configured and also later dismantled. In essence, network slicing is equivalent to operating multiple networks with differing KPI requirements on the same physical infrastructure, as seen in Figure 11.3. This brings in multi-dimensional complexities to 5G network management.

Even at the radio access network (RAN) scope, 5G will be much more diverse and complex than 4G. 5G will occupy spectral bands of a much wider range, from the low 700 MHz bands to mid-range 3.5 GHz and to the high mm-wave frequencies up to 71 GHz (as of 3GPP Release 17). There are distinct variances in the air interface, including the sub-carrier spacing in OFDM, to support these different frequencies and the bandwidths that 5G will adapt. The introduction of very large number of antennas at the base stations (as massive MIMO) is supported in 5G. Also, there are a number of new RAN features supporting the beamforming capabilities of 5G at higher mm-wave frequencies. The frame structure in the time domain is also highly flexible, with the introduction of finer granularity dynamic TDD. It is highly probable that many of these RAN features will come as software configurable options to cater for different RAN slices, making interference prediction and mitigation a very complex issue, for example.

The LTE SON tools will evolve into 5G network management, to provide the necessary levels of accuracy in handling such complexities. One of the main defining features in this evolution will be the use of machine learning (ML) and artificial intelligence (AI) in 5G network automation tools. With the use of AI, algorithms will be able to 'learn' from past data sets (for example, on a certain network functionality like MLB) and make intelligent decisions to alter future behaviour. The basic feature in AI-based algorithms is this learning capability and the ability to adapt a solution to suit the emerging evidence from the data. In the classical SON tools, there are fixed modelling and fixed algorithms (no matter how complex these are) which would give a certain outcome under given conditions. The AI algorithms are very much suited to situations which cannot be understood in a modelling approach and when there is an abundance of data available. In network management, many of the processes generate a lot of data and some of the complex network behaviours to be modelled are intractable. This will be particularly relevant to 5G networks with their complexities, as we discussed above. In these scenarios, AI-based network automation tools will find a lot of traction.

Some of the areas where AI-based tools can be applied, and their current level of maturity, are captured in Figure 11.6 below, from [26]. This reference [26] is also a good introduction to the emerging use of AI in 5G network automation.

The level of reliance on AI-based tools for network operations is an open question, which is under a lot of scrutiny. Whilst AI tools can be very useful in providing solutions formed from learning, the methods of how the solutions are derived can often seem to come from a 'black box'. This can particularly be the case with deep-learning AI algorithms, where training is done on large data sets taking a lot of time but the outputs are derived quickly. Network engineers will thus be aware of the input data sets and the output solutions derived by the AI, but will not be clear on the logics of how the solutions are derived. In situations where these solutions feed onto other more critical functionalities and cascade onwards, such lack of clarity can influence the decisions network engineers have to make. Thus, in the earlier stages of AI adaptation to network management, we may see a nuanced approach where AI is utilised mainly for self-contained configuration and optimisation steps.

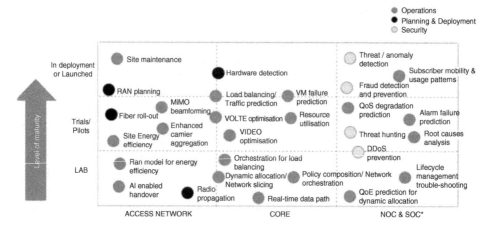

Figure 11.6 Level of maturity of AI-based Network Automation Tools. (*Source* [26]).

11.2.3 The Business Rationale for Network Automation

In the above sections, we looked at how network automation tools can help manage the increasing complexities of mobile networks from a technical perspective. From a business rationale, network automation is not just a cost-saving exercise. LTE SON tools and their evolving 5G counterparts significantly improve QoS, so the operators without this automation are at a competitively disadvantageous position. We will look at some examples of how LTE SON and the newer AI-based automation is helping operators extract the best performance from their networks.

The deployment of a country-wide, all-IP LTE network by Reliance Jio in India in 2016 was a major disruption in the Indian mobile telecoms market. As we detail in Section 11.4.2, the entry of Reliance Jio revitalised the LTE take-up in India and they were capable of capturing a significant market share from the incumbent operators within a very short time period. One of the key reasons for this success, as per the study report by Ovum [27], is that through the extensive use of LTE SON network optimisation tools, the operator was able to provide a far superior service quality at a comparable price. The single vendor network was designed and built in a way that SON optimisation functions like MLB and interference control can easily be carried out based on the dynamic network load conditions, even with an increasing number of subscribers month on month. This higher QoS enabled through SON and the suite of new data services offered in the network attracted and retained millions of new users for Reliance Jio. Their bold decision on upfront investments on an extensive, all-IP network and the associated SON tools like the VoLTE optimiser and Intelligent Traffic optimiser [27] paid off by pushing them to the leading mobile operator position with over 370 million subscribers in India by the end of 2019 [28].

Network automation in the 5G era will become even more critical, as the operator-customer relationships evolve to a new dimension. As discussed above, and in Section 12.2, multiple vertical industries will benefit from 5G and will become 'customers' for 5G network operators. These industries (or businesses) will consume services from respective network slices or augment them and provide them to another consumer down the chain. In this kind of business-to-business (B2B) or business-to-business-to-consumer (B2B2C) relationships, the stakes are much higher and operators simply cannot afford to have suboptimal network conditions which fail to meet the contracted KPI values. Network automation tools, which can actively predict anomalies before their onset and take remedial actions, are highly critical for the successful network management in this set-up. Naturally, 5G automation tools powered by AI, are already beginning to play a significant role.

An early example of how AI-based network automation is benefitting operators is given in the GSMA case study of [29]. SK Telecom (SKT) is South Korea's largest mobile operator with over 40% market share. South Korea has consistently been at the forefront of commercialising new mobile technologies and with 5G there was no exception. At the initiation phase of its 5G network, SKT also introduced a powerful AI-based unified OSS (Operations Support System) called TANGO, which stands for Telco Advanced Next-Generation OSS. TANGO consolidated the OSS of 4G and 5G operations of SKT into a single entity enabling a unified network management, particularly useful when the 5G network is operated in the NSA (Non Stand-alone) mode with the support of 4G LTE core. This single OSS platform makes it easier to create and terminate network slices for different QoS requirements dynamically. TANGO is capable of providing powerful data analytics and optimised real-time automation based on AI capabilities, to execute functions like automatic load balancing in the network. Importantly in the 5G era, TANGO provides the ability to expose network capabilities to third parties via open APIs (Application Programming Interfaces). This has enabled SKT's potential B2B customers in network slices to gain access to the OSS to monitor QoS provision and resource utilisation of the respective slices and also to run analytics on the data. By implementing a unified OSS driven with powerful AI capabilities very early on its 5G launch, SKT has effectively made its network much more attractive to B2B ventures.

We have only given an overview of the network complexities and the drive to automation as a means of managing these complexities in this section. However, this should open up a pathway for more inquiry for interested readers in this very important segment of mobile communications.

11.3 The Need for Greener and Lower EMF Networks

Many hold the view that reducing the environmental footprint of technology is a strategic imperative. Indeed, according to [30], the amount of CO_2 produced in 2002 due to ICT was 151 Mt and is expected to be 349 Mt by the end of 2020, with 51% coming from the mobile sector. The environmental footprint can also impact the application of wireless technology in a positive way, for example, by reducing the footprint in other areas such as transport through optimising journeys. Hence, there is both the direct and indirect impact of wireless technology on the environment. One apparent means of reducing footprint is by reducing transmit power, and hence the electromagnetic field (EMF) of the signal. This is because less power is needed to generate the radio signal and transmitters can be more power efficient. There are also health and safety concerns relating to exposure to RF electromagnetic fields, where concerns diminish as transmit power reduces, hence, there are two advantages of reducing RF transmit power.

Interestingly, figures reported in [30] show that 10% of mobile base stations serve 50% of the traffic; and conversely 50% of base stations only serve 5% of the traffic, hence, there is considerable room for further optimisation of networks. Similarly, base-station power amplifiers are typically 60–70% efficient, with the remaining power being dissipated as heat that may then require active cooling systems. Further, geographic spread of the signal is wasteful, i.e. even if a base station was only serving one device, the signal would be broadcast over an area that is wider and goes beyond the device. Hence, there is potential to only target the device, which would be far less wasteful. The time-domain also has potential to be optimised, for example, with the base station only broadcasting signals when absolutely required.

11.3.1 Greener Mobile Networks

The move towards 'Green' networks started in 2010 with inception of several research projects with this as the focus, as reported in [30]. An example target adopted by several projects was to reduce the energy per bit, E_b, by a factor of 1000. Techniques considered by these projects included cell densification, improved antenna and power amplifier technology and 'lighter' protocols.

Typically, energy cost represents around 20–40% of operator OPEX costs. As we move towards deployment of 5G networks, the necessity for going 'Green' is felt even more strongly. As reported by many expert bodies [31], 5G will likely bring a significant increase in energy consumption, due to its use of wider bandwidths, densified networks and massive MIMO etc. Figure 11.7 showcases how the OPEX cost of energy would surge in 5G networks if any optimisation steps are not taken into practice. The same figure also illustrates the importance of Green technologies to bring down this energy cost significantly. In an environment where the average revenue per user (ARPU) is falling year-on-year in many of the markets (e.g.: Figure 1.6 in Chapter 1), this level of OPEX savings through Green technologies would undoubtedly help operators forge ahead with their 5G roadmaps.

The benefits of Green technology is not confined to 5G. The use of renewable energy sources can have long-term positive impacts for any cellular network. The cost of energy from fossil fuels is highly volatile, driven by geopolitical events. The cost of renewable energy sources like wind and solar power has fallen significantly over recent years, as shown in [31]. Thus, it makes commercial sense for operators to plan forward with renewable options in the energy mix. In some developing countries, the electricity supply from the main grid can suffer numerous interruptions, currently backup diesel generators are used at base stations. Moving to site-generated renewable energy can replace these running costs. Many rural/remote areas in these countries do not have connection to the main electricity grid, so sites run entirely on diesel generators, with constant fuel supply and maintenance needed. Again, site-generated renewable energy can become a viable option as the cost of renewables keeps coming down. Also, as part of the social and environmental responsibility in tackling climate change, it is vital for these big cellular operators to show leadership by moving to Greener energy sources.

Mobile networks can also become Greener by their design. Lowering the power of transmitters whilst reducing the size of cells reduces the overall energy requirement of the network. This is because the power law of signal path loss means that substantial gains can be made in this respect. The path loss power law is a square law in free space and can assume any value between two and four (typically) in cluttered environments such as indoors or surrounded by buildings. To implement denser networks would require careful planning and coordination. It would also require more electronics to be manufactured, transported and deployed. Higher transmitter energy efficiencies can be expected, however, because

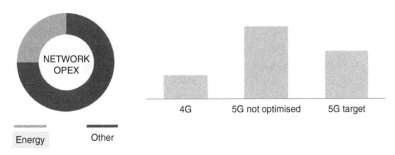

Figure 11.7 Likely energy cost increase in 5G. (*Source* [31]).

the transmitter power amplifiers can be designed to operate with higher efficiencies at lower power levels required for reduced cell sizes. Cell on-off methods are considered as a means to switch off base stations when there are very few or no active users occupying these cells. These schemes are easier to implement with smaller cells. Also, the application of beamforming can really focus the base-station energy to where active users are and hence reduce overall energy consumption.

11.3.3 Green Manufacturing and Recycling

Currently, most technology ends up in landfill at the end of life, with many of the parts being made from polymers and metals which are not biodegradable. There are even materials that are toxic to plants and wildlife such as those in some electronic components. One major contributor is batteries, where the UK alone generates 2000 to 3000 tons of waste from general purpose batteries every year [32]. Thus, with millions of items being disposed of in this way, the environmental contribution through adapting non-toxic and biodegradable materials appears to be substantial. Recycling also plays an important role in reducing the technology footprint on landfill. There are schemes in place for recycling batteries since some of the metals may be reused. Similarly, mobile phones and other technology items have precious metals that can be extracted and reused. There are schemes in place for this in many countries but certainly not in all countries. Even so, substantial material from technology ends up in landfill and hence adapting biodegradable and non-toxic materials would seem to make a great deal of environmental sense. Design and manufacture of mass-scale products (particularly mobile phones) in a manner that is easy to recycle the components is also important in this context.

The environmental impact of primary (non-rechargeable) batteries may be further reduced by replacing them with rechargeable batteries or super-capacitors that are charged from renewable sources. This may be on a large scale for powering base-station equipment or a smaller scale for personal devices. We looked at the Greener technologies of using solar and wind power for base-station equipment above. For smaller equipment such as mobile phones many additional power sources can be used. These include artificial light, kinetic energy in the form of movement and vibrations and electromagnetic energy itself from other radio signals. The crystal radio sets, for example, are powered from the energy captured from the radio waves [33].

The manufacturing processes for telecoms equipment itself can be optimised in a Greener sense. For example, production lines may adopt green energy policies such as using renewable energy and phasing production at off-peak times to limit demand on the electricity grid. Packaging should be carefully chosen to minimise its environmental footprint through reduced use and adopting biodegradable materials. Transportation of components, sub-systems and products is often on an inter-continental basis and so localising supply and manufacturing would reduce this impact.

11.3.4 Applications of Mobile Networks for Energy Reduction

Mobile technology can effectively be used to reduce the environmental impact in many other areas. For example, it may be used to control and optimise transport such that there is reduced energy usage. This could be on a micro-scale through optimising engines; or on a macro-scale for optimising traffic movement. We look at some of these applications in the new 5G automotive vertical area in Chapter 12. With the current lockdown measures due to the COVID-19 pandemic, it has become compulsory for many people to work from home with IT tools and many are finding this as a feasible long-term option. Providing the necessary data connectivity, also aided by other wireless technologies, will have a huge impact in sustaining this trend of working from home, which can reduce CO_2 emissions in the transport sector.

Mobile technology has revolutionised the retail marketing sector, deriving new business models which are also highly energy efficient. For example, it is now commonplace to do shopping on-line on your mobile from sites like Amazon or Boohoo, where goods can come straight from a warehouse, bypassing the energy costs of running a high street store. Deliveries are also highly coordinated to cover many destinations in a single route, so CO_2 emissions of individual shopping journeys are also cancelled out. Similar on-line transactions are possible in many other areas, as with mobile banking and each of these contributes indirectly to reduce the energy consumption and CO_2 emissions.

The potential contribution of these areas may be far more significant than the CO_2 contribution from greener telecommunication networks, although that is not a reason for the telecommunications sector to be complacent.

11.3.5 Electromagnetic Field Exposure and Mobile Networks

Lowering and managing EMF emissions for the purposes of reducing the environmental footprint of wireless equipment has been discussed above. EMF exposure is also a cause of major concern to the public due to potential health concerns and attracts considerable media coverage at the time of writing. It must be noted that the continuing speculation of linking the spread of COVID-19 to 5G networks has no scientific basis or scientific evidence at all. However, increased density of wireless connectivity of people and things raises public concern, even if the increase in density does not necessarily lead to a rise in exposure.

From the point of the mobile industry, stringent regulations do exist that aim to protect people from harmful levels of EMF emissions, see references [34] and [35] for example. According to [36], international authorities, standards' bodies and the mobile industry have cooperated for some time in forming related international regulations. The World Health Organisation and International Telecommunications Union have endorsed the International Commission on Non-Ionising Radiation Protection (ICNIRP) guidelines. Furthermore, standards in most countries are national implementations of these guidelines and some countries have adopted more stringent levels. The levels relate to frequency-dependent maximum permitted exposure levels for parts or whole of the body from any number or type of RF devices, which include mobile devices and base stations.

So, what aspects of a mobile network contribute to the EMF levels, and what can be done to manage these levels whilst maintaining the required QoS? From the analysis in [36], the following points can be highlighted:

- Antennas – For antennas on devices, EM shields have been tried, but these have become impractical. Ferrite absorbers have been tried but this reduces antenna efficiency. For base-station antennas, beamsteering and directional antenna patterns can help by channelling energy where it is required. Integrating beamsteering antennas into form factors available for SCs is more challenging.

- Sleep modes – By only transmitting when required these can help reduce EMF exposure.
- Self-organising architecture – By using a configurable transceiver, a suitable bearer that results in less EMF exposure may be selected and power amplifiers that can quickly switch to support the selected bearer.
- Low-noise amplifiers (LNAs) – Using LNAs with a very low noise floor will reduce the required transmit power, hence reducing EMF exposure. Such an LNA could be a multi-band LNA where the noise bandwidth is reduced compared to that of a wideband LNA. This would also reduce the level of out-of-band interference.
- Radio resource allocation techniques – Power control would have a significant impact on EMF exposure since only the power required to make a link viable would be required. Similarly, managing vertical and horizontal hand-over such that the level of EMF exposure is reduced would also be beneficial.

According to [37], there are additional aspects that need consideration in terms of managing EMF exposure when planning new network roll-out such as 5G.

The cell-densification effect that occurs as a result of managing EMF exposure to be within regulatory limits. This relates to the need to increase the density of transmitters to cover an area whilst avoiding high concentration of EMF in some areas. This requires CAPEX to fund more base-station deployments and may even result in service coverage of some areas being uneconomical.

EMF saturation zones occur when legacy radio equipment is already serving an area. In this case, EMF limits may be saturated making it impossible to install 5G equipment at the same location. Thus, the operator would need to establish a new site at additional CAPEX, which may also incur a reduction in QoS. This possibility may even be exploited by a competing operator in order to deny installation of equipment at a preferred location.

Another possibility is where a minimum distance between a preferred 5G site and a sensitive site such as a school or hospital is required. In this case, the operator would be required to commission an alternative and probably sub-optimal site.

Within the context of 5G networks, what are the specific 5G technologies that may reduce or enhance EMF exposure? Reference [37] has analysed this question and the findings can be summarised as follows.

- Massive MIMO – This can have a positive and negative impact on EMF levels, depending on the configuration and method of determining the EMF levels.
- Beamforming – This can often decrease overall EMF levels in each area since the energy is confined to the target area. In some deployments, it may have a negative impact due to the concentration of energy.
- mmWave communications – Due to the higher path loss at these higher frequencies for a given distance, the EMF level will be less than that experienced at conventional lower frequencies.
- SCs – These will result in a higher density of base stations and thus the EMF levels are expected to generally be lower when compared to conventional cell deployments, but could be higher at close ranges.

- Off-loading – This exploits over-lapping cells and so enables some users being off-loaded from busy cells to neighbouring ones where possible. This means that the busy cell can reduce its power accordingly and hence reduce the EMF level. The overall network EMF level is hence shared across the network and less concentrated.
- Software-defined networking – This refers to the sharing of hardware between network operators. It is expected to result in a large reduction of EMF due to the number of antennas (and hence power) radiated over an area compared to current configurations where multiple base-station systems share a site.
- Mobile Edge Computing – This is expected to result in reducing EMF level since transceivers and equipment (potentially including wireless Fronthaul – transport network options) would be closer to the user. However, the impact depends on the type of service provided by the unit, such as high data rate services that require higher transmit powers.
- Device-2-device (D2D) communications – This may result in a reduction in EMF levels since the data exchanged directly with base stations will be reduced due to the (often) shorter hops between devices requiring less transmit power.
- Sleep modes – Base stations that are not in use may be placed in a reduced power mode until such time as they are needed. Whilst this would reduce EMF levels at a local level, neighbouring cells may need to increase their power in order to provide some coverage.
- Roll-back of legacy networks – 2G and 3G networks may be rolled back or decommissioned with the roll-out of 5G. This would reduce saturation levels and enable better deployments of 5G technology and hence a large decrease in EMF levels.

Reference [37] also provides guidelines that can be followed during network design in order to provide EMF exposure awareness and hence embed the impact of EMF regulation early on in network roll-out. The principal elements of the guidelines are:

- Modelling of 5G radio access technologies, including MIMO arrays.
- Computation of EMF field levels, including specifics relating to mmWave communications.
- Integration of current/future EMF limits, including limits for different countries and building types.
- Modelling the set of candidate sites, including legacy radio equipment.
- Modelling of 5G traffic demands and user QoS, including spatial and temporal fluctuations.
- Validation using realistic scenarios, including sites used by other operators.

When comparing lower EMF and lower energy networks, it is evident that the approaches needed to achieve these are not exactly the same. There has been a lot of theoretical work and practical implementations on developing energy efficiency in mobile networks, whilst the work on achieving lower EMF exposure is not as developed as such. Given the current focus on climate change and the imperative for all industries to play their part to become Greener and finally net zero Carbon emitters, the lower energy networks will be a key theme for the mobile industry. Equally importantly, enhanced efforts to the lower EMF exposure levels in mobile networks will be key to allay health and safety concerns as people and things will become more and more connected through 5G and beyond.

11.4 Covering the Unserved and Under-served Regions

Although the number of mobile subscriptions today exceeds global population number, this does not mean that all of humanity has acceptable levels of mobile coverage. There are huge swathes of rural and far-remote areas that lack access to basic internet, let alone access to broadband. As the latest statistics reveal, only 4.48 billion people or around 58% of the global population have regular access to the internet [38], leaving over 40% without this privilege. Mobile and wireless technologies are seen as formidable tools to overcome this digital divide.

The identification or impetus on mobile and fixed line operators to address this problem has not been a new phenomenon. For these operators, the efforts to cover rural and far-remote areas boil down to a basic business rationale. Does the revenue they receive from a few, basic subscriptions balance out the huge investments needed to extend the coverage to these vast remote areas? In many cases the answer is no, under the current technology and regulatory status and economic frameworks. However, operators continue to try and address this challenge, as highlighted by a recent NGMN study on Extreme Long Range Communications for Deep Rural Coverage [39]. In this study, they identify options to extend current coverage to deep rural areas including, increasing cell radius of current sites and of new sites to fill coverage voids, infrastructure sharing including national roaming between operators, use of relay nodes and the use of wireless backhauling in the form of LTE/5G technologies and satellite links. Interestingly, this study also suggests that users in these deep rural areas should also be provided with low-cost Smartphones with new long-range coverage capabilities.

As 5G dawns on many major commercial and economic hubs of the world, there is a risk that the digital divide will become further entrenched, with most of the population not having access to 5G at least in the short-term. This has mostly been the case when a new mobile generation is introduced, as it takes a number of years for coverage to expand to a truly global footprint. With 5G however, the fixed wireless access (FWA) technology at mmWave frequencies has developed even earlier than the mobile component. This gives 5G a unique opportunity to at least cover some of the non-urban population centres with 5G at an early stage, with FWA.

In the recent past, support from new access technologies and initiatives driven by government policies to provide broadband to everyone have come to the forefront as never before. We will look at examples from these areas as there is now a renewed emphasis to address this perennial challenge.

11.4.1 New Access Technologies

The advent of 5G saw mobile technologies move into the mmWave spectrum (discussed in Section 5.8). A new mobile generation brings in higher speeds and other capabilities but it takes a long time for these capabilities to filter through to the rural communities. However, with 5G and mmWave technologies, a unique opportunity has opened up to provide fixed wireless access (FWA) to these underserved areas as a real alternative to fibre-to-the-home (FTTH) connectivity. With FTTH, it is highly expensive to dig the trenches and connect each of the homes to the fibre network, especially in areas where the planned infrastructure is lacking. The FWA solution provides this 'last mile' through wireless beamforming and an access point located on a lamp-post (for example) can connect many homes through self-aligning CPEs (Customer Premises Equipment). The mmWave bands possess significantly higher bandwidths and it is possible to provide Giga-bits-per-second (Gbps) data rates to match fibre data rates. These data rates are sufficient to give the 5G experience with applications like UHD video or AR/VR in the home.

Samsung has been a pioneer in this area and has developed commercial FWA systems in the 28 GHz mmWave band [40]. In the US, suburban areas in a number of cities have seen FWA deployments with the operator Verizon, using Verizon's early 5GTF standards [41]. There has been a number of similar trials in Europe, targeting the European 26 GHz mmWave band. Samsung has partnered with Cisco and the operator Orange to conduct FWA trials in a remote city in Romania, where aggregated cell throughputs up to 3 Gbps have been recorded [42].

Non-Terrestrial Networks (NTN) can also play a vital role in helping to extend mobile broadband coverage. Traditionally, NTN included satellites only but recent technological advances have seen High Altitude Platforms (HAP) and drones join in as viable NTN options. The NTN solutions are also now in the realm of 3GPP standards. 3GPP is conducting study and work items on how to adapt certain time and delay critical aspects of 5G-NR standards (like random access and HARQ processes) to accommodate NTN communications [43]. Below, we will look at different types of NTN deployments and some of the application areas related to them.

Satellite deployments are the traditional and still the dominant mode of NTN. Communication satellites can now be deployed at much lower altitudes of Low Earth Orbits (LEO) and Medium Earth Orbits (MEO) in addition to traditional Geo-stationary orbits (GEO). The typical altitudes of these satellite types and their coverage beam footprint sizes are noted in Table 11.1 below (from [44]).

The main benefit of satellite access points is the very wide area of coverage they provide, which can easily span several thousand square kilometres. In terms of GEO satellites, they can effectively operate as fixed location base stations (or repeaters), covering the same geographic area. Continuous coverage with MEO or LEO satellites is also possible, with use of multiple satellites (a constellation) in the same orbit and with scheduled handover operations. The main limitations of satellite-based 5G provision are the very high path loss and propagation delays, particularly for GEO satellites. This will limit the 5G eMBB type (unless very high beamforming gains are achieved, which is difficult in mobile environments) and the URLLC type applications. However, there is a scope for providing basic mobile broadband and also supporting wide area mMTC through satellite links. It should be also noted that the propagation delays show wide variances for different types of satellite deployments. For example, the GEO satellites demonstrate more than 250 ms of propagation delay, whilst the LEO satellites at 600 km altitude impose only 2 ms propagation delay. The 3GPP NTN study item [44] is looking at ways of adapting delay sensitive radio procedures like HARQ to reflect these wide variances.

The wide area, reliable coverage of the satellite links make them highly useful for providing wireless connectivity to remote regions of the globe and also for long-distance transport links, like railways and shipping routes. Satellite communications have traditionally needed specialised receiver units, which is a major hindrance in widening their

Table 11.1 Different satellite types, altitudes and coverage footprints. (source [44])

Platforms	Altitude range	Orbit	Typical beam footprint size
Low-Earth Orbit (LEO) satellite	300 – 1500 km	Circular around the earth	100 – 1000 km
Medium-Earth Orbit (MEO) satellite	7000 – 25000 km		100 – 1000 km
Geostationary Earth Orbit (GEO) satellite	35 786 km	Notional station keeping position fixed in terms of elevation/azimuth with respect to a given earth point	200 – 3500 km
UAS platform (including HAPS)	8 – 50 km (20 km for HAPS)		5 – 200 km
High Elliptical Orbit (HEO) satellite	400 – 50000 km	Elliptical around the earth	200 – 3500 km

applicability. With current work in 3GPP [44] to make satellite communication compatible with 5G-NR, so the same 5G handsets can benefit from these satellite (and also HAP, below) links.

The high-altitude platforms (HAP), an emerging form of NTN, are deployed at around 8–50 km altitude. The balloons deployed in the Google Loon project [45], the light un-manned aircraft of the Facebook Aquila project (now concluded) [46] and 'Stratobus' air-ship concept currently developed by Thales [47] are all examples of HAPs. With these solutions, large area coverage can be provided and higher data rates and lower latencies can be achieved due to lower propagation distances. For the balloon and un-manned air-craft deployments, a fleet of units is needed to drift into the serving base station position to cover the same footprint on the ground. It is claimed that the 'Stratobus' airships can stay in position for up to 1 year, in between their service cycles [47]. The Google Loon balloons conduct LTE transmissions in the ISM bands, so the common LTE handsets can be used to receive the basic internet services. These have already been used in disaster and emergency situations in Peru and in Puerto Rico to provide internet access whilst the ground-based networks are rebuilt. In terms of 5G, the HAPs are more capable of meeting some of the basic data rate and latency requirements and also connecting massive numbers of low power devices for IoT type services in remote areas.

Drone-based communications are a third type of NTN, where the dynamics are very different from the above two. Drones can be deployed quickly to cover a small footprint (at 100–1000 m altitude) with very high capacity. They provide flexibility to support ad-hoc events such as airborne base stations or relays but, in many cases, there need to be pre-planned ground networks to receive the drone links either as fronthaul or backhaul. In the context of 5G service provision, drones can support very high data rates and very low latencies, enabling many localised applications from live events and open air concerts to emergency scenarios. Due to battery life constraints, drones would need regular replacements to cover longer events and thus a fleet of drones are needed to provide continuous connectivity. Alternatively, drones in some situations can be tethered to the ground, so the power supply can be uninterrupted. In terms of other relevant use cases, fleets of drones would be highly efficient in capturing mMTC type data periodically from remote areas in applications like agriculture, livestock tracking and forestry and wildlife monitoring.

An illustration of the different types of NTN communication modes is depicted in Figure 11.8 (not drawn to scale).

11.4.2 Initiatives Driven by Government Funding and Policy

Government policy – be it in the form of relaxing regulations on spectrum or other existing telecom infrastructure usage or as financial initiatives to shoulder some of the costs in laying out infrastructure in the under-served areas – can have a major influence in stimulating the coverage expansion of telecom services. We will look at some of the recent examples, which are having some success in bridging the digital divide not just in developing countries, but also in developed countries.

With the second highest population in the world and a vast land mass, India is a challenging country by any means to provide widespread broadband connectivity. The majority of Indians live in rural areas and they lag significantly behind in internet facilities compared to their urban counterparts. According to a reliable report [48], 65% of the urban Indian population had internet connectivity by the end of 2017, whilst this figure was only

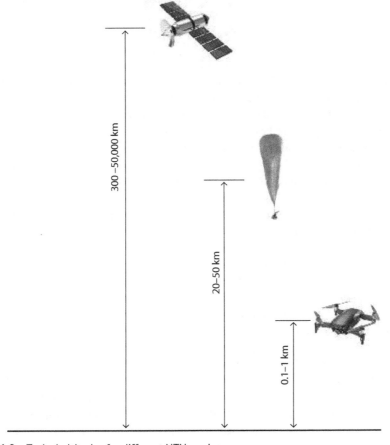

Figure 11.8 Typical altitudes for different NTN modes.

20% in rural India. As a main pillar of the Digital India program, the Indian government launched an ambitious program in 2014 to build a nationwide optical fibre network, to provide all 600000 villages in India with high-speed broadband of at least 100 Mbps speeds. A telecom infrastructure provider, known as BharatNet was created to establish, manage and operate this network. To provide last mile connectivity, up to 700000 Wi-Fi hotspots in these rural areas were planned, with the government hugely subsidising service costs to make these financially viable. By 2019, when the project was due to complete, significant progress had been made. However, the mammoth scale of the task meant that a little less than 50% of the targeted villages were actually connected [49]. BharatNet is, by far, the world's largest rural broadband connectivity program and will be expedited with new funds to meet connectivity targets by 2020.

In parallel to this government drive to digitise India, a new nationwide 4G network was rolled out by Reliance Jio, in September 2016. This new network with 113000 base station sites across India was in partnership with a single vendor in Samsung and used the latest technology from planning, deployment to operations, optimisation and maintenance. This initiative reinvigorated the lacklustre LTE take up in India, reaching 100 million LTE subscribers in just 170 days. This network is considered to be the largest all-IP wireless network in the world. The key to its success was the new exciting range of data services it offered, backed-up with the improved QoS subscribers could experience from the beginning.

Whilst the average broadband speeds and the service take up in the UK were much higher, there was serious underfunding in investments for rural broadband provision. In 2010, there were around 3 million UK households (over 10%) with internet speeds lower than 2 Mbps [50]. The then UK government initiated the rural broadband program with £530 million investment to accelerate the broadband infrastructure development. The next 3 years saw a rapid uptake of faster broadband, with the average UK speeds trebling and 73% of the said premises having access to superfast broadband [50]. The current push is to enable Giga-bits-per-second connectivity anywhere in the UK, with the government committing GBP 200 million to the Rural Gigabit Connectivity (RGC) program [51].

In the US, the regulator FCC (Federal Communications Commission) opened up the use of the 3.5 GHz spectrum band under the CBRS (Citizen's Broadband Radio Service) scheme. Traditionally, this band has been exclusively used by the US military. Under the new CBRS scheme, the incumbent defence services will still have priority access with specific spectrum protection rights at listed sites. There are two new access modes established for civilian use of this spectrum. Organisations can gain priority access licences in the lower 100 MHz of this band for a fee, with the license lasting up to 3 years in a limited geographic area. The rest of the spectrum in this band is open for general use, and a user is covered by a general authorised license. The CBRS scheme has enabled private networks to be established in campuses, hospitals, office premises and other buildings. Also, importantly for our topic, it has enabled fixed wireless networks to be set up to cover residential areas by wireless internet service providers (WISP). Example networks are being deployed in rural Arizona where previously only dial-up internet connectivity was available [52]. These networks currently provide LTE services in this 3.5 GHz CBRS band. Technically, the CBRS band is designated as technology agnostic, so it will be possible to move onto 5G FWA services in this band later.

All these examples show that, with correct technology application and government policies, the digital divide can be effectively bridged. Considering the enormous socio economic benefits that can be borne out of such programs, it should be a priority for all countries in the world.

References

[1] Bell, L., web article. What is Moore's Law? available at: https://www.wired.co.uk/article/wired-explains-moores-law

[2] Ellis, A.D., Suibhne, N.M., Saad, D., and Payne, D.N. (2016). Communication networks beyond the capacity crunch. *Royal Society Philosophical Transactions A* 374: 20150191. http://dx.doi.org/10.1098/rsta.2015.0191.

[3] OFCOM consultation proposal, *Supporting innovation in the 100-200 GHz range*, available at: https://www.ofcom.org.uk/__data/assets/pdf_file/0034/189871/100-ghz-consultation.pdf

[4] Björnson, E., Hoydis, J., and Sanguinetti, L. (2017). Massive MIMO Networks: Spectral, Energy, and Hardware Efficiency. Now Foundations and Trends.

[5] Drysdale, T.D., Allen, B., Chris Stevens, S.J., Berry, F.C., and Smith, J.C. (15 August 2018). How orbital angular momentum modes are boosting the performance of radio links. *IET Map* 12 (10): 1625–1632.

[6] Sasaki, H., Lee, D., Fukumoto, H., Yagi, Y., Kaho, T., Shiba, H., and Shimizu, T., (December 2018), "Experiment on Over-100-Gbps Wireless Transmission with OAM-MIMO Multiplexing System in 28-GHz Band," Proc. of 2018 IEEE Global Communications Conference (GLOBECOM), Abu Dhabi, UAE.

[7] Allen, B.H., Brown, T.W.C., and Drysdale, T.D. (2019), "A New Paradigm for Train to Ground Connectivity Using Angular Momentum," 2019 IEEE 2nd 5G World Forum (5GWF), Dresden, Germany, pp. 185–188, doi: 10.1109/5GWF.2019.8911685.

[8] Nokia White Paper. (2014). Network Sharing: Delivering mobile broadband more efficiently and at lower cost. https://resources.nokia.com/asset/200192

[9] GSM Association. (2017). An Introduction to Network Slicing. https://www.gsma.com/futurenetworks/wp-content/uploads/2017/11/GSMA-An-Introduction-to-Network-Slicing.pdf

[10] IEEE WLAN (Wireless Local Area Network) working group help page. Accessed Dec. 2020. https://www.ieee802.org/11/help.html

[11] CISCO VNI forecast. (2020). Available at: https://www.cisco.com/c/dam/m/en_us/solutions/service-provider/vni-forecast-highlights/pdf/Global_2020_Forecast_Highlights.pdf

[12] Networkworld web article. Viewed Dec. 2020. https://www.networkworld.com/article/3560993/what-is-wi-fi-and-why-is-it-so-important.html

[13] O2 WiFi extra service description. Viewed Dec. 2020. https://www.o2wifi.co.uk/o2-wifi/o2-wifi-extra

[14] GSA report. (Aug. 2020). *5G & LTE in Unlicensed Spectrum.* Available at: https://gsacom.com/paper/5g-lte-in-unlicensed-spectrum-august-2020

[15] Netmania interview with Korea Telecom at MWC. (2015). Viewed Dec. 2020, https://www.netmanias.com/en/?m=view&id=blog&no=7388

[16] Endgadget web article. Viewed Dec. 2020. https://www.engadget.com/samsung-galaxy-note-20-ultra-wide-band-connectivity-151223521.html

[17] Fira (Fine ranging) consortium website. Viewed Dec. 2020. https://www.firaconsortium.org

[18] Pure LiFi website. Viewed Dec. 2020. https://purelifi.com/lifi-technology

[19] NGMN white paper. (Dec. 2008). NGMN Recommendation on SON and O&M Requirements. Available at: https://www.ngmn.org/wp-content/uploads/NGMN_Recommendation_on_SON_and_O_M_Requirements.pdf

[20] 3GPP article on SON. https://www.3gpp.org/technologies/keywords-acronyms/105-son

[21] Osterbo, O. and Grondalen, O. (Sept. 2012). Benefits of Self-Organizing Networks (SON) for Mobile Operators. *Hindawi Journal of Computer Networks and Communications.* Available at https://www.hindawi.com/journals/jcnc/2012/862527.

[22] Hunukumbure, M., Moulsley, T., and Vadgama, S. (March 2012). A Dynamic Resource Allocation Algorithm as a Green Technology for 4G Wireless. World Telecommunications Congress (WTC) 2012, Miyazaki, Japan.

[23] Hamalainen, S., Sanneck, H., and Sartori, C. (eds.). (2011). LTE Self-Organising Networks (SON): Network Management Automation for Operational Efficiency. Wiley Publishers.

[24] 3GPP Technical Specification, TS 32.101. (Jan. 2016). *Telecommunication management; Principles and high level requirements (Release 13)*, v13.0.0. Available at: https://portal.3gpp.org/desktopmodules/Specifications/SpecificationDetails.aspx?specificationId=1838

[25] 5G Americas white paper. Self Optimising Networks in 3GPP Release 11- The benefits of SON in LTE. Available at: https://www.5gamericas.org/self-optimizing-networks-in-3gpp-release-11-the-benefits-of-son-in-lte

[26] GSMA article. (June 2019). AI & Automation: An Overview. Available at: https://www.gsma.com/futurenetworks/wiki/ai-automation-an-overview

[27] Ovum report. (2017). *Jio Is Changing the India Mobile Market*. Available at: https://images.samsung.com/is/content/samsung/p5/global/business/networks/insights/white-paper/jio-is-changing-the-india-mobile-market/global-networks-insight-jio-is-changing-the-india-mobile-market-0.pdf

[28] News article in India Today. (Jan. 2020). Reliance Jio becomes biggest telecom operator in India by subscriber base, Vodafone Idea now second. Available at: https://www.indiatoday.in/technology/news/story/reliance-jio-now-biggest-telecom-operator-by-subscriber-base-1637617-2020-01-17

[29] GSMA Case study on future networks. (Sept. 2019). *Case Study – SKT Tango: AI-assisted Network Operation System*. Available at: https://www.gsma.com/futurenetworks/wiki/case-study-skt

[30] Suarez, L., Nuaymi, L., and Bonnin, J.-M. (2012). An Overview and Classification of Research Approaches in Green Wireless Networks. *EURASIP Journal on Wireless Communications and Networking* 142(2012).

[31] GSMA Future networks article. (May 2019). Energy Efficiency - An Overview. Available at: https://www.gsma.com/futurenetworks/wiki/energy-efficiency-2

[32] http://www2.eng.cam.ac.uk/~dmh/ptialcd/battery/index.htm#Environment

[33] Ajmal, T., Dyo, V., Allen, B., Jazani, D., and Ivanov, I. (2014). Design and optimisation of compact RF energy harvesting device for smart applications. *IET Electronics Letters* 50 (20): 111–113.

[34] ICNIRP. (1998). Guidelines for Limiting Exposure to Time-Varying Electric, Magnetic and Electromagnetic Fields (up to 300GHz). *Health Physics* 74(4): 494–522.

[35] ICNIRP website. https://www.icnirp.org/en/applications/5g/index.html

[36] Tesanovic, M. et al. (June 2014). The LEXNET Project. *IEEE Vehicular Technology Magazine*: 20–28 9(2).

[37] Chiaraviglio, L. et al. (Sept 2018). Planning 5G Networks Under EMF Constraints: State of the Art and Vision. *IEEE Access* 6: 51021–51037.

[38] Global digital population on Statista website. Checked December 2019. https://www.statista.com/statistics/617136/digital-population-worldwide

[39] NGMN study report. (June 2019). https://www.ngmn.org/wp-content/uploads/Publications/2019/190606_NGMN_5G_Ext_Long_Range_D1_v1.7.pdf

[40] White paper. 5G Fixed Wireless Access. Available from: https://images.samsung.com/is/content/samsung/p5/global/business/networks/insights/white-paper/samsung-5g-fwa/white-paper_samsung-5g-fixed-wireless-access.pdf

[41] Verizon https://www.verizon.com/.

[42] ITU news article. https://news.itu.int/orange-romania-5g-fwa-trial-cisco-samsung

[43] 3GPP Article. (January 2018). https://www.3gpp.org/news-events/1933-sat_ntn

[44] 3GPP Technical report. (Dec. 2019). TR 38.821, Solutions for NR to support Non-Terrestrial Networks, version v1.0.0.

[45] Loon project overview. https://loon.com

[46] The technology behind Aquila. Project note 2016. https://www.facebook.com/notes/mark-zuckerberg/the-technology-behind-aquila/10153916136506634

[47] Thales Stratobus project launch. (April 2016). https://www.thalesgroup.com/en/worldwide/space/news/new-video-stratobus-we-go

[48] Mobile Internet Report – India. (2017). https://cms.iamai.in/Content/ResearchPapers/2b08cce4-e571-4cfe-9f8b-86435a12ed17.pdf

[49] BBC reality check on India's rural broadband program. (March 2019). https://www.bbc.com/news/world-asia-india-47053526

[50] https://www.gov.uk/government/publications/2010-to-2015-government-policy-broadband-investment/2010-to-2015-government-policy-broadband-investment

[51] https://www.gov.uk/government/publications/rgc-programme-key-information

[52] Mile High FWA deployments in Arizona. https://www.lightreading.com/mobile/4g-lte/arizona-wisp-plans-lte-expansion-using-35ghz-cbrs-spectrum/d/d-id/755154

12

The Changing Face of Mobile Communications

Over the last few years, there have been fundamental changes in mobile networks and the potential applications it can support. We looked at how the centralisation and virtualisation of the key components in the mobile network are shaping up at a technical level in Section 5.9. In this chapter, we will first analyse the profound changes to the operational and business models these adaptations are causing in the mobile communications ecosystem. We will then investigate some of the vertical industries emerging as potential benefactors from the 5G capabilities of eMBB, URLLC and mMTC. We will also look briefly into how today's Smartphones will evolve within 5G and beyond. Finally, we peek into the 'crystal ball' to identify some of the key enabling technologies that may formulate 6G.

12.1 Changes with Centralisation and Virtualisation of the Mobile Network

With traditional mobile networks, operators have opted to stick with one or two network equipment vendors. The networks would be deployed with specialised hardware components and associated proprietary software. This trend has continued from 2G and the shift to a new generation or even a new release within a generation incurred a major network overhaul at a significant cost. Thus, upgrades within a network would happen in a cautious, step-wise manner and it is not unusual for networks to lag several versions behind the latest finalised standards release.

With the network centralisation and virtualisation trends emerging with 5G, the above operational and business models are set to witness dramatic changes. Some of these changes are necessitated by the nature of the 5G networks themselves. As we will see in the next Section 12.2, 5G networks are set to support multiple industry verticals. These verticals will have distinct KPI (Key Performance Indicator) requirements, such as those unique mixtures of the eMBB, URLLC and mMTC 5G capabilities we looked at in Section 5.8. For a monolithic traditional 5G network, it will be difficult to handle all these different requirements at the same time. After all, traditional networks are designed and built to only support the voice and mobile data-centric consumer sector. The new 5G verticals will need the implementation of end-to-end slicing, where multiple, virtual 'mobile networks' to fit the given

The Technology and Business of Mobile Communications: An Introduction, First Edition.
Mythri Hunukumbure, Justin P. Coon, Ben Allen, and Tony Vernon.
© 2022 John Wiley & Sons Ltd. Published 2022 by John Wiley & Sons Ltd.

sets of KPIs (such as the data rate and latency) can be configured from the same physical network. Virtualised networks provide this capability to configure multiple network slices in a logical manner, which can then meet the required KPIs. Also, some of these slices will not be required all of the time. With NFV and SDN, slices can be created, operated and dismantled without any movement or relocation of physical resources. This will massively increase efficiencies of the future mobile networks, as they will experiment to find the right balance of the resource mix for different vertical industries and also new services in the initial stages. Similarly, with the Open-RAN, the radio access network (RAN) can be virtualised in a cloud native manner and operators will have full control on optimising different RAN slices to meet the different radio layer KPIs to support these vertical industries.

There are huge business implications with NFV, SDN and Open-RAN trends. As the related organisations (mainly ETSI [1], ONF [2] and O-RAN alliance [3]) are striving to build open standards for these concepts, it will be possible for operators to combine the switching, storage, processing and computing elements developed in software on general purpose computing units from different vendors in a seamless manner. This will potentially open up the market for smaller vendors, who would perhaps specialise on a certain component of the NFV/SDN or Open-RAN network architecture. The increased competition would help reduce prices and further enhance the business and technical innovations in this area.

In a fully virtualised network, the upgrades to new versions of a cellular standard would mostly incur software upgrades. The dedicated hardware components of such a network would only consist of remote radio heads (RRH), whilst all other virtualised RAN and core network (CN) components can be upgraded through software. This offers a fast and cost efficient way for virtualised mobile networks to embrace the latest versions of a cellular standard. Thus, with future SDN/NFV and Open-RAN networks, we would see a far closer alignment with the current status of standardisation and the actual networks on the ground.

The ability to virtualise network functions over different hardware platforms means that operators need not own and maintain hardware platforms, which currently incur a very high cost. Instead, they can lease these resources from specialised providers and, using NFV and SDN, map the required functionality and network operational policy frameworks to these. There are many solutions now emerging in this model, termed as Network as a Service (NaaS) or Platform as a Service (PaaS) [4]. These service models take away the hardware ownership efforts and costs from mobile operators. Traditionally, mobile networks are dimensioned to employ resources (as a combination of spectrum and hardware) to meet peak demands, as we studied in Chapter 6. Peak demands only occur a fraction of the time, so there is resource wastage at other times. If the radio access and CN functions are virtualised to cloud servers (through open/cloud-RAN and NFV – as noted in Section 5.9) and these resources are pooled to serve networks across many different time zones for example, peak times can be spread across and hardware resources can be utilised much more efficiently. The PaaS and NaaS solutions also help mobile operators scale their virtual networks to be much in-line with the demand growth cycle (for example, when a new mobile generation is commercialised), avoiding the need for huge up-front costs for network hardware.

Perhaps the biggest transformation in adapting these NFV/SDN and Open-RAN models will come in the form of 'levelling the playing field' for new entrants to the mobile operator domain. Owning and operating hardware infrastructure is the dominant cost for an operator and this includes a significant up-front capital expenditures (CAPEX). The NaaS and PaaS models discussed above remove this CAPEX and replace it with a more manageable operating expenditure (OPEX), which can be scaled in-line with demand for the network services (and thus the revenue). The operator may own and deploy the remote radio heads (RRH) and can densify and/or extend the coverage again in-line with the growth of the customer base. This will leave the spectrum cost as the only major up-front CAPEX that a new operator will have to finance, making new networks more financially attractive.

An early example of this NFV/SDN and Open-RAN based fully virtualised, highly automated 5G network entry come from Japan in the form of Rakuten network [5]. As the fourth mobile operator in Japan, they are poised to disrupt this technologically advanced market by deploying a fully virtualised network, what they claim as the world's first cloud native platform. Being a new entrant to the market with no legacy network, Rakuten has the advantage of designing the entire network with NFV/SDN/Open-RAN functionalities. They obtained 5G spectrum in 3.8 GHz and 28 GHz bands in 2018 with attached conditions to cover 56% of the population in 5 years with a network investment of $1.7 billion. Rakuten is first building their 5G CN, where the commissioning has slipped from October 2019 to the spring of 2020. Initially, the RAN coverage will be provided as 4G, with national roaming agreement involving another Japanese operator (KDDI). The CN is based on a 'microservices' architecture [6] and multiple vendors are involved in designing and building the core and the RAN. With Open-RAN concepts adapted for the 5G rollout, they do not depend on one single vendor system, but integrate components (mostly virtualised) from multiple vendors, including smaller players in the ecosystem. Whilst this first attempt to establish an end-to-end virtualised network is impressive, whether the Rakuten network can be termed as cloud native is discussed in [6]. A conceptual diagram of the planned Rakuten network is shown in Figure 12.1 (from [5]).

With NFV, SDN and Open-RAN future mobile networks will become much more agile, scalable and cost efficient, bringing in significant changes in the business and operational models we discussed above. It should be noted that we are still at the very beginning of commercialising these concepts. It can be envisioned that many more business innovations and disruptions will follow in this domain.

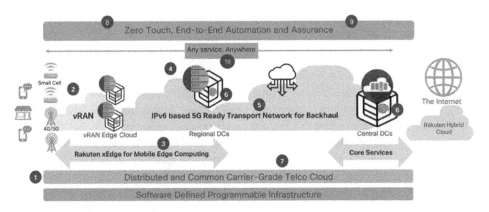

Figure 12.1 The proposed Rakuten end-to-end virtualised 5G network. (*Source* [5]).

12.2 Supporting Multiple Vertical Industries through 5G

So far we have seen that, in all previous cellular generations, the focus was on providing voice or mobile data services to the consumer. The mobile data rates have increased and latencies have gone down over the generations and within their releases, but the primary focus on this consumer sector (be it retail or business consumers) has not changed. We will see a paradigm shift in 5G and generations beyond this. The cellular services will expand to support different vertical industries, in addition to the traditional consumer base. This will be a complex procedure, as these industries are well established with their own demanding requirements. 3GPP and other standardisation, collaboration/advisory bodies have been working over the past few years to identify specific technical, business and regulatory issues each of these specific industry inclusions face. In this section, we will take a detailed look at the context, benefits and challenges in bringing cellular-based services to some of these vertical industries.

One of the key technical enablers in 5G that will facilitate mobile operators to tailor their networks to support these multiple vertical applications is network slicing. Network slicing enables the creation of multiple logically independent networks supporting varied sets of KPIs. All these independent logical networks are supported through a common physical infrastructure, which is the operator's physical network. We discussed this new slicing innovation from a 5G architectural perspective in Section 5.8. A good overview of how network slicing can accommodate different verticals can be found in [7].

Moving cellular applications into these vertical industries requires a thorough understanding of the business needs of each of these unique verticals, in addition to fulfilling the technical requirements. The cellular industry is now working with vertical industry players and other partners to collaboratively develop Business to Business (B2B), Business to Business to Consumer (B2B2C) and other variants of these solutions. Factors such as price points, market penetration to reach a critical consumer mass, ease of use of these solutions and conformance with related regulatory frameworks are just as critical in determining the success of these endeavours. In many of these vertical applications, there will be complex ecosystems with multiple stakeholders and developing successful business models is not a trivial task. In the following examples, we will look at the technical and business context of the emerging key vertical areas.

12.2.1 Automotive Sector

There is a growing recognition of the capabilities that the wireless connectivity can bring to the automotive sector. As the vehicles become more 'intelligent', this connectivity can provide numerous features to enhance the driving experience, improve road safety and reduce environmental impact. Recently, there has been much excitement about driverless cars with the Google Waymo project [8], but, in the wider scale, this should be seen as an end goal for the wireless automotive connectivity rather than a feature available in every new car from the beginning. A more realistic starting point for mass adaptation is assisted and co-operative driving, where wireless technology can provide various indications, warnings and control features to the driver. Another area where wireless connectivity can really make a difference is in tele-operated driving. With this feature, a remote operator can control a vehicle to operate in conditions hazardous to a human driver. Some of the examples will be to remotely operate earth-moving vehicles (bulldozers) in precarious conditions after earthquakes and landslides, to transport coal and other minerals out of deep mines after carrying out explosions and to carry out remote repairs with highly mobile robots in hazardous conditions like chemical or nuclear accidents. Some assisted-driving features regulating vehicle speeds to account for any traffic congestions ahead can remove the need for sudden braking and acceleration and thus reduce the vehicle's environmental impact. If a group of vehicles can be driven together, by maintaining the same distances as a platoon, the aerodynamic impacts on the following vehicles can be reduced, leading to significant fuel savings [9].

The cellular industry is inherently capable of supporting mobility and large area coverage, and hence it is uniquely placed to support the automotive sector. The assisted driving and other related applications need connectivity between vehicle to vehicle (V2V), vehicle to roadside network (V2N), vehicle to pedestrians (V2P) etc. Collectively, all these connections are grouped as vehicle to everything (V2X). The initiative to provide V2X through cellular-is termed as Cellular-V2X (C-V2X). The main technical requirements in C-V2X applications is to ensure ultra-reliability and low latency (URLLC) and many of the solutions discussed today aim to achieve this. Common thresholds being aimed at are referred to as the '5 nines' reliability (99.999%) and sub-millisecond (less than 1 ms) latency.

3GPP has been active in developing the framework for standards to support V2X services for a number of years. The initial standard work for V2X services was completed back in September 2016. The 3GPP V2X support is so far developed in two phases. The V2X phase 1, completed in March 2017, is part of the LTE Release 14. This V2X phase 1 set of standards is sufficient to provide support for basic road safety services. Through this implementation, vehicles can exchange information about their driving status with other vehicles and roadside infrastructure through a sidelink. The V2X phase 2 introduced in 3GPP Release 15 introduces further enhancements to this sidelink with features such as latency reduction. This is again based mainly on LTE and requires the co-existence with Release 14 LTE UEs. The current Release 16 NR V2X work is deemed as phase 3 and will support advanced features with the inherent capabilities of 5G-NR [10]. 3GPP has identified four groups of use cases – vehicle platooning, extended sensors, advanced driving and remote driving – to be supported with the standardisation in this phase 3 [10].

One of the critical steps in building a wider ecosystem to support and develop the 5G Automotive vertical was the formation of the 5G Automotive Association (5GAA) in September 2016. The 5GAA is a global, cross-industry organisation between the ICT and Automotive manufacturing industries. Currently, there are more than 110 members in 5GAA, representing tier 1 mobile product suppliers, automotive manufacturers, mobile operators, chipset and infrastructure vendors. The stated mission of the organisation is to develop end-to-end solutions for future mobility and transportation service sectors [11], where this wider technology area is known as co-operative intelligent transport systems (C-ITS). The 5GAA is of the view that 5G is the best platform to enable C-ITS, mainly through the use of C-V2X. 5G will be able to better carry mission-critical communications for safer driving and further support enhanced V2X communications and connected mobility solutions. The 5GAA is composed of five working groups, looking at use cases and technical requirements, system architecture and solution development, evaluation, test-beds and pilots, standards, spectrum and business models and go-to-market strategies. In mid-2019, the 5GAA and global certification forum (GCF) agreed on a framework to combine their resources to accelerate the certification of C-V2X products [12]. This C-V2X certification, from the initial 3GPP LTE Release 14 (detailed above) compliancy to the latest standards will help to accelerate the 5G product introduction to the automotive vertical, bridging a vital gap in this ecosystem.

A schematic representation of a C-V2X-enabled connected vehicle system is shown in Figure 12.2. The deployment architecture involves connected vehicles, roadside infrastructure and pedestrians (with Smartphones) connected through the cellular RAN and also the ability to directly connect amongst themselves. In 3GPP terminology, these direct connections are enabled by a side-link supported by the standards. The RAN connects to the generic cellular core and a special mobile edge computing (MEC) server located with the core, which runs critical applications some of the time. By having the MEC server at the network edge, some of the latencies in accessing a cloud application server can be removed. The cloud application server is used to support many of the less time critical V2X applications.

This diagram only shows those direct connections the vehicles can have with the network (V2N), other vehicles (V2V), the road infrastructure (V2I) and pedestrians (V2P), as needed for time critical applications. Equally, connections can be derived through the cellular network, for example, vehicle-to-network to infra-structure (V2N2I), for certain applications which are not so time critical.

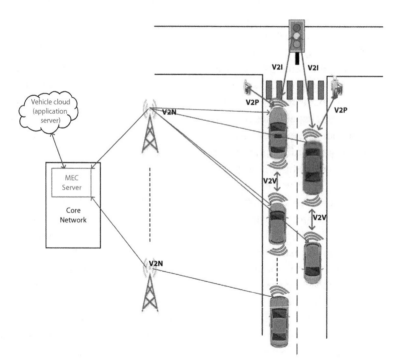

Figure 12.2 A schematic diagram for C-V2X implementation.

Pre-dating C-V2X development, there is another potential technology being touted for V2X applications, based on the Wi-Fi standards version IEEE 802.11p [13]. This technology is known as DSRC (Dedicated Short Range Communications) and, in Europe, it is termed as ITS-G5. Currently, DSRC has an advantage in terms of technology maturity over C-V2X. However, there is no clear global alignment, either with policy makers or car manufacturers, as to which technology is to be adopted. In Europe, both technologies are currently being considered. China has decided to adopt C-V2X, with a commitment to require all vehicles to be equipped with these chipsets by 2025. In the US, and in Japan, DSRC-based safety features have been contained in some of the new built vehicle models since 2017. Both these technologies are targeting the 5.9 GHz ITS band which, in many countries, and regions has now been designated for the use of V2X communications. Many of the chip makers are opting to embed both these technologies in a single chip, but this may drive the costs higher.

Despite this competition between C-V2X and DSRC, and ambiguity in some regions about the technology direction, potential growth forecasts for this sector are significant. One market study predicts that by 2022, the V2X market will be worth $1.2 Billion, with an estimated 6 Million vehicles worldwide equipped with this technology [14]. V2X capabilities will continue to grow and be adapted onto a majority of new vehicles in the coming years, making driving a much safer, enjoyable and less polluting experience.

12.2.2 Smart City

Smart city envisions the application of ICT technologies and artificial intelligence (AI) to solve some of the most critical challenges facing our urban landscapes today. These include reductions in road congestion and air pollution by vehicle emissions, the efficient use of gas, electricity and water utilities in the city, efficient waste management and recycling, using technology for law enforcement and crime reduction, tracking assets moving around in a city etc. Many of these applications rely on large-scale deployment of sensors and regular data collection and related data analytics. In this space, current and future cellular technologies can play a major role. The rollout of Smart metres in the UK to report back on electricity and gas consumption in each household is a recent example [15].

In 5G space, many of these applications fall within the massive Machine Type Communications (mMTC) domain. Individual data rates will be low (with a sporadic transmission period, sometimes called low-duty cycle) and the time latencies required will not be critical, but there will be millions of sensor-type devices sending back data. The baseline 5G target for mMTC is to support 1 million devices per square kilometre area. Thus, the connection density will be very high for mMTC applications. From the device perspective, most of these sensors will run on battery power and extending the battery life through energy efficient transmissions will be a key requirement. Although the battery itself will not be very expensive in a sensor, many sensors will be positioned in difficult to access places and the labour cost to replace these batteries will be high. Typically, in these applications, a 10-year battery life is expected. This type of connectivity is generally termed as Low Power Wide Area (LPWA) communications.

In terms of providing mMTC capabilities, 3GPP has developed two key standards, Narrow Band IoT (NB-IoT) and LTE Machine type (LTE-M). NB-IoT is geared to provide very low data rates per device, Whilst LTE-M can deliver higher data rates, for example, to transfer images and video clips. The NB-IoT standard was first introduced in Release 13, and both these technologies are expected to provide mMTC capabilities well into the 5G era. One of the key technical modifications, in terms of extending the battery life for mMTC applications, is to extend coverage range by narrowing the spectrum band and utilising more robust modulation and coding schemes. Typically, the NB-IoT is designed to operate with 180 kHz bandwidth and can support 20 dB more allowable path loss than the other cellular technologies. The LTE-M will operate in a 1.4 MHz bandwidth (a further variant operating in 5 MHz bandwidth is also available) and will provide 15 dB extra allowable path loss to extend coverage. This additional path-loss headroom will be vital in supporting sensors in low radio visible places like basements and road tunnels in a cityscape. There are also other LPWA standards like LoRa and Sigfox, that are designed to operate in the un-licensed spectrum [16].

Another important requirement for Smart city mMTC is security, particularly for applications involving key utilities, infrastructure and sensitive data. Some security risks identified in a key report on LPWA security are [17]: lack of bandwidth and data rates in LPWA technologies limiting the supported security features, the sporadic connection to the network inhibiting the timely security updates, the need to authenticate devices, subscribers and the network, the overheads to use data encryption and cryptographic keys etc. Cellular-based technologies using licensed spectrum inherently have higher security features and NB-IoT and LTE-M have been designed with this security aspect as a cornerstone.

Both NB-IoT and LTE-M technologies are designed to operate in-band of the current LTE and beyond cellular networks, utilising about 2% to 10% of spectrum or in the guard bands between cellular frequency allocations. The current LTE/5G base stations can be made compatible with both these standards with software upgrades, so there are no huge upgrade costs.

It should be noted that LPWA technologies are not limited to Smart city applications. Applications for rural areas in agriculture, aquaculture, transport and forest maintenance can very well be supported by LPWA. But it is in the Smart city domain that the majority of applications and business traction are envisaged. Reciprocally, Smart cities also employ a full suite of ICT technologies to yield full ICT potential, in dealing with some of the most pressing issues. Some of the early examples include the Smart Nation program in Singapore initiated in 2014, which includes an Intelligent Transport System with real-time traffic information and accident alerts and Smart homes with Smart electricity and water meters in the city state [18]. There is a similar effort in Manchester, UK through the cityverve project, which is building a Smart city demonstrator focussing on the four key areas of health and social care, energy and environment, travel and transport and culture and the public realm [19].

Smart city IoT deployments will be hugely impactful through Smart transport, as a prime example. In many of the world's cities, the main mode of transport is through private vehicles and this creates a lot of resource wastage and environmental impact through congestion and emissions. Some studies have suggested dynamic road pricing for private vehicles, enabled through sensing and predicting the traffic levels in a particular road [20]. Also, Smart sensing technologies can be deployed to better predict and inform the commuters of public transport services of the travel options and schedules, increasing the trust and usage of these services. Effective asset-tracking technologies of green transport options like bicycles, e-bikes and e-scooters can guide potential users quickly to the nearest available item, which significantly increases their usability. In many of the major cities, tourism is an important contributor in terms of revenue and employment generation. Tourism can massively benefit from novel interactive information feed technologies (through AR/VR) and tracking of tourists' visiting, transportation, dining choices (through anonymised data) etc. All these services are just a snapshot of what can be made available within Smart cities. An illustration of various possible services in a Smart city given below in Figure 12.3 (from [21]).

In all these examples, it is clear that multiple stakeholders have to join forces to make them work. The municipal councils, cellular operators, equipment vendors, specific service providers and sometimes specialised data analytics companies all have to play a part to make these successful. In terms of modelling the economic and business viability of these Smart city solutions, both the direct and indirect benefits in the long run have to be considered. For example, an investment now to deploy Smart CCTV with video analytics can cut crime rates and, as an indirect result, may boost tourist numbers 3–4 years down the line. The total return of investment (ROI) should also consider such indirect benefits in the long run. Analysis of the economic viability of Smart city applications has shown that for many use cases, the average payback time for the investment can be around 5 to 7 years [22], which is longer than for many other technology related investments. This is where the policy makers of government and local government sectors need to step in to take on some of these investment risks, for the wider benefit of all city dwellers in the longer term.

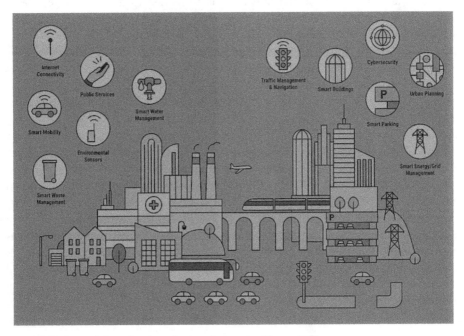

Figure 12.3 Potential service domains in a Smart city. (*Source* [21]).

12.2.3 Industry 4.0

Industry 4.0 refers to the fourth revolution within the manufacturing and wider industrial sectors. In order to understand what Industry 4.0 will bring, it is useful to briefly touch upon the previous three industrial revolutions. The first revolution was from 1760 to around 1840 and came about with the introduction of steam- and water-powered machinery. This enabled many labour-intensive repetitive tasks to be executed with these machinery, increasing productivity of the factories and workshops. The second revolution came about with the invention of electricity, telegraphic communications and widespread use of rail networks. With electric machinery, the forerunners of modern production lines were introduced. Telegraphy helped much quicker communication whilst the railroads enabled faster intake of raw materials and distribution of produced goods. This second industrial revolution lasted till the beginning of the First World War. The third industrial revolution or the digital revolution started around the 1960s. The use of computers and computer-aided machinery, faster communications through telephones and faster goods/raw material transfer through air cargo and container ships were the hallmarks of this revolution, which enabled the massive scale industry productions with global reach we are seeing today.

Industry 4.0, or the fourth industrial revolution, will be characterised by the ability of industrial machines, robots and production processes to communicate amongst themselves and with the wider internet and make automated decisions with AI [23]. This type of extensive M2M communications will enable the production, factory maintenance, supply chain and inventory management and many other processes to be fully automated. With the use of AI analysing the vast amount of collected sensor data in these cyber physical systems, the above processes can be optimised in many different dimensions – in terms of cost, profitability, environmental impact, time to market to name but a few. With Industry 4.0, it is even plausible to predict zero-touch industrial systems in future, where significant human interventions in the processes will barely be needed.

Within this background, it is useful to analyse what kind of value 5G can add to emerging Industry 4.0 'factories of the future'. Currently, many automated functions in the factories are controlled through wired connections. Moving onto 5G wireless connectivity will open up a vast array of new opportunities in this domain. Even highly mobile machines and products, wearable devices, delivery vehicles and forklifts inside and outside of the factory can all be seamlessly connected with a 5G solution. Also, the very high requirements on the reliability, latency, the number of connections and the data rates for these individual service classes can effectively be met by creating different end-to-end network slices supported by 5G. Some of the extreme reliability examples for production line machine parts and robots reach 99.9999% (6 nines) and the latencies reach sub-1 ms levels and these can be met by future 5G-URLLC. Simultaneously, there can be thousands of sensors deployed in the factory to provide regular readings of process monitoring and operational conditions like temperature, humidity, air quality etc. and these will need 5G mMTC. For AR/VR applications related to manufacturing, very high data rates reaching Gbps can be supported by 5G eMBB. All the data collected through the sensors can be fed to application servers in the cloud or edge-cloud servers, where data can be processed with other wider supply/demand parameters (for example) to derive guidelines for production planning. The holy grail of Industry 4.0 will be to seamlessly forecast the production planning (through AI-based techniques), to execute production to match the forecasts through automated factory processes and control supply chains and to manage the automated distribution of goods as inter-connected steps in a single process.

Figure 12.4 (from [24]) illustrates how 5G can help the factory of the future.

Technically, to meet all these challenging KPIs, private dedicated 5G networks covering the factory and its vicinity must be deployed. The coverage, capacity and access rights of these private networks can be tightly controlled to provide guaranteed reliability and availability of the services and also ensure security. We have touched upon service reliability before, but not on service availability. Service availability can be quantified by the percentage of time that the network takes to fulfil all the KPIs demanded by the service. In private factory networks this is easier to manage, as the network can have more control on how its radio resources are deployed and utilised. Also, 5G edge clouds with application server functions can be designed close to the core of these private networks, so that end-to-end latencies for the time critical applications can be reduced. The edge cloud could hold all related applications and relevant data, so decisions driven by AI can be derived on a wide set of industrial procedures as described previously.

In building a wider ecosystem to promote 5G based Industry 4.0 development, the 5G Alliance for Connected Industries and Automation (5G-ACIA) [25] was formed in 2018. It brings together key industry players in the ICT domain and manufacturing and process industries. Born out of an initiative from the German Electrical and Electronic Manufacturers' Association (ZVEI) [26], the 5G-ACIA now has around 50 members from major ICT and manufacturing businesses. The stated mission of 5G-ACIA includes that the

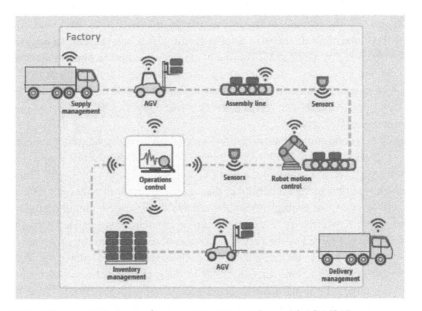

Figure 12.4 5G connectivity for the 'factory of the future'. (*Source* 5G-ACIA [24]).

interests and particular aspects of the industrial domain are adequately considered in 5G standardisation and regulation as well as the on-going developments in 5G are understood by, and transferred to, the industry domain.

The 3GPP has been active in developing the framework to standardise 5G communications for industrial automation. In 2017, 3GPP produced TR 22.804, which looks at system aspects in a study on Communication for Automation in Vertical Domains. More recently, there is an on-going RAN1 study item on channel modelling for indoor industrial scenarios. In November 2018, 5G-ACIA was recognised as a Market Representation Partner (MRP) by 3GPP. This recognition will help collaboration between the two entities to drive future standards support for Industry 4.0 applications. The first steps towards this are seen in Release 17 (with RAN1 work due to start in Q1 2020), where specifications to support Industrial IoT with URLLC requirements will be developed.

A recent report suggests that the Industry 4.0 market will be worth $4 Trillion in year 2020 but at the same time, significant investments will be needed to realise this potential [27]. Considering developments in the market in the recent past, we will see a lot of movement in this vertical domain both in standardisation and deployments on the ground in the next few years.

12.2.4   Critical Communications Sector

Critical communications cover the communications networks to support the blue light services (fire, ambulance and police) and is also known as Public Safety Communications. Traditionally, these emergency communications relied on specialised technologies like TETRA and PS-25, which offered higher reliability and extended coverage (mainly due to the lower frequencies and narrower bandwidths they use) than commercial cellular technologies [28]. However, these specialised networks, collectively known as LMR (Land Mobile Radio) lack the capacity to support very large incidents. This was made evident in the immediate aftermath of the 9/11 attacks in the US, when the emergency networks became overloaded due to the sheer volume of emergency calls. As one of the recommendations from the subsequent inquiries, the US embarked on a long mission to move the emergency networks throughout the country to a single network, underpinned by the commercial LTE infrastructure. The result is the FirstNet network [29], which has now gone live in a number of US states.

Moving critical communications to cellular-based networks also has other benefits. With the rate of development in 3GPP technologies, high-speed data services and advanced Smartphone Apps are now commonplace in LTE and in the upcoming 5G-NR. The specialised emergency networks lag far behind in these capabilities, with only basic data services possible with TETRA, for example. Also, the costs of deploying and maintaining cellular-based emergency networks are less than the capital needed for specialised networks. The economies of scale, enabled by the very high volume of cellular equipment/device vendors and the widespread human expertise available in these technologies play a significant role in enabling this cost reduction. Also, the RAN infrastructure on the commercial cellular networks can largely be re-used for emergency communications. Due to these reasons, many countries are now looking at shifting their entire TETRA-based emergency networks to run on LTE-based commercial cellular infrastructure. The UK is making the first European move here, with their Emergency Services Network (ESN) due to replace country wide TETRA networks by 2024 [30]. The LTE network will have additional base stations and use of additional spectrum to enable the country wide coverage and increased reliability. Also, specialised LTE handsets will be provided to the first responders, that can operate under challenging conditions.

The 3GPP standardisation effort to support critical communications through LTE started with the development of Mission Critical (MC) services from Release 13. A new System Aspects group (SA6) was created to handle the related efforts. The MC work was carried out in several stages and, with each stage, a complete specification was created so the public safety community could make use of these specific services. MC-PTT (Mission Critical – Push To Talk) service specifications were completed in March 2016, as part of Release 13. This MC-PTT service was developed to provide the full capabilities of already existing LMR PTT services. The main features, supported by MC-PTT, include the group calling facility and use of Proximity Services (ProSe or Device to Device communications) where two MC-PTT-enabled devices can communicate directly either with or without network support. The MC-PTT services are continually developed, with enhancements coming in Releases 14 and 15 so far. The MC-data and MC-video services were introduced in Release 14 and further enhanced in Release 15. The MC-data service features in Release 14 included the downlink unicast and multicast, store, forward and reporting functions on and off network. In Release 15, the short data service (SDS), which is confined to 1 PDU is introduced. The MC-video features introduced with Release 15 include the ability to mutually adapt the codec and other parameters of the two devices in a video call, and the facility to generate emergency private video calls. All MC services are protected by security features as specified in the MC-core [31]. There is an emphasis in the MC service development in Release 16 and beyond to include the 5G features as well as to enable inter-operability for the MC services with existing LMR services. A good overview of the MC-PTT services and some of the deployment examples are provided in [32].

Within this context, it is very relevant to see what capabilities 5G can bring to the critical communications sector. The 5G eMBB capabilities can be utilised to provide interactive 3D maps of a building under a fire, to the emergency crew that will enter the building, for example. In the uplink, the emergency crew can upload high-definition video streams from helmet-mounted cameras, which will enable the command and control centres to assess the situation more accurately and can also be a vital part of evidence gathering. The URLLC capabilities can be extremely useful in remotely controlling robots and tele-operated vehicles, which can perform operations under hazardous conditions. Fixing a toxic chemical leak or even a leak in a nuclear power plant remotely are some of the examples where the low latency and reliability of 5G can be very useful. The mMTC capabilities allow extended coverage, typically enabling UEs or devices to operate with narrower bandwidth and utilising very robust modulation and coding schemes (please refer to Chapter 6). The NB-IoT standard we discussed in Section 12.2.2 for example, can tolerate up to 164 dB of path loss, which is typically 20 dB higher than other cellular standards geared for mobile data. This can be very useful in search and rescue missions in mountainous and remote areas, where the network will be able to detect even very low signals from UEs or devices of a missing person.

As seen in many examples with the previous generations, providing new capabilities to the end users result in rather surprising, innovative new applications being developed and this will also be true for critical communications. The mere possibility that 5G-enabled devices of the emergency crew can opt into receive inputs through Apps from the public and the potential victims of an emergency scenario open up so many new opportunities for vital information gathering. The end-to-end slicing capabilities envisaged as native capabilities of 5G networks will also be very useful for critical communications. Network operators will be able to create highly secure and reliable network slices for emergency operations in a dynamic manner (as the ground emergency conditions dictate), which will be much more economical than deploying dedicated hardware resources all the time for critical communications.

As much as the critical communications sector would benefit from early introduction of 5G capabilities, there are many obstacles to overcome. In the early deployment stages, 5G is very likely to be confined to certain capacity hotspots in urban areas. The NSA (Non-Stand-Alone) deployment model discussed in Section 5.8 is primarily aimed at these kind of deployments. The isolated nature of 5G hotspots will seriously inhibit the applicability of such a 5G network to support critical communications. The emergency services would want a particular 5G application to be available across their service areas. For example, a fire brigade would want this 5G connectivity throughout a city or town they are covering. Also, the likelihood that 5G coverage will initially be focussed on urban areas would make search and rescue services typically operating in rural, mountainous and coastal areas out of reach.

However, there are novel approaches to bring 5G to the emergency services with certain coverage guarantees. A potential solution using drones to extend the 5G hotspot coverage across a city area is presented in [33], from the ONE5G EU research project. This solution depends on a network of centralised 4G and 5G small cells (SCs) upgraded to carry the additional fronthaul and backhaul capacity from drone wireless links. The drone network itself is centralised with drone radio heads carrying the minimum radio functions (to be lighter and more power/cost efficient) and the processing carried out in a central BBU (Base Band Unit). Drone links are provisioned to support a minimum data rate of 665 Mbps (for shared use of uplink and downlink in the dynamic TDD mode) and the fronthaul and backhaul links are dimensioned accordingly. A block diagram of this deployment model is shown in Figure 12.5 (from [34]).

Another potential issue when supporting critical communications with cellular (both in 4G and potentially in 5G) infrastructure is that if an eNodeB loses its connectivity to the CN, all radio connections in that cell will also cease to function. It can be common in major incidents (like floods, earthquakes and cyclones) for the wired or wireless backhaul/fronthaul connectivity from the eNodeBs to the core/central processor to get disrupted. Currently, there are many solutions being promoted in the 5G domain for this issue, mainly through use of multi-access edge computing (MEC) techniques. The 5G ESSENCE EU project [35] proposes the deployment of an MEC node as an edge data centre (edge DC) which can run the data plane and control plane of the connected SCs (Small Cells) when required. In case of a loss of backhaul connection to the core, the SCs can be configured to have the control plane orchestrated from these edge servers, thus enabling the SCs to support at least the users connected to this cell. Also, deployment of edge DCs allow the emergency communication applications to be run from these to significantly cut down on the latencies.

The business and economic aspects of transitioning LMR-based critical communications networks to LTE and 5G largely depend on two possible transition models, as noted below.

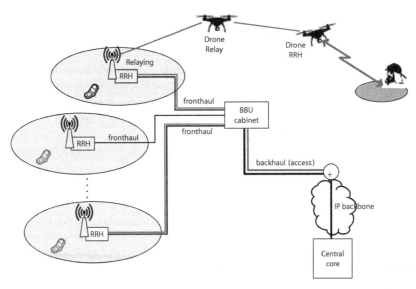

Figure 12.5   A Deployment model for extending 5G coverage for emergency service usage [34].

In the previous examples stated in the US and UK, the entire LMR networks are in the process of being replaced by LTE-based systems. These are massive overhauls, involving a number of government agencies, network operators, specialised device and software vendors and incumbent LMR service operators. The scale and complexity of this overhaul has meant a number of delays in the UK for example. However, the overall economic benefits of these exercises are still significant. The UK Home Office, which oversees transition of the TETRA network to LTE, forecasts net savings of over GBP 200 million per year [30]. Once the painstaking exercise of LMR to LTE transition is completed, the introduction of 5G services on top of this would be relatively straightforward. In another transition model, some countries, like Germany, are planning to introduce new LTE/5G MC services to co-exist with the incumbent TETRA networks for a number of years into the future. This model would mean less disruption and lower transition costs for the network infrastructure in the near term, but the control rooms and dual technology devices would get more complex. Which of the transition models would be more effective is open to debate, but both models will open the way for 4G and 5G technology to enter into this sector.

12.2.5   Other Vertical Areas under Development

There are some other vertical areas where 5G promises to provide enhanced services and improve operational efficiencies. One such area is eHealth. With ageing populations in many countries, governments are having to spend more and more on elderly social care and healthcare. In this context, eHealth and particularly mobile health can improve care quality and, at the same time, reduce operational expenditures. One of the early white papers [36] points out that eHealth and mobile Health (mHealth) solutions can shift the balance of healthcare provision from a very centralised hospital-based model to a more distributed model giving greater emphasis on home care, GP surgeries, day care clinics etc. Importantly, eHealth solutions can allow elderly people to live longer and independently in their own homes, with wearable sensing and monitoring devices to detect any medical emergency or accident. Various health apps on mobile phones are becoming increasingly popular, which can help people track their diet, exercise levels and also key health-related indicators such as heart rate and blood pressure. With non-communicable diseases like diabetes needing long-term care reaching alarming levels in many countries, such apps can keep the population healthier for longer.

In many other vertical areas we looked at before, there is a physical concentration of the service area where targeted 5G connectivity is needed. The eHealth and mHealth applications are most useful when connecting dispersed patients and communities with the required health services. Therefore, 5G will have to provide the specific KPIs needed on a per connection basis, wherever the mHealth application is run. One of the main concerns in this domain is maintaining privacy and security of connectivity and the collected data. In terms of 5G standardisation, 3GPP has been looking at the eHealth requirements (termed as Telemedicine) since Release 14. It was identified as a potential 5G use case in [37].

Smart grid is another vertical area that has attracted attention in the current 5G push. Some publications tend to treat Smart grid within the wider Smart city domain, but there are some significant variations. Smart-grid applications can manage and control electricity demand and supply networks and these applications need very fast response times, in contrast to the normal mMTC/IoT and their Smart-city applications. Through dynamic sensing of demand (network load) the supply can be automatically matched to this demand, and Smart grid can also respond quickly to any abnormalities in network load or supply points. This is vital for managing complex grids with many modes of power generators and a variety of customers cost effectively. Also, many of the future buildings will have the capability to generate green energy, through solar panels or wind turbines. The Smart grid will help these buildings sell the generated excess electricity to the wider network and also buy electricity when its own demand outstrips the internal green electricity supply. This kind of dynamic supply/demand tracking will reduce the need to install expensive, large battery banks and help tackle climate change through greater encouragement for renewable power generation options.

5G, with its capability to create end-to-end slicing to match demanding KPIs, will be an ideal platform to support the connectivity needed for Smart grids. Using a mature (widespread) 5G network to connect the many dispersed distribution nodes and the power generators would be much more cost efficient than deploying a dedicated, private optical network. A commercial feasibility study using 5G network slicing to support Smart grid was carried out in China as reported in [37].

In all these vertical areas we have looked at, the use of 5G will bring tangible benefits right through the value chain down to the end customer. It will also cause significant disruption to the existing operation/business models and it will be a challenging task to manage the transition in these very large scale vertical sectors. Our guess, like many other experts' who have written on this topic, is that 5G will only signal the introductory phase of these transitions. The new challenges identified and lessons learnt from this exercise will help to formulate 6G, which hopefully can complete this mammoth task.

12.3   The Continuous Evolution of the Mobile Device

We have looked at the likely changes in the mobile network in terms of supporting differ-ent verticals in the previous section. Here, we will discuss how the mobile phone will likely evolve in the next few years, as the consumer interface to 5G and 5G Advanced networks. The mobile phone (or the handset) has come a long way from the basic phones supporting only voice and text in the 2G era. The advent of the Smartphone was a critical juncture of this development and it enabled the rapid uptake of mobile data services and mobile Apps. We briefly discussed the impact these Apps are having on facilitating various services on the go in Chapter 1. Let's analyse what the Smartphone may bring in the 5G era.

Today's Smartphones pack more computing power and memory than most of the early laptop models. They provide numerous capabilities such as a digital camera, an audio and video player, a gaming device, a navigator, a mobile payment device etc. Consumers have loved this integration of features into a single device so much that sales of compact digital cameras, mp3 players and in-car navigation units have now declined significantly. The quality and user experience of these features keep advancing year on year from one version of a Smartphone to the next. Today's Smartphone not only provides cellular connectivity, it is also capable of communicating through Wi-Fi, Bluetooth and NFC (near field communi-cations) and even UWB (Ultra Wide Band), making it a multi-RAT device.

A key ingredient of the Smartphone is its mobile operating system, which can manage and optimise the functionalities. This includes managing the numerous Apps that run on the Smartphone and maximisation of the battery life. There are two dominant mobile operating systems today – the Apple iOS and the Google Android, with over 99% of mar-ket share [38]. The iOS is exclusively used for Apple phones and enables seamless con-nectivity with other Apple devices, the Apple App store and iTunes music store. This is a 'walled garden' market model, where a closed ecosystem provides all the hardware and software. The Android OS, on the other hand, is an open system, where the OS is licensed out to many Smartphone manufacturers to develop the hardware products, with a cen-tral Google play App store enabling a market place for Apps. Due to the diversity of the Smartphone manufacturers, the Android phones come to the market at a wide range of price points, from the high-end Samsung phones down to basic Smartphones at a frac-tion of the high-end prices.

Both these OSs have significantly evolved over the years, making the user experience ever more seamless and knocking out a few other competitors along the way. The Windows phone OS and the open source/open community development-based Ubantu Touch are two such OS products that found it difficult to gain traction in this market [39]. As well as the user experience provided by an OS, the acceptance level it will have with the App devel-opers is crucial for its success. Nearly all of the millions of Apps available today are devel-oped either for iOS, Android or both of these systems. Both platforms provide ample support and incentives (mostly with the size of their customer base) to the App developers to remain with them. It is very hard to imagine another mobile operating system coming along and cracking this iOS/Android dominance in the 5G era.

What real breakthrough innovations can be expected in the 5G and beyond Smart-phones? For some time now, the personnel assistants residing in Smartphones has been a feature where you can extract relevant information with a voice command. Now, with the addition of AI to these assistants, much more complex requests can be handled and information can even be made available to suit the user context (including location) even before a request. The Smartphone can gather a plethora of data about user behaviour and preferences (with the user's permission) and use machine learning (ML) and AI algorithms to match user traits with certain choices the user would be likely to make, such as selecting a particular restaurant for dinner. The AI features would enable seamless interactions with the digital assistants, making many of our daily decisions much more convenient.

The emerging 5G phones will undoubtedly offer more appealing applications that will benefit from the higher data rates and lower latencies of 5G. One of the potentially new applications that can benefit from these capabilities is 3D holograms [40]. The 3D image can be created on top of the Smartphone screen. Particularly for 3D hologram videos with smooth transition between frames, peak data rates in the Gbps range will be required, which can push even the 5G Advanced system capabilities to the limit. For a mobile generation, there needs to be a 'killer' application that really tests its performance to the maximum and, in turn, fuels a momentum of growth. For 5G eMBB, the Smartphones with 3D Hologram capability could well be this killer application.

These 5G capabilities also offer the opportunity for more dramatic changes in Smartphone architecture. Just as with the centralisation of the radio network, it may be possible to centralise most of the processing in the 5G Smartphone and run these in the cloud. The physical phone itself can be a 'dumb' device, just an interface to the user. Provided that 5G coverage becomes widespread, it will be possible to operate this phone in the 'Smart' mode wherever the required data rates and low latencies are assured by the network. The cost of the 'dumb' terminals will be lower than today's Smartphones and also the battery life will increase due to less processing being required. A current trend in Smartphones is to double the screen size with foldable devices and this pushes both power consumption and the prices higher. The 'cloud phone' providers will also benefit from economies of scale in processing many phones in a single cloud data centre. One stumbling block, even when 5G becomes widespread, could be the huge data consumption needed per user to operate this model. In 5G however, there are unlimited data plans emerging now and this could remove the burden on the consumer of having to be contained in a fixed data bundle. After years of research and development, Google has just released a cloud-gaming console called Stadia [41], where all the processing happens on the cloud and the user needs at least a 10Mbps connection and a 'dumb' handset for a true gaming experience. This could be a good test case for a future 'cloud phone' running on 5G connectivity.

In terms of the Smartphone market, sales figures of the leading handsets have been falling gradually or becoming stagnant for a good few years now [42]. Smartphone users are not upgrading their phones at the rate they used to at the start of the Smartphone era. One reason being put forward for this trend is that the changes in Smartphones over recent versions have all been evolutionary and incremental. Nothing like a killer new feature has emerged in recent years to capture consumer imagination. Perhaps 5G will provide the right technological conditions to stimulate the next revolution in Smartphones.

12.4   What Will 6G Look Like?

Having come to the end of this book on mobile technology and business trends, we are left with the question, 'What will 6G look like?' Throughout the book, a number of potential candidate technologies and business sectors have been detailed in this futuristic context. Here, we recap some of these and offer further thoughts on what the next generation of cellular networks may resemble. Of course, we must state the caveat that we do not claim to be crystal-ball gazers. Nevertheless, the musings included in this final section have been collated through several years of discussions with academic and industrial colleagues and, thus, we feel we can claim relatively safely that they are fairly representative of the most probable outcome of future cellular developments.

One general forecast that many experts are making on 6G is that it will be a 'fixer' for the nascent capabilities that 5G will introduce, particularly in the URLLC domain. As yet, 3GPP has not addressed the URLLC requirements comprehensively in its 5G-NR releases. Also, the mMTC support is currently limited to what was introduced in Release 14 for NB-IoT and LTE-M. These capabilities will need significant enhancements in 6G, which may even require fundamental changes to the 5G architectural framework. There are some historical similarities of this connotation for 5G and 6G. The basic mobile connectivity was introduced in 1G but it was 2G that really satisfied the user experience. Similarly, 3G introduced mobile broadband, but it needed 4G to fully realise this potential. If the same trend is anticipated, the multi-dimensional communications introduced in 5G will need 6G to fully establish these capabilities.

It will be safe to say that, from the very start (whenever that may be), 6G will have to support predominantly machine-generated traffic over human-generated (through Smartphones or other 'human' consumer devices) traffic. The nature of MTC traffic will be very different to the human-generated Smartphone traffic network operators are used to today. The MTC and URLLC emergence in 5G will give network designers a fairly good idea how networks need to adapt and what KPIs need to be optimised as these traffic flows progressively increase. Thus, 6G networks should be optimised for this machine-type traffic dominance from the very beginning.

Below we will look at some of the emerging technologies that can truly shape the outlook of 6G.

12.4.1   Machine Learning and Artificial Intelligence

ML is a term that encompasses a broad class of supervised and unsupervised learning algorithms. Similarly, the moniker AI has often been used in the context of communication systems to refer to methods of training and using deep-neural networks to achieve some optimisation goal. Tools based on ML and AI have gained considerable momentum in the communications and signal processing research communities over the last two years [43]. A number of mobile vendors and operators have claimed to have an interest in the use of these methods for future transceiver design and network optimisation. At the time of publication, it is fair to say that most major industrial organisations in the telecommunications field have some sort of department or research activity devoted to ML/AI.

Fundamentally, ML/AI can be applied to obtain near-optimal estimates of some desired quantity given sufficient input data is available in the form of training and observations. It should be stressed that the power of these techniques becomes apparent when (a) such data is available and (b) an accurate analytical model of the system in question is not available. To determine the value of ML/AI in the communications and networking context, one must consider whether these two conditions are met.

Let us explore this further in the context of physical layer communications. Much of the work undertaken in this area over the past few decades has been aimed at developing accurate models that can be used for performance optimisation. Hence, condition (b) is often not met in this setting. As a result, to date, very few applications of ML to physical layer communications have been proposed. Example applications that have been considered include the likes of MIMO detection, based on deep learning principles, and OFDM detection with short (insufficient) cyclic prefixes in rich scattering environments. Many studies such as these are concerned with the 'promise' of ML/AI technology rather than actual benefits relative to traditional methods. Indeed, the usual documented drawback of learning-based approaches in the physical layer is high implementation complexity. Other deep learning approaches to physical layer design have also been considered, ranging from modulation classification to treating the communication system as one deep neural network that can be trained in an unsupervised manner to optimise a target objective (e.g. block-error rate). This approach is relatively new and is referred to as 'autoencoding'. Some recent studies have demonstrated that it does not necessarily provide an advantage in some practical cases, and it can be susceptible to adversarial attacks [44]. As security is a major issue in modern communication networks, it remains to be seen whether autoencoding will receive a high uptake.

For network layer practitioners, the use of ML/AI begins to make more sense. Often, the laws of physics cannot be consulted in developing analytical models in these cases. Systems tend not to be stationary, and thus standard statistical approaches do not work well. Yet, data seems to be plentiful. Hence, even engineers that do not have a fundamental interest in learning algorithms have found themselves drawn to the ML/AI world in order to solve their problems. This has led to new thinking with regard to non-stationary systems and how to design ML techniques for them.

12.4.2 Blockchain and the Internet of Things

Blockchain is seen as a secure distributed ledger technology, and as such, is of great interest in high-data IoT networks. It is generally accepted that blockchain will transform B2B transactions [45]. But for technical IoT applications, blockchain will provide a way for sensors to maintain a secure, distributed, causal account of sensor readings and actuator actions, which businesses can see in semi-real time. These could be used for fault diagnoses, process optimisation and automation, as well as security checks. However, blockchain is still vulnerable to cyber-attacks. Indeed, it is impossible, or at least extremely difficult, to modify an entry in the ledger; however, it is easy to add false entries, which might skew statistical measurements taken in sensor networks. Furthermore, the use of blockchain in low-layer communication-related applications is not well understood. Questions around whether blockchain could be useful in low-layer IoT scenarios are starting to be asked, particularly with respect to the (arguably) three most significant IoT sectors: smart energy, health, and transport [46].

With regard to the transport sector, IoT technology has gained interest in the railway industry in recent years. Mostly, the idea is that sensor data (obtained using IoT solutions) can be used for asset management and to reduce OPEX through predictive maintenance and preventative failure prediction. Interestingly, blockchain is also gaining interest, particularly as a solution to managing freight consignments. For example, IoT sensors could be used to track real-time capacity information and log this data in a blockchain for accurate billing. Several issues regarding blockchain-IoT solutions must be overcome before it can be rolled out for transportation. Specifically, trust has been cited as a huge issue in the implementation of blockchain in railway systems. The need for standardisation is also seen to be a significant hurdle, particularly since the rail industry is notoriously slow at rolling out and updating standards. The Blockchain in Transport Alliance (BiTA), which is largely comprised of logistics companies, is driving these standards across the globe [47].

IoT security more broadly relates to the problem of securing IoT devices themselves, which is an important issue, irrespective of whether devices are used in blockchain networks or not. IoT is a nascent market, and developers and product designers tend to forego integrating security features in favour of being first to market. A number of issues inherent to IoT systems cause them to be vulnerable to security breaches. A key problem is the energy constrained nature of IoT devices, which prevents them from running sophisticated encryption/decryption algorithms. Also, most devices are simply set up and left to run until end-of-life; security updates are typically not carried out, or are only implemented for a limited period of time after deployment. A lack of standardisation also prevents a secure IoT infrastructure from being developed. Research into lightweight security measures, such as physical layer security, are currently being developed with a view to improving IoT security.

12.4.3 Evolutions in Cloud and Edge Computing

It has been widely recognised over the past decade that the increasing amounts of data gathered from geographically dispersed sensors and mobile devices will naturally lead to the need for more sophisticated storage and data-processing solutions. These solutions will be required to 1) be energy efficient by not processing data unnecessarily and 2) maximise data value through real-time processing where the generated actionable information has a finite value lifetime. The Cloud paradigm is capable of providing resources that can be used to process these wide and diverse data sources. However, drawbacks exist, with solutions utilising large individual public clouds experiencing limitations once the number and density of data-generating devices grow. These limitations result from communication bottlenecks that manifest when large amounts of data are conveyed from the edge of the network (e.g. the access network) to a centralised single recipient. Growth of this nature is inevitable. This prediction is in line with the IoT concept currently occupying the minds of academics, entrepreneurs, industry leaders and policy makers.

With a view to overcoming the aforementioned problems, one can resort to a broader federated cloud architecture in which, rather than utilising a single public cloud, the data enters a set of inter-operable federated cloud systems providing storage and computing capability with the entry point located nearer to the edge of the network (the so-called *edge cloud*), i.e. close to the devices that generate and use the data. The central Cloud will always be needed for highly complex computational jobs and high capacity storage, but a distributed architecture could reduce latency, improve energy efficiency and increase resilience. Applications for such an architecture include everything from disaster recovery to supporting IoT and cellular services, including those linked to public safety.

The introduction of standards within the cloud computing landscape allows for the creation of borderless federations, with different providers being used, as appropriate for a more nuanced set of requirements by the consumer without worry of alterations to business models and technical interfaces. This also facilitates development of usage or community-specific clouds, their differentiating factors being targeted at higher level services and capabilities. This type of federated cloud model has been successfully deployed, for example, as part of the EGI Federated Cloud, which utilises recognised open standards (e.g. OCCI and CDMI) to enable a smooth transition between providers [48]. As these standards are adopted by commercial providers, auto deployment becomes possible, which would take place across the most advantageous set of resources with a view to fulfilling the complex processing necessary to ensure services are available far beyond what can be supported through single providers. This allows for the exploitation of these technologies in safety critical environments and facilitates deployment of micro cloud services, which support common interfaces with their larger brethren, thus enabling tools and services to be devised independent of the final destination for their data.

With regard to communication and networking aspects of cloud services, a considerable amount of work has been done to develop a concrete notion of a mobile-edge cloud network architecture owing to its origins in the computing and networking communities (layer 3 and upwards). Network function virtualisation has been exploited to invoke edge-cloud services in the access network. It is clear that resource allocation at the lower layers of the protocol stack must also be considered in line with any distributed computing architecture or application. In the context of a distributed or federated cloud network, where the inter-node communication links at the edge are wireless, physical phenomena such as fading, noise, interference and mobility must be treated properly. Recently, compute and wireless resource allocation has been jointly addressed within a structured signal processing formalism through the EU FP7 ICT project TROPIC [49]. However, much more work is required to fully realise the benefits that federated cloud architectures with edge computing capability can offer.

12.4.4 Advanced Hybrid Beamforming

As discussed in earlier chapters, many future cellular systems will leverage millimetre wave (mmWave) transmissions to achieve an optimal balance of rate and reliability. Applications will range from short-range indoor communication to front/backhaul outdoor massive broadband links. It is generally thought that large-scale antenna systems (massive MIMO) will be needed to enable mmWave communication, and transmit/receive beamforming will be applied in order to achieve a high-gain link. In its most general and abstract form, beamforming is achieved by applying optimised phase and amplitude weights to each degree of freedom in the transmitted signal. In traditional

MIMO transmissions, beamforming is performed in the digital domain, before DAC and up-conversion operations are carried out. This approach, by default, requires separate DACs to be used for each transmit branch, which constitutes a substantial portion of the complexity and energy consumption in the transmitter architecture. The other extreme is to apply beamforming weights in the analogue domain. With this approach, a single DAC is required, but flexibility and adaptability in some applications is sacrificed. A more general methodology that has been popularised in the last few years is that of *hybrid beamforming* [50], where the application of beamforming weights is distributed between digital and analogue domains. This method has the advantage that a large number of antenna elements can be employed – thus yielding a large array gain – without sacrificing flexibility. Energy consumption can be traded for rate improvement in a relatively smooth way. With the size of antenna arrays increasing with each cellular generation, it is likely that hybrid approaches such as this will be further developed and implemented in future mobile networks.

12.4.5 New Modulation Schemes

The need for energy efficient modulation schemes that will perform well in wideband channels and at the cell edge (downlink) in future cellular systems has spurred on development of multicarrier/frequency-shift keying (FSK) hybrid techniques. The combination of FSK and OFDM has long been in the making, with early ideas dating back to the 1980s and information theoretical notions going back further to the 1960s. These ideas were eventually applied in the context of ultra wideband (UWB) channels in the early 2000s in what is now known as *multitone FSK* [51]. In this scheme, a fixed number of frequencies are available for transmission, but only a subset of these are used in a given symbol period. Thus, information is encoded in the frequency index set that is employed. These early contributions led to the recent interest in so-called *OFDM with index modulation* (OFDM-IM) [52], whereby a subset of available sub-carriers is transmitted on, and this subset can change from symbol to symbol. Additionally, amplitude and/or phase-shift keying symbols are conveyed on the chosen active sub-carriers in a given OFDM symbol. Hence, with OFDM-IM, information is not only encoded in the usual way (through, e.g. ASK/PSK/QAM), but also through the choice of sub-carrier indices in a given symbol. It has been shown that this approach is capable of enhancing spectral efficiency in some practical system configurations.

Generalisations of OFDM-IM (variable numbers of active sub-carriers and different IM schemes for in-phase and quadrature channels) have been developed in the last few years. Methods of combining OFDM-IM with MIMO have also been devised. A variant of OFDM-IM, known as FQAM, has been proposed as a possible downlink solution for cell-edge communication in 5G networks, and it is thought that variations of OFDM-IM may be employed in 6G systems.

12.4.6 Tera-Hertz (THz) Communications

As radio spectrum becomes scarce for new applications, it has been a common trend to look for spectrum at higher and higher frequencies. This trend will continue into 6G and there has been a steady interest in studying the applicability of Tera-Hertz frequencies (typically classified to be 100 GHz (0.1 THz) and above) for 6G communications [53]. These frequencies do suffer from high attenuation and, being near the visible light spectrum, from shadowing and blocking effects at non-line-of-sight (NLOS). However, there are very wide bandwidths available, and if the RF circuitry can be fine-tuned to work at these frequencies efficiently, there can be many short-range applications. Some of the applications being discussed today include private factory networks which need to cover only short-range work space in factories for industry 4.0 applications. Having wider bandwidths can reduce end-to-end latencies, making THz communications desirable for future URLLC applications, for example. Also, wider bandwidths can support very high data rates, which is appealing for short-range fronthaul and backhaul applications in centralised networks as discussed in Section 5.9.

However, there are significant challenges in making these THz bands workable for mobile and wireless communications. Some specific frequencies in this THz range are highly susceptible to atmospheric attenuation, as the EM energy is absorbed by Oxygen and water vapour at these frequencies. The attenuation peaks shown in Figure 12.6 below [54] indicate the energy absorption effects for water vapour, from a modelling exercise. This indicates a highly frequency selective radio channel behaviour. Also, the design of RF circuit components and RF chips for THz range is still in its infancy. Developing these products cost efficiently, and in tune with mass production procedures, will be key to enable commercial interest in THz communications. Also, air interface design with waveforms and very narrow beamforming capabilities will be needed as pre-standardisation activities. We are still a good few years behind that stage, but seeing how quickly 5G-NR standards moved up the mmWave frequencies, it is reasonable to predict things to evolve quickly. The FCC (Federal Communications Commission) in the US, for example, has now opened up spectrum from 95 GHz to 3 THz for experimental use, offering such licenses for 10 years [55]. There are numerous research initiatives, like the European Teraway project [56], to find solutions to the technical challenges discussed above.

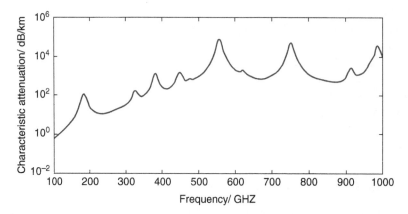

Figure 12.6 The atmospheric attenuation in the THz bands (*Source*: [54]).

12.4.7 Orbital Angular Momentum

Orbital angular momentum (OAM) has been touted by some as a potential disruptive technology for very high-rate applications, such as augmented/virtual reality (AR/VR) and 8k/16k uncompressed video streaming and wireless backhaul. Fundamentally, the technology is very similar to MIMO, in that data can be encoded onto different 'modes', which are theoretically orthogonal [57]. There are some key differences, however. In MIMO, a rich scattering environment is desirable in order to ensure the modes can all support data transmission, which does not lend the technology to line-of-sight (LOS) applications. In contrast, OAM transmitters and receivers should be perfectly matched to ensure all modes are generated and received with high fidelity; this implies LOS is necessary. OAM developments in recent years, particularly by NTT and Tsinghua University, have focused on enhancing data rates by combining MIMO and OAM (NTT demonstrated 100 Gbps over 10 meters in May 2018) and extending the range of OAM transmissions in terrestrial and air-to-ground applications [58]. Due to its LOS nature, OAM may prove useful in next-generation railway communications networks and cellular backhaul links, as described in [59].

12.4.8 Unmanned Aerial Vehicles

Unmanned aerial vehicles (UAVs) have long been popular for military and disaster relief applications. More recently, the advent of small UAV technology has led to a huge interest in using UAVs for agricultural and ecological studies and monitoring. Regulations and policy are lagging, but proprietary systems are being developed and rolled out with great success. For cellular networks, UAVs have been proposed and tested as temporary base stations. This could be useful in scenarios where user densities increase markedly for a short period of time (e.g. at a sporting event) or when fixed base stations require maintenance.

We discussed the potential of drones as an NTN technology in Section 11.4. In early 5G, we see nascent deployments of drones for imaging applications and relaying images back to ground stations in real time, using 5G connectivity [60]. 6G will see a lot more applications of drones and potentially swarms of drones co-ordinating with each other through AI and covering vast swathes of area such as agricultural lands to collect massive MTC data. Many countries are starting to regulate use of drones, to ensure civil aviation safety, for example. But there are many commercial applications coming up, including in communications, where drone manufacturers and vendors are jointly representing for a policy framework to bring these applications to mainstream [61].

12.4.9 Quantum Technology

Although it may not seem to fit within the context of 6G communication networks, quantum computing should be considered when thinking about future mobile telecommunication applications. The simple reason for this is the sheer computational power that quantum computing offers. With a quantum computer, capacity achieving multi-user detection algorithms can be implemented easily in real-time and network scheduling algorithms can be solved optimally. On the downside, state-of-the-art cryptographic codes become breakable.

A huge amount of work is being done in the US, Europe and China to develop quantum technology [62]. Research ranges from quantum cryptography to quantum coding and even quantum estimation (i.e. quantum sensor networks). It is not likely that quantum solutions will be mature enough to play a large part in communication networks during the next decade. However, the promises on offer by this technology will make 6G and beyond an exciting arena, which will bring together academics, technologists, and business leaders with the common purpose of developing revolutionary technology and services for an evolving society.

In this final chapter of the book, we have looked at significant changes that are starting to happen in mobile networks with the advent of 5G as well as likely key technology enablers for 6G. It is evident that these changes, if successfully implemented, will only make mobile communications an even more vital and essential component of the digital society. The mobile industry has withstood many challenges over the last 30 years and has always shown the ability to adopt technological advancements to support its growth. It will undoubtedly do the same as it navigates the complex ecosystems presented by 5G and later on by 6G.

References

[1] ETSI (2021). Website. https://www.etsi.org (accessed 19 June 2021).

[2] ONF (2021). Website. https://www.opennetworking.org (accessed 19 June 2021).

[3] O-RAN (2020). Website. https://www.o-ran.org (accessed 19 June 2021).

[4] CISCO (2008). Why should you consider Network As a Service? https://www.cisco.com/c/en/us/solutions/enterprise-networks/network-as-service-naas.html (accessed 19 June 2021).

[5] Sultan, O. (2019). Cisco blog article. Rakuten Cloud Platform is a Blueprint for the Future. https://blogs.cisco.com/sp/rakuten-cloud-platform-is-a-blueprint-for-the-future (accessed 22 June 2021).

[6] Pooley, M. (2017). Rakuten's network: is it really cloud-native, multi-vendor and fully-virtualised? https://stlpartners.com/virtualisation-nfv-sdn/rakutens-network-is-it-really-cloud-native-multi-vendor-and-fully-virtualised (accessed 19 June 2021).

[7] GSMA (2017). Smart 5G networks: enabled by network slicing and tailored to customers' needs. https://www.gsma.com/futurenetworks/wp-content/uploads/2017/09/5G-Network-Slicing-Report.pdf (accessed 19 June 2021).

[8] WAYMO (2019). The Google Waymo project. Technical details. https://waymo.com/tech (accessed 20 June 2021).

[9] CORDIS (2009). Safe road trains for the environment. https://cordis.europa.eu/project/id/233683 (accessed 20 June 2021).

[10] 3GPP (2019). 5G V2X with NR sidelink. https://portal.3gpp.org/ngppapp/CreateTDoc.aspx?mode=view&contributionUid=RP-190766 (accessed 20 June 2021).

[11] 5G Automotive Association (2019). Website. https://5gaa.org/about-5gaa/about-us (accessed 20 June 2021).

[12] 5G Automotive Association (2019). Press release. https://5gaa.org/news/5gaa-and-gcf-announce-collaboration-on-the-certification-and-testing-of-c-v2x-technologies (accessed 20 June 2021).

[13] IEEE (2010). IEEE 802.11p-2010. https://standards.ieee.org/standard/802_11p-2010.html (accessed 20 June 2021).

[14] SNS Telecom (2019). The V2X (Vehicle-to-Everything) Communications Ecosystem: 2019 – 2030 – Opportunities, Challenges, Strategies & Forecasts. https://www.snstelecom.com/v2x. (accessed 20 June 2021)

[15] OFGEM (2000). Transition to smart meters. https://www.ofgem.gov.uk/gas/retail-market/metering/transition-smart-meters (accessed 20 June 2021).

[16] Mekki, K., Bajic, E., Chaxel, F. et al. (2019). A comparative study of LPWAN technologies for large-scale IoT Deployment. *ICT Express* 5: 1–7.

[17] Franklin Heath (2017). LPWA Technology Security Comparison. https://fhcouk.files.wordpress.com/2017/05/lpwa-security-white-paper-1_0_1.pdf (accessed 20 June 2021).

[18] GSMA (2017). Maximising the Smart cities Opportunity. https://www.gsma.com/iot/wp-content/uploads/2017/05/Smart-Cities-Report-web.pdf (accessed 20 June 2021).

[19] University of Manchester (2015). Manchester Cityverve project. http://www.digitalfutures.manchester.ac.uk/case-studies/the-cityverve-project/ (accessed 22 June 2021).

[20] LocalGov (2018). Implementing road pricing systems. https://www.localgov.co.uk/Implementing-road-pricing-systems/44791 (accessed 20 June 2021).

[21] CB Insights (2020). What are smart cities? https://www.cbinsights.com/research/what-are-smart-cities (accessed 20 June 2021).

[22] European Commission (2017). Creating smart cities together. https://eu-smartcities.eu (accessed 20 June 2021).

[23] Marr, B. (2018). Forbes. What is Industry 4.0? Here's a Super Easy Explanation for Anyone. https://www.forbes.com/sites/bernardmarr/2018/09/02/what-is-industry-4-0-heres-a-super-easy-explanation-for-anyone/#167aaeff9788 (accessed 20 June 2021).

[24] 5G-ACIA (2019). 5G for Connected Industries and Automation. https://www.5g-acia.org/fileadmin/5G-ACIA/Publikationen/Whitepaper_5G_for_Connected_Industries_and_Automation/WP_5G_for_Connected_Industries_and_Automation_Download_19.03.19.pdf (accessed 20 June 2021).

[25] 5G-ACIA (2019). Website. https://www.5g-acia.org (accessed 20 June 2021).

[26] ZVEI (2021). German Electrical and Electronic Manufacturers' Association. https://www.zvei.org/en/association/about-us(accessed 20 June 2021).

[27] KPMG (2017). Beyond the hype. https://assets.kpmg/content/dam/kpmg/xx/pdf/2017/05/beyond-the-hype-separating-ambition-from-reality-in-i4.0.pdf (accessed 20 June 2021).

[28] Tait Radio Academy (2008). Introduction to Unified Critical Communications. https://www.taitradioacademy.com/topic/pros-and-cons-of-land-mobile-radio-lmr (accessed 20 June 2021).

[29] First Net (2019). Transforming Public Safety Broadband. https://www.firstnet.gov/newsroom/resources/reports (accessed 20 June 2021).

[30] GOV.UK (2018). Emergency Services Network: Overview. https://www.gov.uk/government/publications/the-emergency-services-mobile-communications-programme/emergency-services-network (accessed 22 June 2021).

[31] 3GPP (2016). Mission Critical Video over LTE. https://portal.3gpp.org/desktopmodules/Specifications/SpecificationDetails.aspx?specificationId=3018 (accessed 20 June 2021).

[32] GSMA (2020). Network 2020 – Mission Critical Communications. https://www.gsma.com/futurenetworks/wp-content/uploads/2017/03/Network_2020_Mission_critical_communications.pdf (accessed 20 June 2021).

[33] ONE5G (2019). Deliverable D2.3. Final system-level evaluation, integration and techno-economic analysis. https://one5g.eu/documents (accessed 20 June 2021).

[34] Hunukumbure, M. and Tsoukaneri, G. (2019). Cost Analysis for Drone based 5G eMBB Provision to Emergency Services. *IEEE Globecom conference proceedings*, Waikoloa, USA (9–13 December 2019). IEEE.

[35] 5G ESSENCE (2020). Embedded network services for 5G experiences. http://www.5g-essence-h2020.eu (accessed 20 June 2021).

[36] 5GIA (2016). 5G and eHealth. www.5g-ppp.eu(accessed 20 June 2021).

[37] GSMA (2018). Smart Grid Powered by 5GSA-based Network Slicing. https://www.gsma.com/futurenetworks/wp-content/uploads/2020/02/5_Smart-Grid-Powered-by-5G-SA-based-Network-Slicing_GSMA.pdf (acessed 20 June 2021).

[38] Statcounter (2019). Global Mobile OS statistics 2019. https://gs.statcounter.com/os-market-share/mobile/worldwide (accessed 20 June 2021).

[39] Pocket-lint (2018). Windows Phone: 10 moments that defined the life and death of Microsoft's mobile platform. https://www.pocket-lint.com/phones/news/microsoft/142501-windows-phone-10-moments-that-defined-the-life-and-death-of-microsoft-s-mobile-platform(accessed 20 June 2021).

[40] Computerworld (2018). The future of 3D holograms comes into focus. https://www.computerworld.com/article/3249605/the-future-of-3d-holograms-comes-into-focus.html (accessed 20 June 2021).

[41] Techradar (2020). Google Stadia review. https://www.techradar.com/uk/reviews/google-stadia (accessed 20 June 2021).

[42] The Telegraph (2019). Smartphones face 'biggest ever decline' in sales. https://www.telegraph.co.uk/technology/2019/09/26/smartphones-face-biggest-ever-decline-sales (accessed 20 June 2021).

[43] Jiang, C., Zhang, H., Ren, Y. et al. (2017). Machine Learning Paradigms for Next-Generation Wireless Networks. *IEEE Wirel Commun* 24(2): 98–105.

[44] Goodfellow, I., Bengio, Y., and Courville, A. (2016). *Deep Learning*. Cambridge, MA: MIT Press.

[45] The Economist (2015). The great chain of being sure about things. https://www.economist.com/briefing/2015/10/31/the-great-chain-of-being-sure-about-things (accessed 20 June 2021).

[46] Banafa, A. (2018). *Secure and Smart Internet of Things (Iot): Using Blockchain and AI*. Gistrup, Denmark: River Publishers.

[47] BiTA (2017). Blockchain in Transport Alliance. https://www.bita.studio (accessed 20 June 2021).

[48] EGI (2020). The EGI federated cloud. https://www.egi.eu/federation/egi-federated-cloud (accessed 20 June 2021).

[49] Tropic (2012). EU project Tropic. https://wwwicttropic.webs.upc.edu/index.php?option=com_spcom&Itemid=12 (accessed 22 June 2021).

[50] Ahmed, I., Khammari, H., Shahid, A. et al. (2018). A Survey on Hybrid Beamforming Techniques in 5G: Architecture and System Model Perspectives. *IEEE Commun Surv Tutor* 20(4): 3060–3097.

[51] Luo, C., Medard, M., and Zheng, L. (2005). On approaching wideband capacity using multitone FSK. *IEEE J Sel Areas Commun* 23(9): 1830–1838.

[52] Basar, E. (2016). Index modulation techniques for 5G wireless networks. *IEEE Commun Mag* 54(7): 168–175.

[53] Xing, Y. and Rappaport, T.S. (2018). Propagation Measurement System and Approach at 140 GHz–Moving to 6G and Above 100GHz. *2018 IEEE Global Communications Conference*, Abu Dhabi, UAB (9–13 December 2018). IEEE.

[54] Jing, Q. and Liu, D. (2018). Study of atmospheric attenuation characteristics of Terahertz wave based on line-by-line integration. *Int J Commun Syst* 31(12): 33718.

[55] FCC (2019). FCC takes steps to open spectrum horizons for new services and technologies. https://docs.fcc.gov/public/attachments/DOC-356588A1.pdf (accessed 20 June 2021).

[56] Teraway (2020). Terahertz technology. https://ict-teraway.eu (accessed 20 June 2021).

[57] Drysdale, T., Allen, B., Stevens, C. et al. (2018). How orbital angular momentum modes are boosting the performance of radio links. *IET Map* 12(10): 1625–1632.

[58] Lee, D. et al. (2018). An Experimental Demonstration of 28 GHz Band Wireless OAM-MIMO (Orbital Angular Momentum Multi-Input and Multi-Output) Multiplexing. *2018 IEEE 87th Vehicular Technology Conference*, Porto, Portugal (3–6 June 2018). IEEE.

[59] Allen, B., Simmons, D., Drysdale, T.D. et al. Performance analysis of an orbital angular momentum multiplexed amplify-and-forward radio relay chain with inter-modal crosstalk. *R Soc Open Sci* 6(1): 181063.

[60] 5G World Pro (2019). Website. https://www.5gworldpro.com/5g-news/77-kt-uses-ar-to-promote-5g-network-and-drones-to-expand-5g-services.html (accessed 20 June 2021).

[61] Drone Alliance Europe (2016). Pursuing a bold vision for Europe's done industry. http://dronealliance.eu (accessed 19 June 2021).

[62] Kaye, P. (2007). *An Introduction to Quantum Computing*. Oxford, UK: Oxford University Press.

Index

1G 8–9, 18
2G 9–11, 18–19, 118–121, 124, 169–170
3G 10–13, 15, 18–19, 124–133, 169–170
3GPP 12–17, 20, 28, 76–77, 79, 86, 88–99, 109,124–129, 130–134, 148–154, 165–166, 169, 319–321, 324–325, 329, 335–337, 339, 352–353, 355–356, 358–359, 369–370, 380–383, 388–389, 392, 395
3GPP2 12, 15, 86
4G 13–16, 18–19, 133–150, 192–194, 199, 213, 345, 348–350, 354–355, 358–359, 361, 372, 379, 390–391, 395
5G 16–20, 24–25, 31, 35–41, 53–55, 58, 62–63, 66–69, 76–77, 80, 94–98, 150–169, 200, 207–209, 213–215, 350–352, 354, 359–363, 366–370, 379–394
   5G New Radio (5G-NR) 45, 55, 95, 155–162, 169, 369–370, 381, 388, 395, 400
   5G NodeB (gNB/ gNodeB) 142–154, 156, 160–161, 164, 214
   Air Interface 158–160
   Bandwidth Part (BWP) 158–159
   Energy Efficiency 54
   Frame structure 158–159
   Non-Stand-Alone (NSA) deployment 96, 151–153, 155, 169, 361
   Protocol Layers 155–158
   Reference Signals 161–162
   Mini-slot 156, 158

   Service Based Architecture (SBA) 95, 162–163
   Stand-Alone (SA) deployment 96, 151–153
   Sub-Carrier Spacing (SCS) 158–159
   Synchronisation Signals (PSS, SSS, SSB) 160–161
   System Architecture 153–154
5G Automotive Association (5GAA) 381–382
5G Alliance for Connected Industries and Automation (5G-ACIA) 387–388
6G 170, 395–402

Adaptive modulation 296
Additive white Gaussian noise (AWGN) 257
Airtime Provider 56
Antenna 109, 144, 155, 161, 174–175, 178–179, 181–182, 186–187, 194–195, 208–210, 212–214, 320, 323, 334, 336, 3339, 348–349, 359, 365–366, 398–399
   Antenna gain 52, 178–179, 185
   Directional 178
   Isotropic (unity gain) 174–175, 178, 286, 314
   Omni-directional 174, 178, 187, 210, 323, 334
   Planar array 209, 213
   Radiation Pattern 52, 179, 208–210, 212
   Sidelobe Response 52

The Technology and Business of Mobile Communications: An Introduction, First Edition.
Mythri Hunukumbure, Justin P. Coon, Ben Allen, and Tony Vernon.
© 2022 John Wiley & Sons Ltd. Published 2022 by John Wiley & Sons Ltd.

Artificial Intelligence/ Machine Learning (AI/ ML)   40, 166, 383, 386–387, 394–396, 401

Asset depreciation   43–44

Average Billing per User (ABPU)   60

Average Revenue per User (ARPU)   19–20, 43, 60, 355, 363

Backhaul   87, 95, 102, 164–165, 173, 187, 199–208, 213–215, 318–319, 322, 327, 332–335, 340–342, 345, 370, 390–391, 398, 400–401
  23/38 GHz Fixed Links   50
  Dark fibre   200
  Fibre backhaul   200–201, 206–208
  Integrated Access and Backhaul (IAB)   200, 208
  Microwave Links   50, 199–200
  Rain Fade   50
  Self-Coordination   50
  Self-backhaul   200
  Self-similarity   205–206

Baseband Processing Unit (BBU)   46, 164–166, 214, 320, 390–391

Base Station Subsystem (BSS)   49, 118–119

Base Station Rigging Costs   44

Broadcast Channels   54, 104, 129, 137, 145, 160

Business Support System (BSS)   167–168

Capacity   260–262

CAPEX (CAPital EXpenditure)   43, 56, 366, 379

CDMA 2000   12–13, 86–87, 89–90

Cell Radius, cell range   54, 105, 179, 184, 186, 195–198

Cellular Vehicle to Everything (C-V2X)   150, 381–383

Central Office (CO)   350

Centralised Unit (CU)   153–154, 350–351

Channel coding   243, 246–251
  Channel coding theorem   247
  Linear block codes   248–249
  Trellis codes   250–251

Channel estimation   280–282

Channel Quality Indicator (CQI)   108, 195

Commercial-off-the-shelf (COTS)   31, 166

Control Plane Service-Oriented Core (CP-SOC)   350

Core Network (CN) or Core 46, 101–102, 110–11, 116, 118–120, 124–126, 132–138, 152–154, 156–157, 162–163, 165, 168–169, 187, 199–201, 203, 207   215, 346–347, 349–352, 360–361, 378–379, 382, 387, 389–391
  2G core   119–120
  3G core   124–125
  5G Core (5GC)   152–154, 156–157, 162–163, 215, 351–352
  Evolved Packet Core (EPC)   46–47, 134–135, 152–153, 201, 203, 206, 215

Cost-Plus Business Model   43

Data Session Routing Area   45

Demodulation   242, 258–260

Detection   260
  Maximum a posteriori (MAP)   260
  Maximum likelihood (ML)   260

Distributed Evolved Packet System (DEPS)   49

Distributed Unit (DU)   154, 164–165, 350

Diversity   297–307, 179, 184–186, 195
  Cyclic delay diversity (CDD)   304–305
  Diversity order *or* diversity gain   299, 179, 185–186, 195
  Frequency diversity   300
  Maximum ratio combining (MRC)   298, 301
  Maximum ratio transmission (MRT)   302–303
  Repetition coding   298
  Selection combining   301–302
  Space-time codes   302, 305–307
  Spatial diversity   300–307
  Time diversity   298–299
  Transmit antenna selection   304

Drone links (drones)   208, 369–370, 390–391, 401

Enhanced Data for GSM Evolution
(EDGE)   10–11, 86–87, 120, 124, 126
Enhanced Mobile Broadband (eMBB)   16,
19, 151–152, 351, 354, 359, 369, 377,
384, 389, 394
Equalisation   264–272
BCJR algorithm   264
Frequency-domain equalisation   267–272
Maximum likelihood sequence estimation
(MLSE)   264
Orthogonal frequency-division
multiplexing (OFDM)   267–270
Time-domain equalisation   264–267
Single-carrier with frequency-domain
equalisation (SC-FDE)   271–272
Estimated Mobile Network Cost   46
European Telecommunication Standards
Institute (ETSI)   12, 27, 83–86, 96–98,
166–167, 169, 378
EV-DO   12–13, 15, 90

Fading   287–293
Fast fading   184, 292–293
Fast Fade Margin (FFM)   184, 186, 215
Frequency-selective   291
Large-scale, shadowing   182–183, 213, 288
Log-Normal Fade Margin (LNFM)
182–184, 186–187
Nakagami   290
Rayleigh   184, 289
Rician   290
Slow fading   183–184, 187, 291–292
Small-scale   289
First-time Installation Expenditure   44
Fixed Wireless Access (FWA)   155, 368–369,
372
Frequency Allocation Table (FAT)   222
Primary   222
Secondary   222
Fronthaul   164–166, 173, 200, 207–208,
213–214, 341, 390–391, 400

Gateway GPRS Support Node (GGSN)   45,
120, 126, 132
General Packet Radio System (GPRS)   10–11,
111–112, 118–120, 124–125

GPRS Tunnelling Protocol (GTP)   77,
111–112, 136, 138
Groupe Spéciale Mobile (GSM)   9–12,
116–125, 194
Base Transceiver Station (BTS)   118–119,
123
Base Station Controller (BSC)   118–119,
123, 135
Gaussian Minimum Shift Keying
(GMSK)   11, 122–124
Transcoder Rate Adapter Unit
(TRAU)   118–119
GSM Association (GSMA)   3, 5, 35, 83–85,
93–95, 97, 131, 361

Heterogeneous Networks (HetNets)   39, 211,
213, 318, 320–322, 327, 330–332,
337–338, 354–357
High Speed Packet Access (HSPA)   12,
87–92, 109, 124–125, 132–133, 148
Home Subscription Server (HSS)   47,
135–136, 163

IMT-2000 10–12, 86–87, 89–91.   125
IMT-2020   16–17, 95–96, 152
IMT-Advanced   15, 90–91, 134
Infrastructure Vendor Selection   44
Initial Network Planning   44, 208–211
Interleaving   184, 278, 295,297, 300
Interference   106–107, 121–123, 127–128,
131, 139–142, 149, 159, 173–175,
177–179, 194–195, 203, 210–211, 242,
256, 261, 264, 266, 270, 274–278, 285,
287, 307, 309, 313, 320–321
Co-tier interference   321
Cross-tier interference   321
Interference Margin (IM)   177, 186
Inter-Symbol Interference (ISI)   139, 141,
285
Interference Coordination   322–324
Successive interference cancellation
(SIC)   159, 278, 322
Fractional frequency reuse (FFR)   323
Inter-cell interference coordination (ICIC)
323–324, 357
Enhanced ICIC (eICIC)   149, 325, 357

Almost blank sub-frames (ABS)   149, 325, 357–358

Reduced power sub-frames (RPS)   326–327

Cell range expansion (CRE)   326, 357

International Roaming   9, 57, 78–79, 83–85, 92, 94, 96–98, 101, 110, 112, 119, 136, 163

International Telecommunications Union (ITU)   13, 15–17, 75–77, 86–87, 89–91, 95–97, 125, 151–152, 220

Internet of Things (IoT)   43, 155, 351–352, 370, 384, 388–392, 396–398

Investment/ Investors   25, 30, 34, 38–39, 90, 95, 345, 361, 368, 372, 379, 385

IP Multimedia Sub-system (IMS)   88–89, 93–94, 125–126, 130–131, 135

IPv6 QoS Parameters   44

IS-95 (CDMA One)   11–12

Local Data Centre (LDC)   350

Long Term Evolution (LTE)   4, 15–16, 106–109, 116–117, 133–158, 160–163, 178, 185–190, 194–195, 197, 199–203, 206–207, 215, 319, 329–330, 336, 339, 350, 352–353, 355–358, 361, 368, 370, 372, 381–382, 388–391

Evolved NodeB (eNodeB)   45, 134–138, 141, 145–147, 149–150, 164, 200–201, 205, 356, 358

Evolved Packet System (EPS)   47, 79–80, 89

Evolved UTRAN (E-UTRAN)   46, 79, 135, 154

Femto cells   318–319, 324

Frame, sub-frame   142–143, 186, 194–195, 199, 202

HARQ   147–148, 369

Handover   146–147, 329–330

Handover failure (HOF)   330

Hard Handover   45, 130, 146

Hysteresis margin (HM)   322

Late path switch   138, 147

Radio link failure (RLF)   330

Time to trigger (TTT)   322, 331

LTE-Unlicensed (LTE-U)   150, 353

Mobility Management Entity (MME)   46, 116, 135–138, 147, 163

MIMO, MIMO-B   148, 195–198, 200, 202, 207

OFDMA   131, 139–142, 146

Power control   146

Random access   145–146

Reference signals   144

SC-FDMA   139, 141–142

Slow Power Control   146

Synchronisation signals (PSS, SSS)   143, 145

System Architecture   134–136

Licence (spectrum)   224, 228

Licence Exempt   232

Licenced shared access   231

Local Area Network (LAN)   150, 219, 320

Low Power Wide Area (LPWA)   383–384

LTE-M   149–151, 383–384, 395

Machine to Machine (M2M)   23–24, 30, 33–34, 40, 317–318, 386

Management and Orchestration (MANO)   95, 167–168

Marketing-as-a-Service (MaaS)   57

Massive Machine Type Communication (mMTC)   16, 19, 149, 151, 156, 359, 369–370, 377, 383–384, 386, 389, 392, 395

Millimetre wave (mmWave)   27, 95, 154–155, 158, 207, 213, 347, 352, 366–369, 398, 400

Mission Critical Communications (MCC)   150, 389, 391

Mobile Network Operators (MNO) or Operators   2–4, 7, 10–19, 74–80, 82–97, 125, 130–131, 133–134, 142, 152, 155, 163–164, 166, 168–170, 187–188, 199, 345–346, 350, 353–355, 358, 361, 363, 367, 377–378, 380–381, 385, 390–391, 395

Mobile Licence Coverage Requirements   45

Mobile Switching Centre (MSC)   45, 118–119, 123, 126

Mobile Virtual Network Enabler (MVNE)   57

Mobile Virtual Network Operator
    (MVNO)   56, 350
Mobility Management   45, 116–117, 135,
    137, 163
Modulation   242, 251–255
    Amplitude shift keying (ASK)   253
    Band-pass versus baseband   252
    Constellation   254–255
    Line codes   253
    Orthogonal frequency-division
        multiplexing (OFDM)   267–270
    Phase-shift keying (PSK)   254
    Quadrature amplitude modulation
        (QAM)   254
    Single-carrier with frequency-domain
        equalisation (SC-FDE)   271–272
Modulation and Coding Schemes
    (MCS)   105, 108, 138, 162, 195–199,
    202, 211
Multi-Operator Radio Access Network
    (MORAN)   350
Multipath   287
Multiple access   272–279
    Code Division Multiple Access
        (CDMA)   272–275
    Frequency Division Multiple Access
        (FDMA)   116, 121, 139, 275
    Non-Orthogonal Multiple Access
        (NOMA)   277–279, 348
    Orthogonal Frequency Division Multiple
        Access (OFDMA)   15, 98, 275–276,
        131, 139–142, 145–146
    Single Carrier Frequency Division Multiple
        Access (SC-FDMA)   139, 141–142,
        276–277
    Time Division Multiple Access (TDMA)
        86, 116, 121–122, 124
Multi-access/ Mobile Edge Computing
    (MEC)   97, 166, 169, 382
Multiple-input multiple-output
    (MIMO)   109, 143–144, 148, 307–314,
    334, 348
    Beamforming   310–311
    Capacity   308–309
    Massive MIMO   214, 314

MIMO-B   195–196
Maximum likelihood (ML) detection   311
Minimum mean-square error (MMSE)
    detector   312
Multi-user MIMO (MU-MIMO)   109, 144,
    148, 313–314
Spatial multiplexing   311
Sphere decoding   312
Zero-forcing (ZF) receiver   312
Multiple SIM Data Use   43

National Roaming   368, 379
NB-IoT   149, 151, 383–384, 389, 395
Negative Asset Value   44
Network as a Service (NaaS)   378–379
Network Slice Controller   350
Network Slicing   163, 350–351, 359–360,
    380
Network Function Virtualisation (NFV)   49,
    166–168, 341, 349, 377–379
Next Generation Mobile Network
    (NGMN)   95, 201, 203–204, 206–207
NodeB (3G)   45, 126–132
Noise   123, 158, 162, 173–174, 175–174,
    176–178, 185–186, 215, 247, 278–282,
    257–258, 260–262, 265–266, 272–275,
    288, 293–294, 296–299, 308, 311–312,
    366, 398
    Additive White Gaussian Noise (AWGN)
        257–258, 260–263, 274, 281, 288, 296
    Shot Noise   257
    White Noise   176
Nordic Mobile Telephone System (NMT)   9,
    57

Open Systems Interconnection
    (OSI)   113–116
OPerational EXpenditure (OPEX)   43, 44,
    51, 56, 363, 379, 397
Operations Support System (OSS)   110,
    167–168, 355, 358, 361–362
O-RAN Alliance   166, 378
Orthogonal frequency-division multiplexing
    (OFDM)   267–270
Over The Top (OTT)   19, 31, 37, 131

Packet Data Network (PDN) 45, 103, 120
Path loss 106, 130, 146, 149–150, 174, 180,
      183, 214, 347–348, 363, 366, 369, 384,
      389
PDN-Gateway (P-GW) 136
Personal Communications Network
      (PCN) 49
Phase-locked loop (PLL) 280
Pre-pay/Post-Pay Revenue per
      Subscriber 43
Primary Synchronisation Signal (PSS) 143,
      145, 160–161
Policy Control and Charging Rules Function
      (PCRF) 47, 135–136, 163
Profit/Loss Equation 57
Protocol Stack/ Layers 112–117, 136–138,
      155–157
   2G 116–117
   4G 116–117, 136–138
   5G 155–157

Quality of Experience (QoE) 30, 360
Quality of Service (QoS) 44, 105–106, 112,
      130–131, 136, 147, 157, 188–190, 201,
      207, 318, 322, 338, 348, 350, 352–353,
      356, 360–362, 365–367, 372
Quality of Service Class Indicator
      (QCI) 190, 195, 350
Quantisation 245

Radio Access Network (RAN) 45, 77, 85,
      96–98, 101–106, 109–110, 118, 122,
      124–125, 131–133, 137, 144, 146,
      152–154, 156–158, 160, 163–166,
      168–169, 173, 199, 207, 214, 359–360,
      378–379, 382, 388
   Centralized RAN (C-RAN) 164–166, 207,
      214
   Cloud RAN 165, 207
   Distributed RAN (D-RAN) 164–165
   Open RAN (O-RAN) 166, 378–379
Radio over Fibre 48
Radio Resource Control layer (RRC) 116–117,
      137–138, 155–156, 164, 166
RRC connected state 105, 136–137, 146, 156

RRC idle state 137, 156
RRC inactive state 156
Radio Resource Management (RRM) 45,
      116, 119, 126, 135, 154, 164
Regional Bodies (Spectrum Regulation)
      221
   APT 221
   ASMG 221
   ATU 221
   CEPT 221
   IATC 221
   RCC 221
Regional Data Centre (RDC) 351
Remote Radio Unit/Head (RRU/RRH) 46,
      48, 164, 214, 378–379, 391
Return of Investment (ROI) 385

S1 Link 46, 135, 200–201, 203
Sampling 243
   Sampling frequency 243
   Sampling interval 243
Satellite links 40, 368–370
   Geostationary (GEO) 370
   Low Earth Orbit (LEO) 370
   Medium Earth Orbit (MEO) 370
Self Organised Network (SON) 40, 355–361
Service Provider 56, 372, 382
Serving Gateway (S-GW) 46, 135–136, 147
Serving GPRS Support Node (SGSN) 45,
      118, 120, 126, 132, 136, 349
Signal to Interference plus Noise Ratio
      (SINR) 108, 144, 177, 195, 202–203,
      210, 278, 322, 332, 337–340, 348
Signal to Noise Ratio (SNR) 51, 105, 126,
      177, 179, 185, 189, 195–196, 202–203,
      207, 211, 215, 258, 274–275, 280, 285,
      289–295, 299–302, 304–306, 309, 312,
      332
SIM (Subscriber Identity Module) 56
Single-carrier with frequency-domain
      equalisation (SC-FDE) 271–272
Site Build Costs 53
Small cells 149, 182, 199–200, 207–208,
      211–215, 318–322, 332–342
Smart Antennas 48

Smartphones   4, 14, 16, 18–19, 34, 87–88,
    125, 149, 317, 352, 354, 368, 377, 382,
    393–395
Software Defined Network (SDN)   39,
    166–168, 331, 341
Source coding   243, 245–246
    Huffman coding   246
    Source coding theorem   246
Spectrum assignments   222–223, 225–227
    Auctions   227
    Beauty contest   227
    Market based mechanisms   225
Spectrum management   219
    Allocation   220, 222, 225
    Allotment   222
    Geneva 2006 Agreements   220
    Monitoring   224
    Radio Regulations   220
Spectrum Re-farming   149, 155, 228

T1 Transmission Link   50
Tail insertion loss   48
TERA MVNO analysis tool   59
Terrestrial Trunked Radio (TETRA)   388–389,
    391
Titan MVNO modelling engine   59
Total Cost of Ownership (TCO)   54
Transmission Control Protocol (TCP), TCP-IP
    114–115, 190
Transmission Costs   49–50

Universal Mobile Telecommunication System
    (UMTS)   12, 15–16, 85–87, 89–90, 92,
    125–128, 130–131, 134, 136, 178
    Code orthogonality   131
    Direct Sequence CDMA
        (DS-CDMA)   127–128
    Fast power control   106, 129
    Near-far effect   129, 131

Node Base-station (NodeB)   126, 128–130,
    132
Orthogonal Variable Spreading Factor
    (OVSF)   128, 131
Radio Network Controller (RNC)   45, 126,
    130, 132, 135, 138
Soft handover   45, 130, 135
Wideband CDMA (W-CDMA)   12–13, 15,
    86–87, 90, 124, 126, 128–129, 131, 133
UMTS Terrestrial Radio Access
    (UTRA)   46, 125–126, 128–129
Ultra Reliable Low Latency Communication
    (URLLC)   16, 19, 151, 156–158, 168,
    351–352, 354, 359, 369, 377, 381, 386,
    388–389, 395, 400
Ultra Wide Band (UWB)   354
User Datagram Protocol (UDP)   115, 138
User Plane Service-Oriented Core
    (UP-SOC)   350

Value-Added Server   44
Vehicle to Everything (V2X)   150, 165,
    381–383
Vendors (Equipment manufacturers)   13, 16,
    31, 74–76, 78, 80, 83, 88–89, 91, 95–96,
    98, 113, 131, 166, 167–168, 170, 355,
    377–379, 381, 385, 388, 391, 395, 401
Voice Call Location Area   45
Voice over IP (VoIP)   15, 19, 111, 130, 133,
    138, 147, 170, 189–193, 198

Water-filling   296–297, 309
Wi-Fi   90–93, 101, 11, 113, 139, 150, 219,
    352–353, 372,
WiMAX   15–16, 90–92
World Radio-communication Conference
    (WRC)   152, 220, 222, 235

X2 Links   46, 135, 138, 147, 200–201, 203